Ad Hoc Networks: Current Status and Future Trends

Ad Hoc Networks: Current Status and Future Trends

Edited by
Tyler Ward

www.statesacademicpress.com

Published by States Academic Press,
109 South 5th Street,
Brooklyn, NY 11249, USA

ISBN: 978-1-63989-014-9

Cataloging-in-Publication Data

   Ad hoc networks : current status and future trends / edited by Tyler Ward.
      p. cm.
   Includes bibliographical references and index.
   ISBN 978-1-63989-014-9
   1. Ad hoc networks (Computer networks). 2. Wireless communication systems.
   3. Computer networks. I. Ward, Tyler.
TK5105.77 .A33 2022
004.6--dc23

For information on all States Academic Press publications
visit our website at www.statesacademicpress.com

Contents

Permissions

List of Contributors

Index

Preface

The wireless network which is set up for a single session and does not need a wireless base station or a router is termed as an ad hoc network. It does not rely on a pre-existing infrastructure, rather each node participates in routing by forwarding data to other nodes dynamically. On the basis of the applications, ad hoc networks can be classified into vehicular ad hoc networks, smartphone ad hoc networks, iMANETs, wireless mesh networks, army tactical MANETS, etc. These are mostly wireless local area networks. This book outlines the current status and future trends of ad hoc networks in detail. It will also provide interesting topics for research which interested readers can take up. The book is appropriate for students seeking detailed information in this area as well as for experts.

The researches compiled throughout the book are authentic and of high quality, combining several disciplines and from very diverse regions from around the world. Drawing on the contributions of many researchers from diverse countries, the book's objective is to provide the readers with the latest achievements in the area of research. This book will surely be a source of knowledge to all interested and researching the field.

In the end, I would like to express my deep sense of gratitude to all the authors for meeting the set deadlines in completing and submitting their research chapters. I would also like to thank the publisher for the support offered to us throughout the course of the book. Finally, I extend my sincere thanks to my family for being a constant source of inspiration and encouragement.

Editor

Efficient trust management with Bayesian-Evidence theorem to secure public key infrastructure-based mobile ad hoc networks

Janani V S[*] and Manikandan M S K

Abstract

In mobile ad hoc networks (MANETs), the reliability of nodes, quality of data and access control cannot be achieved successfully for various network functionalities through traditional cryptographic security, which makes MANET vulnerable to illegitimate node behaviour changes. These node misbehaviours, referred as soft security threats, need to be detected and prevented in order to protect against the accumulation of false measurements with selfish and malicious intentions. Trust has been employed as a powerful tool to handle the soft security threats and to provide security among uncertain and dynamic nodes effectively in MANET. Therefore, it is of great importance that efficient trust management mechanisms should be developed in a public key infrastructure (PKI), in order to verify the identities on the ad hoc networks for reliable and secure group communication. However, the independent nature of nodes and the computational complexities make the trust management a challenging one in MANET. In this paper, we present an efficient distributed trust computation and misbehaviour verification method with Bayesian and Evidence theorem, on hexagonally clustered MANET. Besides, a secured PKI system is designed in the paper by applying the proposed trust management scheme in terms of certificate revocation, which is an important functionality of PKI cryptosystem. The uncertainty impacts the node's anticipation of neighbour's behaviour and decisions during communication; we include uncertainty in the trust management system. An efficient method to reduce the uncertainty is to exploit the mobility characteristics of MANET that accelerates the trust convergence. The simulation results reveal a better performance against adversaries in creating considerable untrustworthy transactions with a mobility-aware cluster guarantee. Moreover, the proposed trust application shows its betterment in the revocation process in terms of revocation rate and time. Thus, the proposed scheme provides an effective security solution that incorporates the optimistic features of trust mechanisms and hierarchical Voronoi clustering.

Keywords: Bayesian and evidence theorem, Certificate revocation, Clustering, MANET, Public key infrastructure, Trust computations, Verification, Security

1 Introduction

In mobile ad hoc network (MANET), malicious attacks on different layers have been identified and analysed by researchers over several years. Several routing protocols were introduced in order to secure routing and forwarding packets in MANET from malicious attackers. Most of these conventional routing protocols rely on a centralized public key infrastructure (PKI) to detect and secure malicious

misbehaviours using hard security or cryptographic mechanisms. However, these solutions provide only a partial security in the initial stages of managing mobile nodes, where malicious nodes can affect the credibility of the network. In certain cases, nodes may be vulnerable to the behaviour changes with the influence of attacker participants, even if they behave as legitimate nodes initially in a secure group communication and therefore pass the hard security checks. However, these unauthorized nodes may become selfish or malicious nodes and report false information with the intention to damage the reliability

* Correspondence: jananivs1987@gmail.com
Thiagarajar College of Engineering, Madurai-15, India

of group communication. The traditional cryptographic mechanisms cannot detect and prevent these continual changes in the node behaviour. In other words, the reliability of communication, the quality of data and access control cannot be achieved fully with the hard security techniques. Therefore, a security mechanism is required to defend against the node behaviour changes commonly referred as soft security threats and assure integrity, reliability and access control to the group communication in MANET. Consequently, an effective distributed and self-organising mechanism quantified with trust to identify and secure the misbehaviour in ad hoc network should be established.

While considering the PKI security system, the necessity of centralised or distributed certificate authority (CA) is of greater importance. The PKI system manages trust in communication conducted by the nodes, over the network. The vital elements used for trust management are the certificates and security protection in the environments of the different participants involved. The CA controls the entire certificate and public key management in which trust plays a vital role. These elements are derived by a trust management mechanism for the communication purpose of the exchanges, associated with the public-private keys. In the PKI domain, to establish a distributed trust relationship, the public key needs to be imported and afeguard its integrity, communication or storage to other entities.

The researches on distributed trust systems in MANET require the nodes to be organized with some hierarchical security methodology to achieve performance guarantee, especially when applied to emergency communication. To manage the uncertain mobile nodes, various clustering techniques have been introduced as a hierarchical architecture for scalability issue in wireless networks. A cluster structure manages network functionalities with efficient spatial reuse, in order to deploy the PKI based security in MANET, over a finite network region. The self-organization property should be combined with distributed clustering architecture to coordinate and collaborate the dynamic nodes in MANET. This eliminates the single point of dependency and failure that occur in every traditional centralised methodology and provides a PKI framework with self-healing, self-configuration and self-management features to adapt the frequently changing network conditions. This can be successful only when the nodes behave in a trustworthy manner. Trust management encounters these network challenges in order to develop an optimized distributed and self-organized security system. The trust in ad hoc networks is the subjective evaluation on the node behaviour of its neighbouring nodes. It reflects the belief and expectations on the credibility of behaviour and information sends by any node.

In spite of that, there are several pitfalls in establishing a self-organizing and distributed trust-based PKI security

system with partitions in ad hoc networks. Some of them are as follows:

- Maintaining trustworthy cluster members and headers increases computation and communication complexities.
- The traditional centralized trust block may depend on a single point for functionalities and requires more computational and infrastructural cost.
- Most of the recommendation-based trust management works on the assumption that the belief is of equal weight, which is prone to attackers.
- Mobility oblivious PKI system in MANET weakens the trust computation as it is hard to find the behaviour as the nodes moves dynamically.
- The trust measurements in the traditional trust methodologies are instantaneous and not precise.
- The data sharing between nodes in a cluster greatly depend on the location of mobile nodes. Therefore, the distance between the nodes should be computed accurately for any group communication, which is hard to attain with the traditional clustering techniques.
- It is difficult to develop a complete security system with underlying distributed trust-based clustered MANET where link failures occur frequently.

Therefore, it is comprehensible that the drawbacks of the widely known trust management techniques should be minimized to make the PKI-based security flexible for group security communication. In this pursuit, the proposed work focus on developing a distributed and self-organized security solution for PKI framework, which quantifies nodes behaviour in the form of trust.

The most influential complication in distributed trust management is how to collaborate the observations from multiple sources to calculate the trust of any node. The primary intention of the proposed work is to adapt the dynamic topology with a hybrid trust management mechanism. The trust establishment maintains the self-organizing property with no trust agent involved in trust calculation. This is attained by incorporating the direct trust measurements and the recommendations obtained from the cluster members. The direct trust is evaluated and verified using Bayesian theory, whereas the indirect trust is calculated by the Dempster-Shafer (DS) evidence theory which combines recommendations obtained from various neighbouring nodes. The observations in the proposed scheme are taken as evidences on the node behaviour. We make use of the well-established mathematical structure called Voronoi diagram to overwhelm the neighbour-searching problem and to reduce the distance computation complexities. Unlike the traditional circular clusters, a regular hexagonal shape is constructed with improved spatial reuse to group the MANET area into adjacent, non-overlapping clusters of nodes. The proposed trust-based clusters guarantees improved performance with dynamic re-configurability, scalability and security.

2 Related works

Over the past several years, there has been a large amount of researches on security protocols and their implementation in a PKI-based MANET security system. Most of these researches focus on the routing protocols, medium access and data forwarding algorithms. Distributed communication is important to be achieved for MANET-based sensing and scrutiny applications. The communication will be effective only if all the nodes follow a trustworthy behaviour [1–3]. The MANETs is established in unconstrained environments with no centralized controlling authority, where the node compromising and attacks happen at higher probability. These unique features make constraints on the nodes to be prudent for a secure communication, predominantly in the PKI framework. Therefore, it is important to quantify the behaviour of each participant in such collaborative communications. This can be achieved by deploying trust as a system of measuring the node behaviour, where the mobile nodes are grouped into clusters in order to maintain scalability and reduce frequent link failures during a secure group communication.

2.1 Trust management in MANET

Numerous trust models have been proposed for secure node communication based on sharing group recommendation to establish cooperation in computational networks [4–7]. The trust can be defined as the degree of individual belief on the behaviour of any participant node [8]. In [9], the trust management was distinguished from other security services to provide and manage security policies and relationships. In MANET, trust management is applied to evaluate the belief level of information and nodes, to detect intrusions and to provide security services including key management, authentication, access control and node revocation [10–14]. On that account, certain computational methodologies should be utilized at regular interval to assess the trust level. Unlike a wired network, in a dynamic mobile network like MANET, the trust computation can be made only with many numbers of such periodic observations. The trust computation is, however, a challenging task because of random node mobility and the lack of central authority. The surveys of trust management in MANET [15–17] give a summary of various techniques for trust calculations. The formalising trust method in [18] made a contribution to many later on schemes to consider the neighbour opinion along with the direct interactions in decision making. In [19], the trust of each node is calculated with two schemes, namely the reputation framework and trust establishment. A direct observation and further distribution of information is done in reputation framework. Whereas, in trust establishment, direct observations and opinion from one-hop neighbours are combined for evaluating the trust relations. In [20–23], the concept of combined trust computation is presented in which direct trust is computed with direct observations and indirect trust is computed from recommendation. The misbehaviour verification in trust computation for non-cooperation is presented in [24]. In contrast with wired networks, to estimate trust in a fully distributed network is demanding to attain [25, 26]. A mathematical model with the Bayesian theory was introduced in [27–30], to update the reputation from direct observations. In [31–38], various trust models in a public key infrastructural network are discussed. These trust models are developed on a clustered mobile node network where security enhancement is certainly important. On the other hand, these existing trust models for computing the trust level of each node in MANET multiplied the computation as well as communication complexities.

2.2 Trust for MANET scalability

A Cluster based Trust-aware Routing Protocol (CBTRP) to protect packets from the attackers was proposed in [39]. With the aim to provide security, trust-based security systems were presented in different network architectures [40–43]. To overcome the drawbacks of conventional security systems, the uncertainty reasoning has been assessed as the probabilistic technique with trust in MANET where mobility is considered with greater importance [38, 39]. In most of such uncertain management methodologies, the distance of the nodes is calculated with respect to Euclidean distance. However, this distance calculation suits only for a specific distance function [44, 45]. To handle this distance computation issue in uncertain space, Voronoi diagrams have been introduced by [46, 47]. This computational geometric structure is applied for decomposition of network space into polygonal regions, to evaluate the distance distribution [48, 49]. The distribution of mobile nodes with increased network capacity and throughput in hexagonal structures was introduced in [50]. The regular hexagonal partitions have proven its flexibility to form non-overlapping clusters in large ad hoc networks [51]. In order to secure network functionalities, trust management has been widely applied in ad hoc networks [52–57]. These methodologies prevent various attacks that might affect the system passively or actively.

By taking everything into account, it can be stated that trust has been employed as a powerful tool to handle the soft security threats and to provide security among uncertain and dynamic nodes effectively in MANET. The trust computation of a node has a high impact on the reliability and quality of any secure communication, particularly for public key distribution. The PKI requires a chain of trust to verify the identities on the ad hoc network. Therefore, it is of great importance that efficient trust calculation and management mechanisms should be developed in a PKI-based ad hoc network with efficient clustering model (Table 1).

3 Motivation of proposed work

With the comparison of the related works, the advantage and disadvantage of trust management and its application are analysed and incorporate the best suited techniques to implement PKI system in MANET. Providing a distributed hybrid trust mechanism for MANET security is difficult

Table 1 Comparison of different trust mechanisms

Authors and year	Context in use	Advantages	Disadvantages
Trust management in MANET			
Li et al. 2008 [19]	A reputation based on direct observations	Certain attacks such as selective misbehaving, bad mouthing and On off attacks are reduced	The ratio of trustworthiness over reputation is based on direct observations
Hui Xia et al., 2013 [22]	Novel on-demand trust-based unicast routing protocol for MANETs to provide a suitable approach to select the shortest route for secured data packet transmission	Black hole attack and gray hole attack are reduced with the proposed protocol	Trust is derived only based on direct observations
A.M Shabut et al. 2015 [23]	Proposed a recommendation-based trust model with clustering technique to dynamically filter out attacks related to dishonest recommendations	Tested under several topologies and route changes	The work does not consider the past node behaviour
S. Marti et al., 2000 [24]	Proposed a reputation-based trust management system	Node behaviours are monitored by watchdog and collect the reputation with pathrater	Trust evaluation is based only on direct observations
C. H. Ngai Edith and R. Lyu Michael 2004 [31]	Presented a secure PKI-based trust model to prevent false key propagation	Trust is calculated based on direct monitoring and recommendation to prevent attackers	This work does not consider the effect of mobility and distance between the nodes on trust management
Trust for MANET scalability			
Cho et al. 2013 [40]	Past experiences and current behaviour are combined to estimate trust using the Bayesian approach	No single point failure	No precise trust measurements
Cho, J. H. et al.. 2011 [16]	Trust is calculated based on packet forwarding behaviour	Can be applied to any wireless networks	Trust has instantaneously calculated based on individual nodes
R. H. Jhaveri and N. M. Patel 2016 [42]	A trust model is integrated with an attack discovery technique	Earlier detection and elimination of adversaries	No trade-off between security levels and energy consumption
H. Safa et al. 2010 [39]	Organizes the network into disjoint clusters and elects cluster head with the most qualified and trustworthy nodes	Ensures the trustworthiness of by replacing malicious cluster heads	Load balance clustering is a dynamic optimization problem
J. M. Nichols and J. V. Michalowicz 2017 [44]	Distance statistics for mobile ad-hoc wireless network have focused on the three-dimensional spatial cases	High network reliability quantified with distance distribution	Distribution is performed with Euclidean distance
Kao. B. et al. 2010 [45]	Propose pruning techniques that are based on Voronoi diagrams to reduce the number of expected distance calculations	Reduces the computation of expected distances between uncertain objects and cluster head	The complexity of the UK-means are not reduced by the proposed pruning techniques
X. Xie et al. 2012 [46]	Voronoi diagram is used for uncertain spatial data for evaluating nearest-neighbour queries	Support probabilistic nearest-neighbour queries execution	It is computationally infeasible to create and store UV partitions
Matthew L.et al. 2017 [47]	Finds the Voronoi neighbours directly from inter-object distances, before assigning coordinates	Effectiveness in the presence of noise	Increased computational complexity
Fan P. et al. 2007 [49]	The probability density function of the distance between two nodes is derived using the space decomposition method. The node degree is calculated with a simple path loss model	Efficient node degree distribution and maximum flow capacity of the network	Limitation with multi-hop networks
Fei T et al. 2016 [51]	A probabilistic distance-based model is presented for nodal distance distribution over a finite network	Extended to the networks with the shape of one or multiple arbitrary polygon	Trust metrics are not considered as functions of the distances among interfering nodes

to establish, in the presence of differing topology. An efficient security solution for this issue should combine the beneficial features of trust and Voronoi that are partitioning for managing MANET nodes, which is still problematic. Such an optimal solution is presented in this paper for providing PKI security in MANET by resolving the drawbacks in the existing mechanisms. With this objective, we make the following contributions in this paper:

1. In this paper, we propose a novel trust management strategy which combines two prominent theorems: the Bayesian and Evidence theorem to compute the trust level directly and indirectly for use in ad hoc networks in order to reduce the complexity of managing the underlying PKI-based security framework.
2. To reduce the nearest-neighbour finding (NNF) problem in the conventional clustering mechanisms, the uncertain nodes are grouped by Voronoi geometric patterns.
3. To be inconsistent with the cellular clustering structure with highly overlapping partitions, the mobile nodes are grouped into Voronoi polygons with hexagon structure to reduce the cluster construction complexities. Further, the proposed scheme shows resilience to many attacks, mainly recommendation attacks.

Even though the idea of using the Dempster-Shafer theory of evidence for trust management is familiar as presented by [56] and [57], the proposed work introduces certain novel features as follows:

(a) Misbehaviour detection and isolation model
(b) Hexagonal-Voronoi clustering model to form non-overlapping spatial reuse clusters
(c) Case study, i.e., application of cluster-based trust methodology in PKI security system for certificate revocation
(d) Security-related simulation parameters such as security level, attack model, the rate of detection, revocation time and revocation rate

In this paper, the self-organized security system is developed with trust as the quantifying factor on node's behaviour. To manage the challenges with node cooperation and security, hybrid trust management is proposed, where cluster heads (CH) are elected with low uncertainty level and high trust level. The novelty of the proposed work incorporates Voronoi clustering and Bayesian-Evidence trust management to predict the distributed security solution. The trust level of the neighbouring nodes is estimated with hybrid trust that combines direct and indirect trusts. This trust management

is validated to adapt the dynamic mobility of MANET nodes.

This paper is structured as follows. In section 2, the related works on trust management and its application in MANET are discussed, followed by the motivation in section 3. The system architecture for deploying trust scheme in MANET is mentioned in section 4. Section 5 describes the proposed mathematical model for trust management scheme. The proposed misbehaviours evaluation methodology is described in section 6 with the Voronoi clustering scheme in section 7. The case study of the proposed scheme is explained in section 8 followed by the attack mitigation model in section 9. The performance evaluation and simulations are illustrated in section 10 and the concluding remarks appear in section 11.

4 System architecture

The MANET functionalities are performed in a distributive manner due to lack of infrastructure. A two-dimensional bounded space is assumed to set for our dynamic and distributed trust computation, so that the nodes move freely and randomly around the network. The transmission ranges and the location of each node denote the neighbours within which the nodes perform their communication directly. Whereas, the communication, exterior to the transmission range are forward through intermediate nodes. It is difficult to obtain a completely authenticated public key pair in MANET even in the presence of various conventional authentication metrics.

The invasions from the adversaries make a node misbehave or malicious at any time during communication. Considering this as a significant issue, we propose a trust management and clustering model to enhance the security of PKI infrastructure in MANET. Apart from providing security, Voronoi diagram-based clustering improves the efficiency of trust model as well. The entire MANET region is clustered into a set of non-overlapping reliable and scalable hexagonal clusters with CH elected based on highest trust value by the members. The CH performs the complex functionalities and processes data in a collaborative fashion. With this cluster-based MANET model, monitoring and availability of each introduced node can be ensured in the network. Moreover, a misbehaviours evaluation methodology to analyse the direct observations and indirect evidences is proposed. The entire proposed system is secured with an attack-mitigating model which provides a defense mechanism for selfish and malicious node activities. We consider the selfish behaviour of the node as dropping packets in a group communication transmitted among the cluster nodes. Thus, even if the nodes behave selfishly, it cooperates to perform the public key management operations. The energy level of each node is set to its status.

The node's trust is assessed with direct and indirect information, where the indirect measurements are obtained from the one-hop neighbouring nodes of the target node called the recommenders. In our scheme, the recommenders are selected based on their trust level. We consider two main hypotheses for hybrid trust management. First, with the direct observations that revoke the untrustworthy node, the probability of selecting a trustable recommender gets higher. Second, the selection of higher trust recommenders conveys that those recommenders have participated constantly in group communication and are therefore familiar with the target node. However, the trustable recommenders are randomly selected to avoid undetected compromises which may dominate the communication of recommendations. The proposed system model is shown in Fig. 1: system architecture. The architecture includes the step by step processes of the proposed trust management, clustering and its application to construct a secured PKI framework in MANET. Initially, the MANET nodes are computed for its trustworthiness in terms of direct and indirect trust methods, i.e., with the Bayesian and Evidence theorems, respectively. The hybrid trust values are then combined with the Dempster-Shafer (DS) theorem. During this phase, the nodes are categorised into trustworthy and untrustworthy from which the trustable nodes are chosen

and forwarded for other network functionalities. The untrustworthy nodes are thus isolated and revoked from the system. The trustworthy nodes are grouped into hexagonal clusters in which the node with the highest trust value is elected as a header node in the next phase. To adapt mobility node registration and resignation, procedures are carried out whenever nodes join or leave the MANET clusters. Finally, the clustered trust platform is applied for public key functionalities in order to secure the MANET environment.

5 Proposed trust management method

This section describes the distributed trust computation method to adapt the active topology and to secure PKI-MANET system. The proposed trust methodology is assumed to deploy in the clustered environment with header and members nodes. Generally, the trust of a node can be defined as the probability of belief of a trustor (m) on a trustee (n), varying from 0 (complete distrust) to 1 (complete trust). The probability of trust and distrust of the trustor on information (i) send by the trustee with context to belief (b) is given as:

$$Trust\ Degree,\ TD(m, n, i, b) \tag{1}$$
$$= P[belief\ (m, i)|made\ By(i, n, b) \wedge beTrue(b)\ ]$$

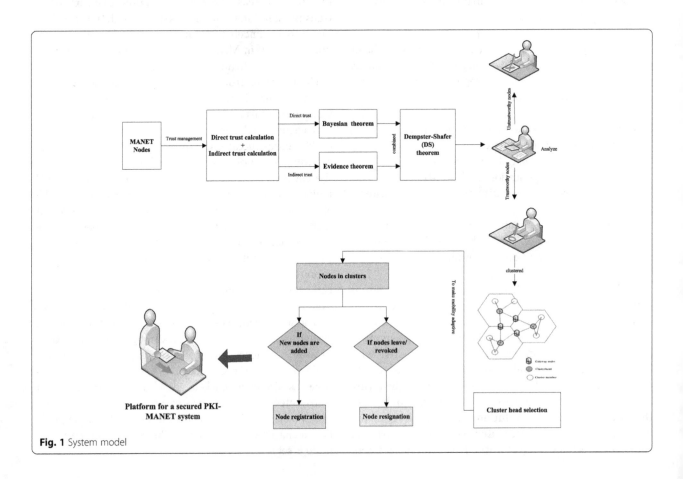

Fig. 1 System model

$$\begin{aligned} &Distrust\ Degree, DTD(m,n,i,b) \quad\quad\quad (2)\\ &= P[belief\ (m, \neg i)|made\ By\ (i,n,b) \wedge beTrue(b)\ ] \end{aligned}$$

5.1 Distributed trust management

The distributed trust is computed based on a hybrid method which combines the direct and indirect trust values. The direct trust is based on direct observations obtained by sending *SENSE* beacon constantly to the neighbouring nodes and evaluating these observations. Whereas, recommendations from the one-hop neighbour contributes to the indirect trust computation. The hybrid trust is computed by combining the direct as well as the indirect components. Unlike a centralized trust calculation, here, each node computes its own trust value on its neighbour. The trust computation of trustor x on trustee y, ($T_{x,\ y}$), by hybrid mechanism is given in Fig. 2: hybrid trust method. It is calculated as:

$$T_{x,y} = (1-f)\ T_{x,y}^{D} + f T_{x,y}^{ID} \quad\quad\quad (3)$$

where f is the trust component; $0 \le f \le 1$
$T_{x,\ y}^{D}$ is the direct trust made by m on n; $0 \le T_{x,\ y}^{D} \le 1$
$T_{x,\ y}^{ID}$ is the indirect trust made by m on n; $0 \le T_{x,\ y}^{ID} \le 1$

The direct trust computation is performed with the direct observations of x on y at time t is given by (4). The trust may decay with the change in the time (t_1), represented by the fading component δ.

$$T_{x,y}^{D} = \begin{cases} T_{x,y}^{D}(t) & ; \text{if hop count} == 1 \\ \\ \delta\ T_{x,y}^{D}(t-t_1) & ; \quad\quad \text{else} \end{cases}$$

$$(4)$$

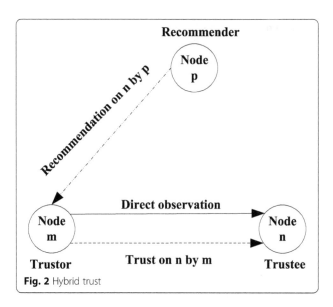

Fig. 2 Hybrid trust

The indirect trust evaluated by x on y with respect to the recommendation from one-hop neighbour of x (node k), at time t is given by (5). The trust reduces with t_1 when y receives false recommendations from a recommender (say node p) located within an appropriate trust length from y.

$$T_{x,y}^{ID} = \begin{cases} T_{k,n} & ; |TR| > 0 \\ \\ \delta T_{x,y}\ (t-t_1); & \text{else} \end{cases}$$

$$(5)$$

where TR is the set of true recommendations received from x's one-hop neighbour (i.e., k). When $TR > 0$, x appoints those neighbouring nodes to evaluate the trust indirectly. On the other hand, if $TR = 0$, y uses its previous trust value $T_{x,\ y}\ (t - t_1)$, since it received no true recommendations.

5.2 Direct and indirect trust management

Uncertainty is an unresolved problem in MANET, especially while evaluating the trust of the network. With the uncertainty, the nodes may misbehave due to selfish or malicious attackers. In each cluster, the cluster heads are authorized to monitor the misbehaviours locally and to collaborate the cluster members to further investigate the effect of misbehaviour on the network. When a cluster head detects a sign of misbehaviour from any node (say node x), it first evaluates the credibility of the message. Subsequently, the *CH* requests the cluster members, especially the one-hop neighbours of the suspicious node to share their individual observations about x. We consider these observations as evidences which are assembled to evaluate the evidence trust factor ($\mathbb{E}^x(e)$). Furthermore, the *CH* monitors the rate of misbehaviour by directly observing the node x as ($\mathbb{E}^x(d)$). The trust management systems combines these direct observations and the evidences obtained from the one-hop cluster members to evaluate the trustworthiness of x.

The trust management becomes more complex when the observing node (called recommender) itself behaves untrustworthy, which contributes false evidences. Such system makes MANET trust evaluation impracticable especially in detecting which recommender is untrustworthy. Therefore, we make use of the well-known Dempster-Shafer (DS) evidence theory, where the uncertainty of nodes is represented using belief functions. The main idea of the DS theory is that a recommender attains a certain degree of belief on a hypothesis based on the subjective probability. DS theory provides an appropriate mathematical model for MANET, to combine distributed information gathered from different sources.

5.2.1 Trust verification with the Bayesian theory

We consider that the *CH* monitors the packet forwarded by the suspected node and compare them with the original packets send directly to the node, in order to identify the misbehaviour nature of the node x. Let consider a node x maintains for its neighbouring node y. Then, for a set of nodes N, the *CH* supervises the packet ratio as in (6):

$$\sum_{x \in N} S_{xy} = \sum_{x \in N} F_{xy} \tag{6}$$

where S_{xy} is the number of packets forwarded to node x by the neighbouring node y and F_{xy} is the number of packets forwarded by node x. If the packet ratio is not equal, a misbehaviour is identified by the *CH*, i.e., if $\sum_{x \in N} S_{xy} \neq \sum_{x \in N} F_{xy}$, it is understandable that node x is misbehaving either due to selfish or malicious attackers.

Thus, the *CH* directly evaluate the misbehaviour and calculates the trust factor of its cluster members with a Bayesian inference, where the unknown probabilities are hypothesized using observations. The measure of belief about a hypothesis shall be represented by the well-known Baye's theorem:

$$P(i|j) = \frac{P(j|i)P(i)}{P(j)} \tag{7}$$

where $[i|j]$ is the measure of belief about the hypothesis (i) on the subject of the evidence (j)

$P[i]$ is the belief about a in the absence of j

In MANET, the higher the probability of any misbehaviour, the more likely it is that the misbehaviour will occur. Therefore, the Baye's theorem may be expressed in terms of probability distributions as:

$$P(\delta|data) = \frac{P(data|\delta)P(\delta)}{P(data)} \tag{8}$$

where $[\delta|data]$ is the posterior distribution for the parameter δ, $P[data|\delta]$ is the sampling density function, $P[\delta]$ is the prior distribution and $P[data]$ is the marginal probability function of data.

From (8), we shall modify the misbehaviour verification as:

$$P(\delta, a|b) = \frac{f(b|\delta, \alpha)P(\delta, \alpha)}{\int_0^1 f(b|\delta, \alpha)P(\delta, \alpha)d\delta} \tag{9}$$

where degree of belief and $0 \leq \delta \leq 1$, b is the rate of correctly forwarded packets by a node, α is the rate of packets received by the node, $f(b|\delta, \alpha)$ is the probability function that follows a binomial distribution given by

$$f(b|\delta, \alpha) = \binom{\alpha}{b} \delta^b (1-\delta)^{\alpha-b} \tag{10}$$

To describe the initial knowledge concerning probabilities of success, we use the beta distribution to the Bayesian approach and hence the prior distribution $P(, , i)$ can be stated as:

$$ʒ(\delta; \alpha, \beta) = \frac{\delta^{\alpha-1}(1-\delta)^{\beta-1}}{\int_0^1 f(b|\delta, a)P(\delta, a)\,d\delta}$$

where $\alpha, \beta > 0$, is the power function of ⊣ and b.

The mean and variance of the beta distribution function is given as:

$$M(\delta|\alpha, \beta) = \frac{\alpha}{\alpha + \beta} \tag{12}$$

and

$$V(\delta|\alpha, \beta) = \frac{\alpha\beta}{\alpha + \beta + 1} * \frac{1}{(\alpha + \beta)^2} \tag{13}$$

In our scheme, the trust factor represents the behaviour which grows feebly, thereby giving more impact on the misbehaving rate in Bayesian networks. The trust factor for misbehaviour verification is given as:

(12)⇒

$$M(\delta|\alpha, \beta) = \frac{\alpha}{\alpha + \alpha^x \beta} \tag{14}$$

The beta distribution is well suitable for the random behaviour of proportions. While considering the event history in the Bayesian framework, the expected value of beta distribution can be written as

(14)⇒

$$M(\delta|\alpha, \beta) = \frac{\alpha_t}{\alpha_t + \alpha^x{}_t \beta_t} \tag{15}$$

where

$$\alpha_t = \alpha_{t-1} + i_{t-1}$$
$$\beta_t = \beta_{t-1} + b_{t-1}$$

and with the prior probability distribution, we assume no observations are made initially and so α_0, $\beta_0 = 0$. Therefore, the direct trust factor that quantifies the behaviour of node x is deduced from the above calculations as:

$$T_x{}^D(t) = (\mathbb{E}^x(d)) = M(\delta|\alpha, \beta) \tag{16}$$

The accuracy of the proposed direct trust evaluation is improved by calculating the rate of correctly forwarded packets (b), which is incremented by one for each successful transmission. If the rate b is not increased, either due to unreliable network conditions or packet lifetime, the packets are considered as dropped and so discarded from the communication. Algorithm 1 describes the accuracy of direct calculation trust in the Bayesian framework.

Algorithm 1: Direct Trust Computation

1:    if the trustor node m, observes its neighbour trustee node n receives packets then
2:        the rate of packets received is incremented by 1
3:        if node m identifies packets forwarded by node n is correctly done then
4:           rate of correctly forwarded packets (b) is incremented by 1 else
5:           if the packet lifetime is set to 0
6:           then
7:           rate of correctly forwarded packets (b) is decremented by 1
8:           end if
9:        end if
10:     end if
11:    Compute and update the direct trust with $T_x^D(t)$ with equation (16)

5.2.2 Misbehavior verification with evidence theory

This section describes the misbehaviour verification with respect to the recommendations for the suspicious node x from the one-hop neighbours within each cluster. The cluster head requests the one-hop neighbours of x referred as recommenders, to verify the misbehaviours based on their independent observations, as shown in Fig. 3: indirect misbehaviour verification. The recommendations called evidences received from the cluster neighbours give assistance in evaluating the trust value of x. The DS theory is used in practice with uncertainty or ignorance to evaluate the value of trust. This theory utilizes a belief function to combine the indirect evidences, which reflects the subjective probabilities.

The probabilities which are mutually exclusive and exhaustive are computed as a set of functions with Φ

as a frame of discernment, in the DS evidence system. By including all the probabilities of the hypothesis called focal values P_k as a function of m, we consider a power set 2^{Φ} and satisfy the conditions as follows:

1. The probability value of the null set is zero, i.e., $\mathcal{M}(\delta) = 0$.
2. The sum of all elements in the power set is 1, i.e., $\sum_{P_k \subseteq \Phi} \mathcal{M}(P_k) = 1$

The belief function of subjective probabilities shall therefore be defined as

$$F(x) = \sum_{P_k \subseteq x} \mathcal{M}(P_k) \tag{17}$$

In the proposed trust management scheme, we consider two node behaviour states, i.e.,{accept, evict} represented with the DS theory. Using this, the frame of discernment is included with a set of probability pair regarding the behaviour of any random node. That is $\Phi = \{trust, distrust\}$; where '$ust$' represents the trustworthy behaviour of the node and '$distrust$' represents the misbehaving node state which occurred in the presence of selfish and malicious attackers.

On considering Fig. 3, the neighbours node A, B and C of the suspicious node x at a hop distance equal to 1 shares their evidences with the CH, as a subset of Φ. We interpret the power set with three probability forms of proposition, i.e., proposition $T = \{trust\}$, proposition $M = \{distrust\}$ and finally proposition $H = \Phi$, which represent the uncertainty state where node x is uncertain whether to include as acceptable or misbehaving state. The neighbours provide recommendations as evidences by sharing its belief over Φ.

Consider an example, if node A believes that node x behaves trustworthy, then $\mathcal{M}_A(T)$ is $\mathbb{E}^x(A)$ and therefore $\mathcal{M}_A(M)$ is 0. The evidence from node A can be stated as:

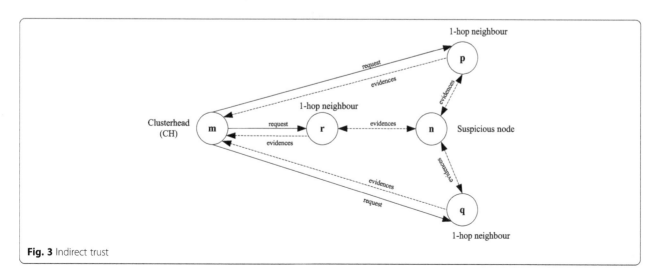

Fig. 3 Indirect trust

$$\mathcal{M}_A(T) = \mathbb{E}^x(A)$$
$$\mathcal{M}_A(M) = 0 \qquad (18)$$
$$\mathcal{M}_A(H) = 1 - \mathbb{E}^x(A)$$

Likewise, if node B believes that node x misbehaves, its recommendations favours the evict function as follows:

$$\mathcal{M}_B(T) = 0$$
$$\mathcal{M}_B(M) = \mathbb{E}^x(B) \qquad (19)$$
$$\mathcal{M}_B(H) = 1 - \mathbb{E}^x(B)$$

5.2.3 DS theory of combining evidences
In the proposed trust management scheme, the DS theory combines all the recommendations of one-hop neighbours with the condition that the recommendations are independent. Suppose $F_1(x)$ and $F_2(x)$ are belief functions of two independent recommending nodes, over the same suspicious node, then the orthogonal sum of these belief functions is given and represented as:

$$F(x) = F_1(x) + F_2(x)$$
$$= \frac{\sum_{j,k,P_j \cap P_k = x} \mathcal{M}_1(P_j) * \mathcal{M}_2(P_k)}{\sum_{j,k,P_j \cap P_k \neq \Phi} \mathcal{M}_1(P_j) * \mathcal{M}_2(P_k)} \qquad (20)$$

where $P_j, P_k \subseteq \Phi$.

With reference to Fig. 3, the belief of node A and B is calculated as

$$\mathcal{M}_A(T) \oplus \mathcal{M}_B(T)$$
$$= \frac{1}{I}[\mathcal{M}_A(T)\,\mathcal{M}_B(T) + \mathcal{M}_A(T)\mathcal{M}_B(H) + \mathcal{M}_A(H)\,\mathcal{M}_B(T)]$$
$$\mathcal{M}_A(M) \oplus \mathcal{M}_B(M)$$
$$= \frac{1}{I}[\mathcal{M}_A(M)\,\mathcal{M}_B(M) + \mathcal{M}_A(M)\mathcal{M}_B(H) + \mathcal{M}_A(H)\,\mathcal{M}_B(M)]$$
$$\mathcal{M}_A(H) \oplus \mathcal{M}_B(H) = \frac{1}{I}[\mathcal{M}_A(H)\,\mathcal{M}_B(H)]$$

$$(21)$$

where

$$I = \mathcal{M}_A(T)\,\mathcal{M}_B(T) + \mathcal{M}_A(T)\mathcal{M}_B(H)$$
$$+ \mathcal{M}_A(H)\,\mathcal{M}_B(H) + \mathcal{M}_A(H)\,\mathcal{M}_B(T)$$
$$+ \mathcal{M}_A(H)\,\mathcal{M}_B(M) + \mathcal{M}_A(M)\,\mathcal{M}_B(D)$$
$$+ \mathcal{M}_A(M)\mathcal{M}_B(H) \qquad (22)$$

We assume the rate of acceptance of the probability of node A and B as 0.8 and 0.7, respectively, and thus,

$$F(T) = 0.94$$
$$F(M) = 0$$
$$F(H) = 0.6$$

Thus, we shall conclude the acceptable behaviour rate from the indirect evidences with DS theory is 0.9. By combining all the belief values, we get

$$T_x^{ID}(t) = \mathcal{M}_A(T) \oplus \mathcal{M}_B(T) \oplus ..... \oplus \mathcal{M}_N(T) \qquad (23)$$

where nodes A, B, ...N are one-hop recommenders of node x.

Therefore, the evidence trust value obtained from the recommendations can be computed as

$$T_x^{ID}(t) = (\mathbb{E}^x(e)) = F(x) \qquad (24)$$

The indirect trust evaluation with Evidence theory and DST is depicted in Algorithm 2.

Algorithm 2: Indirect Trust Computation

```
1:   if the trustor node m and trustee node n has more than one recommenders,
2:   then
3:       Calculate T_x^{ID}(t) from equation (23)
4:   else
5:       set T_x^{ID}(t) = 0
6:   end if.
```

6 Proposed misbehaviours evaluation methodology
Unlike other hybrid trust computation methodologies, to improve the precision of measurement this section evaluates the misbehaviours obtained from the direct and indirect trust mechanisms as follows.

Due to the unique characteristics of MANET, nodes move independently without restrictions. In such environment, misbehaviour is more likely to appear due to selfish or malicious nodes. The selfish nodes are characterized by their disinclination to spend resources to cooperate on a group communication. On the other hand, the malicious nodes attack the availability of the network through flooding, wormhole, black hole, rushing and denial of service (DoS). The misbehaviour verification process of the proposed scheme includes two main phases: evaluating and revocation. In the first phase, the hybrid trust values of the misbehaving nodes are evaluated with a vector model. During this detection phase, the misbehaviours are classified into selfish or malicious based on their characteristics. In the next phase, the misbehaving nodes are revoked based on the analysis.

To detect and isolate a misbehaving node, we use a trust evaluation vector (TEV) to configure the mobile nodes, which is given as:

$$TEV(A \rightarrow B) = [D_{AB}, ID_{AB}] \qquad (25)$$

where D_{AB} and ID_{AB} are direct and indirect trust

evaluation of node A on node B. In order to normalize the value of TEV, we define

$$|TEV(A{\rightarrow}B)| = \mathbb{W}_A \otimes TEV(A{\rightarrow}B)$$
$$= [\mathbb{W}_D, \mathbb{W}_{ID}] \otimes [D_{AB}, ID_{AB}]$$
$$= \mathbb{W}_D * D_{AB} + \mathbb{W}_{ID} * ID_{AB} \quad (26)$$
$$= T_{A,B}$$

$$(\{D_{AB}, ID_{AB}\}) \in [0,1], \{\mathbb{W}_D, \mathbb{W}_{ID}\} \in [0,1]$$

where $\mathbb{W}_A$ is the trust vector of node A and $T_{A,B}$ is the trust value of node A on node B.

The direct trust value of any suspicious node is evaluated as:

$$D_{AB} = \frac{\overline{PC_B}}{PC_B} = \frac{PC_B^{out} - PC_{B,A}}{PC_B^{in} - PC_{A,B}} \quad (27)$$

where PC_B is the total packet count that node B have to forward,

$\overline{PC_B}$ is the total packet count that node B actually forwarded,

PC_B^{in} is the total packets forwarded to node B,

PC_B^{out} is the total packets forwarded by node B,

$PC_{B,A}$ is the total packets forwarded from node B to node A

and

$PC_{A,B}$ is the total packets forwarded from node A to node B

Now, the indirect trust value of the suspicious node is evaluated as:

$$ID_{AB} = \frac{\sum_{R \in \mathrm{g}} |TEV(A{\rightarrow}R)| * |TEV(R{\rightarrow}B)|}{\sum_{R \in \mathrm{g}} |TEV(A{\rightarrow}R)|} \quad (28)$$

where R is the recommender node which is an element of the set of recommenders represented by g. In MANET, the cluster membership changes dynamically whenever a node is added to evict from the cluster. The new nodes are added and registered into the cluster with trust verification, whereas the evicted nodes are deleted from the cluster. This is to maintain the forward and backward secrecy of the mobility aware cluster. Another significant challenge that MANET faces with this membership reformation is the re-evaluation of trust within each cluster. Let us consider initially, at time t, the node A places a trust $T_{A,B}(t)$ on its neighbouring node B. With the change in mobility, at time t_1, the node B leaves the current cluster and joins an adjacent cluster. The node B is now resigned from the particular cluster. With the progress in time and mobility, the node B may re-join the home cluster of node A during which eventually decays the trust value $T_{A,B}(t)$. This time and mobility dependent trust value can be evaluated as:

$$T_{A,B}(t_1) = T_{A,B}(t) * e^{-\left(T_{A,B}(t) \triangle T\right)^2} \quad (29)$$

where $\triangle T = t_1 - t$ and x is an integer, where $x \geq 1$.

Let S be the event that a suspected node is selfish and $\overline{S}$ be the event that the node is normal with density function $P(x|R)$ and $P(x|\overline{R})$. By Baye's theorem, we compute a prior probability function as:

$$P(S|x) = \frac{P(S)P(x|R)}{P(R)P(x|R) + P(\overline{R})P(x|\overline{R})} \quad (30)$$

while considering the ratio of prior probabilities which is written as:

$$\mathbb{p} = \frac{P(S|x)}{P(\overline{S}|x)} \quad (31)$$

If the ratio of probabilities is less than one, i.e., $\mathbb{p} < 1$, the nodes are considered not to be normal than to selfish. Additionally, in the proposed trust management scheme, a malicious node test is incorporated to detect the malicious activities in the clustered MANET. Using the Baye's theorem, we calculate the malicious events as:

$$P(M|p) = \frac{P(M)P(p|M)}{P(M)P(p|M) + P(\overline{M})P(p|\overline{M})} \quad (32)$$

where M be the event that a node behaves malicious, $\overline{M}$ be the event that a node behaves normally and p be the event that malicious test is positive. If the value of $P(M|p) \geq 0.5$, it is concluded that the suspected node is more likely not to be a malicious node.

Thus, the misbehaviour is detected by evaluating the hybrid trust value with the trust evaluation vector method. This detection mechanism shall be effectively integrated into the hexagonal clusters in order to secure the PKI framework. The mechanism detects and classifies the misbehaviour, either selfishness or malicious, to take revocation actions on those nodes.

7 Proposed clustering methodology

This section describes the distributed trust-based clustering framework to adapt the active topology and to secure MANET. An efficient clustering scheme is designed with the ad hoc environment to form stable clusters for the underlying network operations. To adapt the dynamic mobility of MANET, the diameter of the cluster should be flexible, and so herein, we use hexagonal shape non-overlapping clusters. In the proposed scheme, each cluster has exactly one CH elected based on trust value as shown in Fig. 4: hexagonal clusters. The nodes in the boundary region and within the transmission range of any two CH

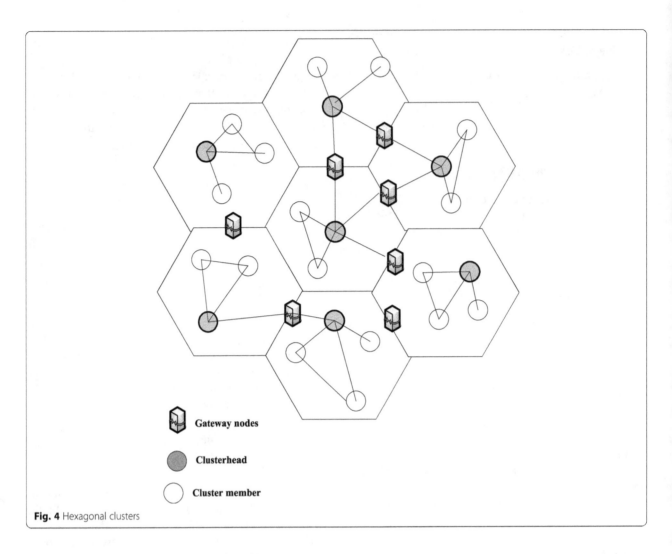

Fig. 4 Hexagonal clusters

are considered as gateway nodes, which handles cluster-to-cluster operations. The CH monitors its neighbour nodes with their trustworthiness, within each cluster. We assume all the nodes communicate through bi-directional channels so that each node can forward as well as hear from its neighbouring nodes.

In an ad hoc uncertain clustering (UC) model, it has been assumed that a node n_i should be located inside a region with a probability density function (PDF) to describe the distribution of nodes within a region. To compute the closeness of the node and the cluster representative, different methods based on mean, Euclidean distance and probability have been in practice. However, these traditional clustering techniques of uncertain nodes increase the computational complexities and communication cost in mobile environment, especially in mobile ad hoc networks. To construct a highly desirable uncertain clustering cell in MANET, we propose to use Voronoi diagrams (VD) based clustering in which the clustering issues are managed considering the drawbacks of existing UC methods.

Voronoi diagrams are applied for wireless application to compute the Voronoi region of each node. To increase the spatial reuse, the network areas are clustered into congruent polygons with Voronoi geometric features. A hexagonal spatial geometric distribution of nodes is utilized in order to increase the network capacity and throughput of the network. It was proven the regular hexagons have the flexibility to be partitioned into smaller hexagonal shapes and grouped together to form larger ones.

In MANET, VD is used to partition network into clusters based on Euclidean distances to nodes in a specific subset of the plane. A Voronoi diagram represents the region of influence around each of a given set of nodes. This geometric structure partitions the entire plane into polygon cells, called Voronoi polygonal, formed with respect to n nodes in a plane. In recent years, this structuring concept is widely used for exploring location and routing based issues. The Voronoi partition or cluster for a given set of nodes is unique and produces polygons which are route connected. A Voronoi polygon is, traditionally, constructed as follows

$$V_{(n_i)} = \left\{y \middle| d(n_i, y) \le d\left(n_j, y\right); i \ne j\right\} \quad (33)$$

where $V_{(n_i)}$ is the Voronoi polygon of n_i, n_i is the node and y is the set of points closer to n_i, $d(n_i, y)$, distance from point y and n_i and (n_j, y) is the distance from point y and n_j.

7.1 Cluster construction

Consider N as the number of nodes distributed independently and uniformly in a regular hexagon with distance between them as d, radius of the hexagonal cluster as r and $R \in E^2$, where E^2 denotes the 2D Euclidean space and R denotes an arbitrary point in the hexagon. The probability distribution of d is given as $P(d \le r)$.

In the first step, Voronoi clusters (VC) are constructed on a set of nodes $N = \{n_1, n_2 \ldots \ldots n_k\}$ with a distance function $d : S^m \times S^m \to S$ (m-dimensional space) giving the distance $d(x, y) \ge 0$ between any nodes $x, y \in S^m$. The VD partitions the space S^m in k cells with cluster representatives $C = \{c_1, c_2 \ldots \ldots c_k\}$ with the property as:

$$d(x, c_i) < d(x, c_j) \forall x \in V(c_i), c_i \ne c_j$$

In the second step, the distance between the nodes and a cluster node is calculated. The Voronoi partitioning of a network can be of any polygonal shape and for its beneficial geometrical characteristics, we assume that the uncertainty region of N_i is a regular hexagon with nodes whose centers are equidistance to each other by distance d and radius r, where $r > 0$. The hexagonal clustering partitions a larger area into adjacent, non-overlapping areas and can be subdivided into smaller hexagons. Nodes join to form hexagonal clusters, and each cluster consists of CH and cluster members (CM). The distance $d(a, b)$ between nodes in MANET plays an important role in determining the network performance. We shall assume that the nodes of the ad hoc network are independently and randomly distributed in the hexagonal structure. The edges of the hexagonal polygon are perpendicular to the line joining a node with another in N. Considering the radius for any query point, $\in S$, $d(x, c_i)$ can be written as:

$$d(p, c_i) - d\left(p, c_j\right) = r_i + r_j \quad (34)$$

If two nodes overlap, the distance $d(n_i, n_j) < r_i + r_j$ and (34) become unreal, which means the edges cannot be found, and we consider the cluster as empty. The hexagonal cluster construction in the MANET as shown in Fig. 5 is illustrated in Algorithm 3.

Algorithm 3: Proposed Cluster Construction

Input: Nodes $N = \{n_1, n_2 \ldots \ldots \ldots n_k\}$

Output: Clusters $C = \{C_1, C_2 \ldots \ldots \ldots C_k\}$

1.   for each $n_n \in N$ do ;

2.   The VD for cluster construction consider an expected region of node n_i and the neighbouring region of VC edge $E_n(m)$. The expected region of n_i, denoted by E_{r_i} is the intersection of all the internal regions. ie.,

$$E_{r_i} = \cap_{j=1 \ldots |E| \wedge j \ne i} \overline{X_n(m)} \quad (35)$$

where the neighbouring region, $X_n(m)$ is the region on one side of the cluster cell edge $E_n(m)$ and $|E|$ is the empty set.

3. $E_{r_i} \leftarrow S^m$; initialize expected region

4.   for each $n_m \in N \wedge m \ne n$, do

5.   The clustering polygon can be generated by excluding all the neighbouring regions from the domain space. The overlapped regions are reduced to generate the expected region E_{r_i}.

6. $E_n(m) \leftarrow$ VC edge of n_n ; compute edge of Voronoi cluster

7. $N_n(m) \leftarrow$ neighbour of $E_n(m)$;compute the neighbour

8. $E_{r_i} \leftarrow E_{r_i} - N_n(m)$;reduce overlap

9.   end for

10. For each node n_j, we verify the expected region lie inside a Minimum and Maximum Region Bounding (MinMax-RB) of the domain space.

11. if $E_{r_i} \subseteq$ MinMaxRB, do

12. Let us consider six equilateral triangles in a regular hexagon. For the calculation we take a single equilateral triangle $\triangle OAF$. A circle with center c_n and radius r_n is assumed to intersect the $\triangle OAF$.

13. $C_n \leftarrow E_{r_i}$;assign expected region as cluster

14. Considered as neighbouring regions $N_n(m)$ and the region where the area of the circle and the neighbouring region overlap as overlap region O_i (ie., $O_i(x, y) = O_1 + O_2 + O_3$).

15. Calculate probability of the expected region E_{r_i} in a hexagonal cluster with area A and (x, y) as co-ordinates of any random node is given as

$$P_{E_{r_i}} = \frac{1}{A^2} \iint [\pi r_n^2 - \Sigma_{i=1}^6 O_i(x, y)] \, dx \, dy$$

$$P_{E_{r_i}} = \frac{\pi r_n^2}{A} - \frac{6}{A^2} \iint O_i(x, y) dx dy \quad (36)$$

16. end if

17. end for.

7.2 Cluster head selection

In MANET, the nodes join or leave the cluster dynamically, and thus, the CH selection is difficult. We consider a distributed cluster head selection procedure with n nodes, which are of h hops distance within a cluster. It is much easier to select an efficient mechanism to establish security, if the trust relationship among the nodes is obtainable for every cooperating node. Hence, to provide a secured communication amongst cooperative nodes, it is important to calculate the trust and distrust levels of nodes in the network.

In order to measure the trust level explicitly in an ad hoc environment, we present a trust calculation method with uncertainty level. With this, a high level of trust can be achieved for a secured communication. The certainty of nodes in MANET is considered as the summation of trust and distrust levels. Consequently, thus, the uncertainty level (UL) is defined as

$$UL(m, n, i, b) = 1 - certainity \ of \ nodes \quad (37)$$

The uncertainty impacts the node's anticipation of neighbour's behaviour and decisions during communication;

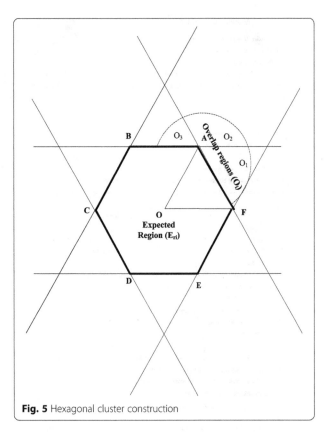

Fig. 5 Hexagonal cluster construction

$$UL(t,s,i,b) = 1 - \left[M(\delta|\alpha,\beta) \; * \; \frac{\sum_{x=1}^{n} d_p(x)}{H_e} + (\mathbb{E}^x(e)) \; * \; \frac{\sum_{x=1}^{n} d_n(x)}{H_e} \right]$$

$$(39)$$

The degree of successive encounter $'x'$ made be trustee on trustor may be either positive (represented as $d_p(x)$) or negative (represented as $d_p(x)$). Here, to evaluate the trust, we consider three cases of uncertainty level, i.e., =0, $0 < UL < 1$ and $UL = 1$. When the uncertain level is low ($UL = 0$), the nodes are highly trustable. This highly certain case shows that the trustor is very much confident with the trustee. If the uncertain level varies from low to high ($0 < UL < 1$), the trustor may not have sufficient confidence with the trustee. On the other hand, a highly uncertain case occurs when the uncertain level $UL = 1$. At this state, the trustor may be completely unknown about the trustee.

The nodes with the highest trust level, i.e., $UL = 0$ and $TL = 1$, is considered as CH, initially at time T_1. As time progresses, the topology changes frequently in a MANET that varies the cluster nodes and the cluster heads. Hence, the cluster head selection procedure is adaptable for the change in topology. The trust value of each node is recomputed and the CH is selected, comparing the current CH (CH_c) with the previous CH (CH_p) and location (L_p).

The nodes with trust level between 0 and 1(i.e., $0 < UL < 1$) have undergone a distrust test to reduce the rate of risks. In comparison with the trust level and the distrust level of such nodes, they are either revoked or considered as cluster members, i.e., the nodes with the highest distrust level ($DL = 1$ or $DL > TL$ and $UL = 1$) are revoked and the remaining nods are assigned as CH. This trust-based cluster head selection as shown in Fig. 6 eliminates a certain amount of risk in communication within the network. To perceive the exact location information of any node, each node in the network is enabled with a position identification system. Our proposed scheme makes use of the clusters as well as the location information intensively. To construct a mobility adaptive MANET, nodes are either registered or resigned whenever the cluster membership changes.

8 Case study: application of cluster-based trust in PKI MANET systems

The PKI-based security architectures are being actively investigated to ensure the integrity of node-to-node messages. The basic strategy in PKI-based security is to equip nodes with asymmetric cryptographic key pairs (public key, private key) and certificates issued by a trusted certification authority (CA). The certificates are used to authenticate the genuine nodes for communications. The other desirable property of the PKI-based security scheme is certificate revocation. That is, the

we include uncertainty in the trust management system. It represents whether a trustor node collected the required information from past communications with a trustee and its confidence in that communication. An efficient method to reduce the uncertainty is to exploit the mobility characteristics of the MANET. The node mobility can increase the propagation of direct and indirect measurements and hence accelerates the trust convergence.

An important factor that affects the trust level of a node is the history of events (H_e), which specifies the number of successive interactions between the trustor and the trustee in a network. Initially, we assume H_e as greater than or equal to 0. The trust and the distrust level of any node can be measured with the relation as shown in (38).

$$TL(m,n,i,b) = M(\delta|\alpha,\beta) \; * \; \frac{\sum_{x=1}^{n} d_p(x)}{H_e}$$

and

$$DL(t,s,i,b) = (\mathbb{E}^x(e)) \; * \; \frac{\sum_{x=1}^{n} d_n(x)}{H_e} \qquad (38)$$

Therefore, (37)$\Rightarrow$

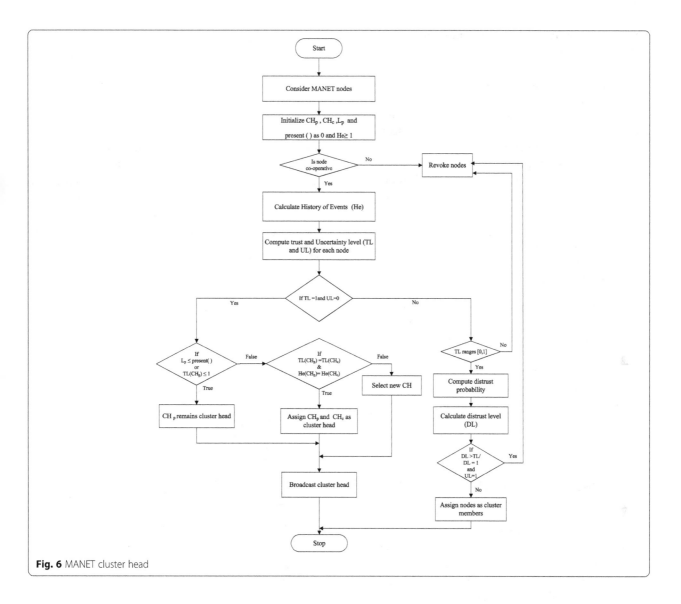

Fig. 6 MANET cluster head

certificates of a detected attacker or malfunctioning vehicles can be revoked. The most common way to revoke certificates is the distribution of CRLs (Certificate Revocation Lists) that contain the most recently revoked certificates. The nodes in a secured group communication in ad hoc networks participate until the certificates are valid. A certificate is said to be valid if it has not expired and it is not revoked by the CA. Checking the revoked status of any certificate involves acquiring the CRL corresponding to that certificate (i.e., the CRL with the CRL series number specified in the certificate). When transmitting a message, the sender appends to the message the following: (a) the sender's certificate and (b) the signature of (the hash of) the message using the sender's private key. When receiving a message, the receiver (a) verifies the validity of the sender's certificate and (b) verifies the signature on the message

(using the sender's public key that is a part of the sender's certificate) before accepting it.

In traditional PKI system, single CA maintains the certificate authorization and complete CRL list for the entire network. Such a structure can be delayed prone and also maintaining such an infrastructure that is a high-speed wired connection from CA to cluster heads and then headers to the nodes may add up the infrastructural cost to a large extend. Revocation checking can be problematic in these structures, and since all the revoked certificates in the entire network are listed in a single CRL, the number of entries on that CRL can become quite large. A large CRL takes significant bandwidth as well as computational resources to check the revocation status of a particular node also, and the amount of revocation information that can be stored at a CH is limited by the memory available at the CH.

Therefore, it is clear that the complexity of the PKI system should be minimized in order to make the PKI-based security viable for node to node security deployment. In this pursuit, we propose a trust-based certificate revocation for use in ad hoc networks with significant reduction in the cost. In addition, we ensure the infrastructural complexity does not grow further in order to improve the performance of the PKI-based security framework; in particular, it reduces the load on the wireless communication medium for disseminating the certificates and CRLs. The network is initialized as follows:

1. CA chooses a secret polynomial function F_i and private key K_s, where $F_i = \sum_{x=1}^{t-1}(F_x i^x) \ mod \ m$, with coefficient F_x and variable i^x.
2. CA computes a secret share key for group communication and broadcasts through secured channel to the group members as $K_i = F(n_i)$, where n_i is the identity of group members.
3. CA constructs a polynomial function f_x^m by interpolation of points for each clusters, to determine the public information. The polynomial is constructed as $f_x^m \Rightarrow (dk_{i \mapsto m}, Encrypt_{H_1(ek_i \ A_m)}(ek_m)$
4. Each group node computes its share key as $k_{s_i} = \sum_{i=1}^{n} f_x^m (I_i)G$, where $f_x^m(I_i)$ is the encrypted subshare with $I_i = H_1(n_i)$ and G be the generator of G', an additive cyclic group of order q.
5. Each node verifies the integrity of the secret value as $K_i \Theta \ G = \sum_{y=0}^{x-1}(x^y F_y)$.

The revocation process with hybrid trust is performed within each cluster whenever misbehaviours are identified. It is important to evaluate the trust to authenticate and manage certificates in PKI system. Therefore, the application of proposed hybrid trust management in the public key functionalities is significant to provide soft security for a secured group communication. The node's trustworthiness determines the revocation rate. The revocation rate depends on the number of revocations made against node n_i, as well as the number of attacker node n_i made. If a number of uncertainty states are made against a member, it is likely that this member might be a misbehaving node. During such cases, the certificate of the accused node is revoked by the CA and the revocation information is distributed within each cluster. This paper presents an efficient method of revoking certificates by quantifying the trustworthiness of nodes to construct trust framework in PKI without assessing the PKI structure. Compared with the conventional methods, this scheme has lower revocation time and higher revocation rate in order to guarantee a secured MANET framework. Now, the revocation list cost of single cluster is given as:

$$Cost_{revoke} = \frac{Q}{T} * \frac{\frac{3\sqrt{3}a^2}{2}}{A} * L_{revoke} + \left(1 - \frac{\left(\frac{\sqrt{3}}{2}a - x\right)^2}{a^2}\right) * \frac{\frac{3\sqrt{3}a^2}{2}}{A}$$
$$* T_{x,y} * L_{revoke}$$
(40)

where Q is the estimated number of certificates that will eventually be revoked prior to expiration, T is the number of time slots for which a certificate is issued.

$\frac{Q}{T}$ is the average number of certificates revoked per time slot, L_{revoke} is the length of the revoked message corresponding to each revoked certificate.

$\frac{A}{\frac{3\sqrt{3}a^2}{2}}$ is the number of hexagonal regions with area of overall region as A.

The revocation mechanism is described in the Algorithm 4 as:

Algorithm 4: Revocation model

1. When a node is accused as misbehaving, the CA performs as revocation coordinator to evaluate and isolate the misbehaving nodes.
2. The CA broadcasts the revocation request '$REVOKE_{req}$' to all the cluster members. The $REVOKE_{req}$ packet includes the revoked node identity, its certificate, time stamp and public parameter, k_{pub}^{CA}, where $k_{pub}^{CA} = K_s * G$.
3. The cluster members on receiving the $REVOKE_{req}$ message initially verify the signature of CA using its public parameters and check the revocation time stamp to ensure the freshness of operations. Each reviver node verifies the hybrid trust value of the accused node to check its trustworthiness.
4. Each node replies CA with '$REPLY_{revoke}$' with its identity, certificate, time stamp and public parameter $A_i = a_i G$, where random number $a_i \in Z_q^*$, when the accused node is found untrustworthy.
5. When the CA receives $REPLY_{revoke}$ from a member, it verifies the trustworthiness of the sender node with its trust table. The nodes whose $TL \geq threshold \ (TL_{th})$ are allowed to contribute in the revocation and if $TL < threshold \ (TL_{th})$, those nodes are excluded from the revocation.
6. The CA constructs revocation function with all the $REPLY_{revoke}$ obtained from the trusted members using Lagrange Polynomial Interpolation as: $F_i = \sum_{x=1}^{t}(revoke)_{sign_{K_i}} \prod_{k \neq x}^{t} \frac{i-k}{x-k} \ mode \ m$.
7. The CA broadcasts the revocation information $REVOKE_{info}$ within the corresponding clusters. The $REVOKE_{info}$ includes the revoked node identity, its certificate and time stamp.
8. Any member receiving $REVOKE_{info}$ verifies the time updates to ensure the freshness of revocation.

9 Attack mitigation model
The attacker capabilities that affect the system are enumerated as follows:

- Attackers can control the group communication between the nodes and CA
- Attackers can modify/alter the message in group communication.
- Attackers can remove or add messages, shared among the group members.
- Attackers can be an identity spoofing, node cloning, reply or an unauthorized access.
- Attackers can remotely access CA for altering the shared parameters.
- Attackers can flood the packet to consume larger resources.

- Attackers falsely send recommendations to create an untrustworthy network.

We consider the following attacks that affect the trust computation

- *False recommendation attack* falsely sends recommendations to include an untrustworthy node in the cluster functionalities. The hybrid trust calculation we used measures the direct trust from direct observations, in addition to the indirect trust obtained in the form of recommendations. This direct trust value gives higher importance for analyzing the trustworthiness of any node, which degrades fake recommendations.
- *Impersonation attack* can be an identity spoofing, node cloning, reply or an unauthorized access. However, the attackers fail to pass the source and location authentication as well as integrity check.
- *Packet dropping attack* interrupts the service availability of the nodes. The attackers deactivate nodes from their cluster by making a connection failure or cluster disconnection. The *SENSE* beacon send by the *CH* during node missing, re-establishes the connection with the deactivated node, after verification process.
- *Flooding attack* resends replicate of packets received previously from the node members. This flooding consumes larger bandwidth and power that might terminate network functionalities.
- *Sybil attack* can break down the security, when a node in the network claims multiple identities. The integrity check of the node gets rid of such attackers, where the honesty of that node is proved. Also the *CH* records the location, history of each node, which aids it to detect the attacker node with multiple identities and same location particulars.

Besides, we consider the attacks that generate with the malicious and selfish node behaviours, such as flooding attack, wormhole attack, black hole attack, rushing attack and denial of service (DoS). These attacks are mitigated with the misbehaviour evaluation mechanism explained in section 6. The potential countermeasures proposed to isolate these attacks are as follows:

- *Black hole attack/wormhole attack*: By ensuring trust-based secure packet transmission in group communication selects reliable routes that mitigate black hole attacks. This authenticated routing protects routing messages from unauthorized modifications.

- *Impersonation*: To prevent identity theft in the PKI MANET system, an effective access control mechanism is provided by hybrid trust, by which stronger authentication and authorization is achieved.
- *Dropping attack*: The two-level security, i.e., cryptographic and soft securities, provided in the proposed scheme detects and prevent the packet drop attacks. By monitoring the packet send and the packet delivery ratio, the presence of attackers is identified here.
- *Flooding attack*: The proposed distributive self-organised scheme runs the trust management code in cooperative fashion to identify and isolate flooding attackers in PKI MANET system. By categorising the nodes as {trust, distrust, uncertain}, the probability of malicious behaviour is identified in which the packets from distrust nodes are isolated. To prevent packet flooding, a threshold level is set by each node to accept packets from its neighbours.
- *Sybil attack*: It is detected by cooperative monitoring of MANET nodes. With authorised certificates, the integrity of nodes can be monitored for determining the attackers, whenever packets are transmitted. The possibility theory applied in trust computation detects Sybil attackers by logically evaluating the node behaviour and assigning trust value. Accordingly, the proposed system identifies the node behavioural discriminations caused by Sybil attackers.

The final trust level of any node is the comprehensive value of both direct and indirect trusts. This direct-indirect trust calculation followed by the misbehaviour verification is explained in the previous section. Despite that, an attacker neighbouring node can provide fake recommendations to mitigate the indirect trust value. To reduce such fake recommendations, an attacker defence scheme is proposed as given below in Algorithm 5.

Algorithm 5: Attack defence model

1: Find the common neighbours between the trustor and trustee with their ID.

2: Verify the table of trust maintained at each node.

3: If direct trust value is above the desired limit (say 0.5), ie., $T_{m,n}^{D} > 0.5$ then;

4: Trustor node broadcasts the request for recommendation to the trustee enlisted.

5: Identify the sender node when a recommendation reply is received.

The node can be identified as three types: trustable, suspicious and newcomer depending on the history of recommendations or communications between the node and the trustor. The trustor accepts the recommendations from trustable and suspicious node and eliminated those from the newcomer nodes.

else;

6:    Set $T_{m,n}^{ID} = 0$

7:    End if;

8:    End.

By executing the Algorithm 1, a final trust value can be evaluated by mitigating fake recommendations. Consequently, a secure communication can be achieved between the trustor and the nodes with higher trust value.

10 Performance analysis

10.1 Simulation analysis

To evaluate the performance of the proposed method, we have developed a MANET environment in QualNet 4.5 simulator. The node behaviour comprises the packet sending and forwarding, observations as well as recommendation broadcasting. The simulation platform is setup in such a way to monitor the neighbour's behaviour and to categorise it into trustworthy and/or untrustworthy actions, with a time gap exponentially distributed between successive actions. We consider a 50 number of nodes simulated at a time of 500 s. A MANET environment is configured with many mobile devices (mobile phones, laptops, etc.) which move randomly to communicate among their neighbours in the network of transmission range 250 m.

The nodes are assumed to move randomly at different node mobility from 5 to 25 m/s over network traffic of constant bit rate (CBR) that is applied between the sender and receiver nodes. The probability of selecting a new node as CH is set to 0.3. The nodes follow a random way point (RWP) approach, where the speed and the direction of each node are chosen randomly and independently.

When the simulation starts, each node chooses one location randomly as the destination within terrain of 1000 by 1000 m terrain in QualNet simulator for 802.11b and ad hoc on demand routing protocol over the simulation field. The nodes then moves with constant velocity chosen uniformly and randomly in a range $[0, V_m]$, where 'V_m' is the maximum range of velocity that a node travels. When the node reaches its destination, it halts for a time period, referred as halt time 'T_{halt}'. If $T_{halt} = 0$, a continuous mobility is experienced. However, when the 'T_{halt}' expires, the nodes again move randomly in the simulation field. The performance of the proposed THCM is evaluated by varying the two parameters 'V_m' and 'T_{halt}' for topology alterations (i.e., if 'V_m' is less and 'T_{halt}' is high, a relatively stable topology is achieved, while a highly dynamic topology is obtained if 'V_m' is high and 'T_{halt}' is less). Each data point in the simulation was limited to 10 observations for trust value calculation during simulation. We analyze the node behaviour by sustained monitoring system that includes two parts: monitoring phase and calculation phase. In the monitoring phase, the *CH* closely monitors its members and indicates the probability of behaviour changes if any.

The higher the probability rate the more will be the accuracy. In the second phase, the trust value of each node is evaluated based on the set of observations obtained previously.

10.1.1 Direct and indirect trust for different nodes

Figure 7: direct trust for different nodes shows the direct trust calculated for random node 5, 20, 30 and 40 at a maximum time period of 500 s. From the figure, it is clearly shown that the nodes 30 and 40 misbehaved and so the trust value that is calculated directly by observing node 30 and 40 is gradually decreased to zero, whereas the other two nodes show an increased trust level with their trustworthy behaviour. The indirect trust for nodes 5, 20, 30 and 40 are calculated and plotted in Fig. 8: indirect trust for different nodes with a simulation time set at 300 s. The trust value for nodes 5 and 10 is greater than 0.5, which show higher node cooperation for trust value. On the other hand, the indirect trust of nodes 30 and 40 degrades below 0.5 due to misbehaviour observed using the Bayesian-Evidence theorem.

10.1.2 Performance metrics

10.1.2.1 Network complexity The network complexity greatly depends on the convergence time. The convergence time is the time period required to achieve a trust convergence. The trust convergence of a node can be defined as the difference between the variance of two continuous trust values above a predefined trust threshold of 0.5. With the increase in the number of nodes, the convergence time increases, which in turn contributes to network complexities. From Fig. 9: convergence time, we compare the proposed trust methodology with CTrust schemes in [58] for various ɗ, where ɗ represents the node degree. In both schemes, the convergence time multiplies gradually with the growth in the network size.

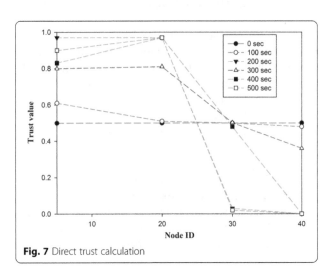

Fig. 7 Direct trust calculation

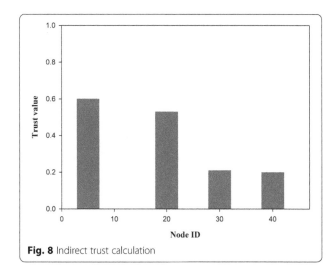

Fig. 8 Indirect trust calculation

the network from the increase in convergence time, even if the number of recommenders and their evidences increases. Therefore, on comparing with the CTrust method, the proposed shows a better performance marginally by decreasing the convergence time that results in controlling the further rise in network complexity. This distinctly shows the scalability feature of the proposed trust management scheme.

10.1.2.2 Communication overhead The average communication overhead occurs during the trust computation per node in a cluster is shown in Fig. 10: communication overhead. The proposed scheme reveals a reduction in overhead in communication by using the mathematical theorems compared to CTrust scheme [58]. For each recommendation request, each node receives recommendation reply only for its one-hop neighbours which lower the redundant accumulation of packers that urges in overhead reduction.

10.1.2.3 Trust accuracy It measures the inferred trust computations with its attacker mitigating property. Compared to CTrust, the accuracy is maintained above 92% in all the cases except when ḏ = 30, where almost the

This is because of the false trustworthiness values computed by the recommender which is high in the existing methodology that increases the convergence time.

With the increase in the node degree, the measurements from the recommenders gets increased which further increases the network complexities. The misbehaviour verification algorithm in the proposed scheme safeguards

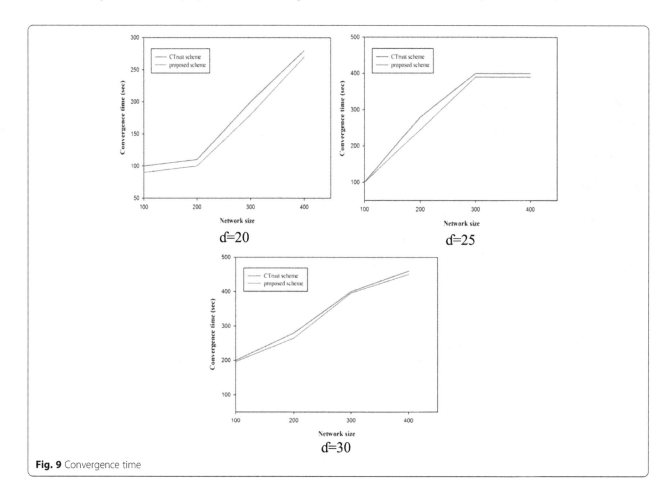

Fig. 9 Convergence time

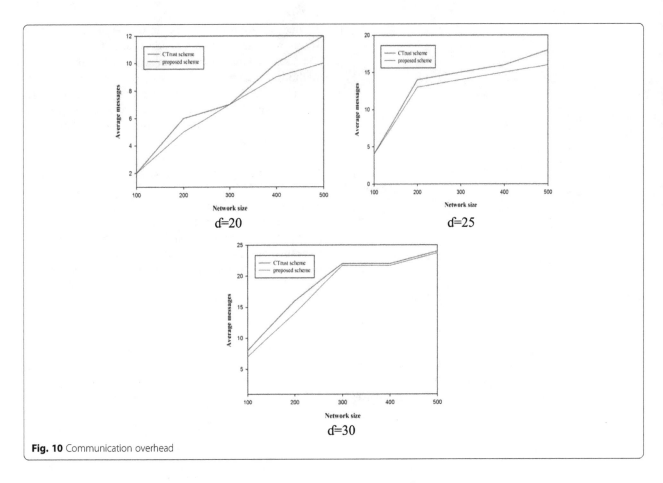

Fig. 10 Communication overhead

same level of accuracy is achieved for larger network size in both the schemes as shown in Fig. 11: trust accuracy. Though the network complexity is lowered, the MANET shows a high accuracy rate that makes our scheme more advantageous.

10.1.3 Mobility factors

This section discusses some factors that affect the cluster property with respect to the mobility of MANET nodes, namely, cluster size, node's probability in a cluster and average cluster head changes as shown in Figs. 12, 13 and 14. We compare the proposed scheme with the established existing protocols such as 2ACK [59] and CBTRP [39].

10.1.3.1 Cluster size with node mobility With the increase in the node velocity in MANET, the size of clusters varies. The network performance may get interrupted with the traffic overload, when the cluster size increases. Therefore, the cluster size should be maintained from increasing to achieve favourable clustering scalability. Figure 12: cluster size shows the mobility influential clusters for the existing 2ACK, CBTRP with the proposed trust-based scheme. The result demonstrates how each methodology accepts the

cluster changes whenever the membership alters. When the node speed is increased as high as 25 m/s from a lower speed of 5 m/s, the cluster size get reduces from 25 to 7 nodes in the proposed scheme. This makes the proposed method more suitable for packets to establish and maintain routes. On the other hand, the existing schemes present a higher size of clusters with different increased node velocity. This further increases the cluster communication as more data need to be transmitted among the CH and the cluster multi-hops.

Simultaneously, the communication from cluster members to the CH drops significantly, since the less number of CH is present. This is because, if these protocols does not restrict the cluster size, a less number of clusters results in high intra-cluster communication overhead with the increase in the single size of clusters. It is clear that all the schemes construct large clusters with low mobility of nodes and smaller clusters over higher mobility. The efficient hexagonal clusters with Voronoi geometric patterns divide the network area into regular clusters with the shortest distance and expected number of transmissions computed between each node and the corresponding CH. The proposed scheme, thus, maintains appropriate clusters of optimal size with effectual mobility adaptiveness.

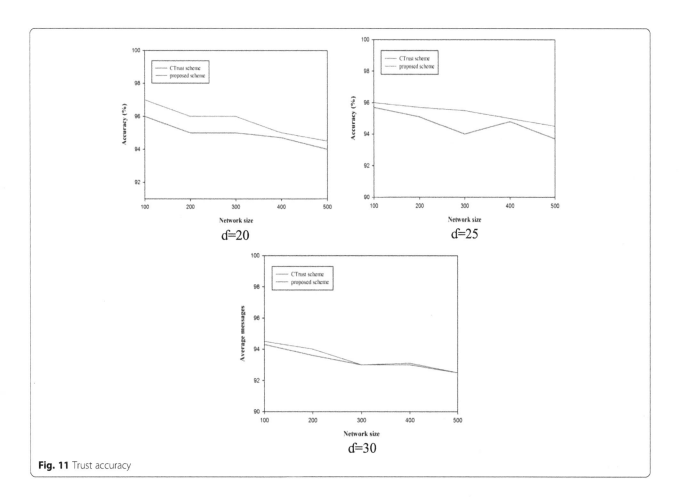

Fig. 11 Trust accuracy

10.1.3.2 Node probability Figure 13: node's probability, illustrates the probability that each node is available in the clusters with respect to the mobility. The efficiency of any scheme depends on the high probability of the node that remains in the clusters which greatly depends on the clustering parameters. In the proposed scheme, the nodes remain clustered every time, which is greater than 0.9 even in the presence of large mobile nodes at a speed of 25 m/s. Whereas, the existing schemes show lesser probability on nodes being clustered compared to the proposed methodology. This beneficial feature of the proposed scheme is attained only because of the Voronoi clustering technique, where the nearest neighbour problem is solved greatly on non-overlapping

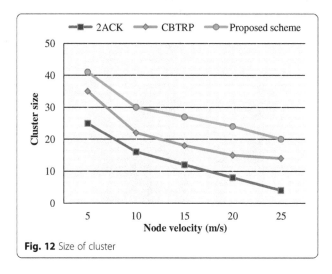

Fig. 12 Size of cluster

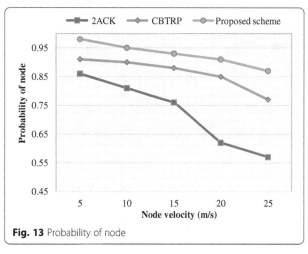

Fig. 13 Probability of node

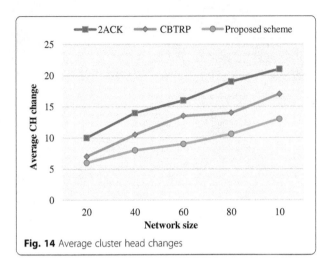

Fig. 14 Average cluster head changes

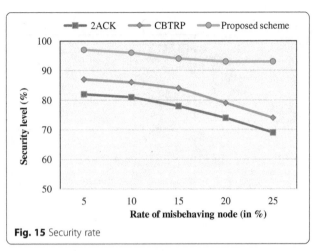

Fig. 15 Security rate

partitions so that each node remains in the cluster region. The simulation result thus shows the desirable property of the proposed scheme that the probability of nodes being clustered is high even in the presence of greater node speed.

10.1.3.3 Cluster head change with mobility Figure 14: average cluster head change with mobility demonstrates the CH age of the proposed scheme against existing schemes. The CH duration is measured as the average time that a CH is active at each time instants. This factor indicates the cluster stability, i.e., with more change in the CHs the lesser will be the cluster stability. For a stable cluster construction, the CH duration should be relatively lesser with high trust level. As expected, the proposed scheme performs better than the existing methodologies as the former exclusively uses higher trust level and the latter identity and node degree information to form the cluster structure. The CBTRP scheme also incorporates trust metric in cluster construction, thereby undesirably influencing the cluster stability; also, as the size of cluster increases, it is more predictable to appeal to re-clustering due to nodes mobility. The proposed scheme provides better results, as the CH depends on the node mobility with hybrid trust. Compared to the existing schemes, the proposed mechanism has lesser CH age, even at higher rates of node mobility. The result also shows the advantages of the proposed scheme in the reaffiliation rate, which represents the average CH change and its affiliation with rate of change of mobility. The proposed scheme presents a higher probability of reaffiliation that remains its CH for a longer time. This advantage of the average CH change for the proposed scheme is because of the lower link formations and failures in the cluster construction.

10.1.4 Security level with mobility

Figure 15: security level demonstrates the level of security, which is one of the significant factors for measuring the security strength of the proposed scheme. The Hackman tool integrated with the QualNet network simulator analyses different attackers at periodic time intervals. The Block Cipher Cryptography Class (BCCC) interface with Hackman tool enabled with Hackman SDK. The tool in the simulator tries to break the data packets and calculates the packets that are hacked successfully for evaluating the security level in percentage. The proposed scheme presents a higher security level to different selfish and malicious attackers compared to other existing schemes. An overall security of 93%is attained by the proposed scheme in the presence of different misbehaving activities, at larger node mobility. Whereas, the existing schemes such as 2ACK and CBTRP shows lower security level of 69 and 74%, respectively.

Besides, the attacks that generate the malicious and selfish node behaviours, such as flooding attack, black hole attack, wormhole attack, impersonation attack, packet dropping attack and Sybil attacks, are managed

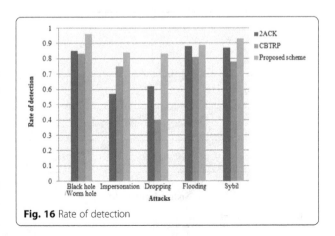

Fig. 16 Rate of detection

by the proposed trust management system. In order to reduce the false recommendation attack, the proposed system undergoes the misbehaviour verification procedure. We have selectively chosen the above mentioned attackers to represent how effectively they are detected and revoked. The detection rate of various attackers for different scheme varies. Figure 16: rate of detection shows the detection rate for all the schemes and for each attacker. The attack with the highest rate of detection for the proposed scheme is malicious attackers namely black hole and wormhole attacks. This shows the resistance of the proposed scheme to the malicious activities that can collapse the entire MANET functionalities, unlike the selfish behaviour. It can also be seen that with the proposed trust scheme, it performs well than the other scheme for some attackers.

10.1.5 Cost of cluster formation

The benefits of clustering comes with cost-effectiveness of the proposed hybrid trust-based clustering scheme that aims at minimizing overheads incurred in reducing control traffic and communication, enhancing the cluster stability with no prolonged cluster head resistance time. Figure 17: cluster overhead increases gradually to 72% in the presence of 25% of attacker nodes in the proposed scheme. Whereas, the overhead increases greatly in the existing scheme due to the flat network architecture that floods the cluster formation packets throughout the network region.

Figure 18 shows the cost of cluster formation of different schemes compared with the proposed scheme. The cost of clustering is a crucial issue to evaluate the scalability and effectiveness improvement of a cluster structure. By validating the cost of clustering for different qualitatively or quantitatively characteristics, its usefulness can be specified. The proposed methodology shows lower cost for constructing the hexagonal clusters and

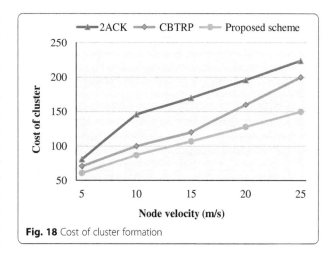

Fig. 18 Cost of cluster formation

re-construction. The cost of re-clustering is minimized due to the mobility aware cluster construction presented in the Voronoi clusters.

The proposed scheme has reduced the amount of message exchanged in the cluster construction. The communication complexity for re-clustering in the cluster formation phase may be equal to the cluster maintenance. An important factor that increases the cost is the rate of overlapping clusters in the MANET region. If the clusters are highly overlapping, the average number of clusters increases. All the clustering schemes are active with explicit control message among the MANET nodes for clustering. In 2ACK scheme, the mobile nodes are unable to elect the CH until an acknowledgement is received from the cluster members. The number of rounds for cluster construction is equal to the clusters formed, which represents that only one CH is elected in each round. However, the cluster construction is performed in parallel to the PKI functionalities and the cluster formation rounds should be less. The proposed scheme maintains the cluster architecture

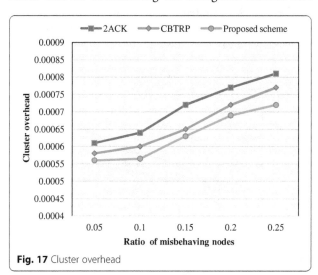

Fig. 17 Cluster overhead

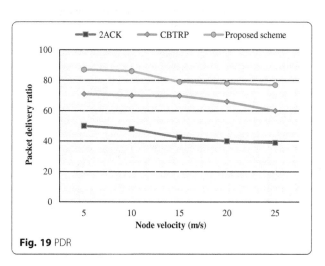

Fig. 19 PDR

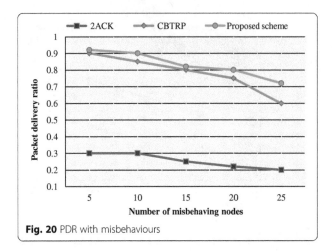

Fig. 20 PDR with misbehaviours

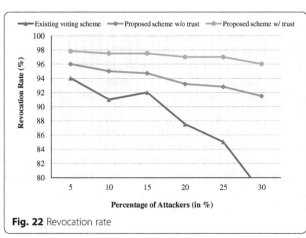

Fig. 22 Revocation rate

well throughout the functionalities and effectively lengthens the lifetime of the clusters under a dynamic mobile environment. The scheme can outperform the existing scheme in terms of cluster stability and overhead since it provides guarantee with no ripple effect of re-clustering. Hence, the proposed scheme is more feasible for a large dynamic scenario, where nodes are highly connected.

10.1.6 Packet delivery ratio

Figure 19: packet delivery ratio with node velocity, and Fig. 20: packet delivery ratio with misbehaving nodes, represent the efficiency of the proposed scheme in packet delivery ratio (PDR) while participating a secure group communication. Figure 19 shows the impact of node mobility in 25 nodes MANET. It is observed that, as the node velocity increases, the PDR drops gradually. This is due to the higher node speed with may increase the packet dropping. However, the proposed scheme delivers a higher ratio of packets compared to the existing one. In Fig. 20, it is clear that the PDR is maintained with a higher percentage of misbehaving nodes in the proposed scheme than existing schemes. This is because of the trust-based misbehaviour calculation of selfish

and malicious nodes. The results demonstrate that the scheme with indirect and direct observation has the highest PDR among the other two schemes. The PDR of all the schemes reduces gradually with the increase in the number of nodes. This is due to the packet collision or packet dropping that occurs either due to the frequent node movement or with the influence of misbehaving nodes. In the proposed scheme, the packet dropping attack is handled effectively by detecting and isolating the attackers that initiate the attack and therefore the packets can be delivered successfully to the destination node while carrying out a secure group communication. The packet dropping in the existing schemes is higher due to the inefficiency in handling the dropping attackers.

10.1.7 Certificate revocation with hybrid trust

Revocation time is a crucial factor for estimating the performance of revocation strategy. Revocation time is defined as the time for which the rate of nodes revoked per second. Figure 21 shows the advantage of a trust-based mechanism in terms of revocation time compared to the trust-less strategy. To analyse the impact of attacker nodes on revocation, we deploy 100 nodes in the network, whereas the attacker nodes ranges up-to 30%.

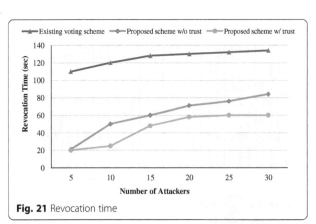

Fig. 21 Revocation time

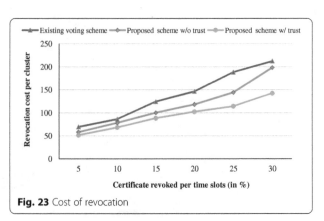

Fig. 23 Cost of revocation

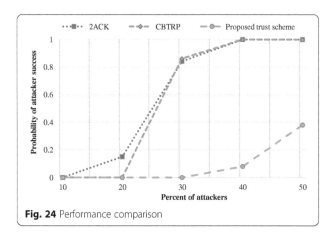

Fig. 24 Performance comparison

compared with the trust-less strategy and existing voting scheme. The performance of the proposed trust-based scheme is evaluated with various existing schemes for its efficiency to resilience against attackers. Figure 24 provides insight on the effect of the probability of success of attackers against various attacker ranges. We assume the attackers might report false events with the aim to interrupt the functionalities by trustable nodes. The average trust value of attackers (0.8) is considered as higher than the average trust value of trustable nodes (0.6). From the figure, it is clear that the existing schemes are less resilient to attackers and the proposed trust-based scheme is the most resilient among the existing methods. We also evaluated the proposed trust scheme for various performance parameters as rate of detection, false alarm, detection method and attacks analysed as given in Table 2.

11 Conclusions

In the dynamic environment of MANETs, trusting the neighbours for secure communication is strenuous to achieve. Traditional cryptographic schemes do not contribute a complete solution to detect and secure the ad hoc nodes from various attacks. An efficient tool to manage this drawback in MANET is the establishment of trust among nodes. The proposed trust model successfully secures the communication in the clustered network that confirms trust among the participant nodes. Additionally, the trust recommendations and trust computation reduce the chances of attackers in a large amount with mobility adaptive and stable clusters. The theoretical bases for trust computation in this paper also provide a platform for practical implementation in a MANET to provide an efficient public key infrastructure (PKI)-based security framework. Finally, a simple analysis to highlight the benefits of the proposed strategies was presented. From the analysis, we can observe that in the trust-based certificate management strategy, the increases in revocation time, revocation rate, cost or CRL list is almost maintained at constant, and hence, the system is scalable. In the future, we plan to analyze the performance of the proposed strategies assuming node mobility across geographic clusters, taking into account the overhead incurred in obtaining new certificates, and the corresponding region-specific CRLs.

Figure 21 shows the change in the revocation time with the increase in attacker nodes, between the proposed scheme (with and without trust) and existing voting scheme [60]. It is clear that the voting scheme requires a longer time for revocation compared to the other two schemes. On the other hand, the proposed trust-based scheme maintains a beneficial and steady revocation time, even with a higher percentage of attackers. When the revocation is performed without trust, the time of operation increases since there required more verification steps. Whereas the revocation time gets reduced in a larger amount when a certificate assignment is performed in a trust-based scheme, which is shown in Fig. 21. The rate of revocation for different number of attackers is shown in Fig. 22. A revocation rate can be defined as the rate of rate of attackers revoked before launching the attacks. It is noted that the rate of revocation improves with the increasing number of attackers for the proposed trust-based scheme. Even though the rate gets down a little for some attacker percentage, it gradually increases for larger number of attackers.

Another important factor that shows the efficiency of any certificate revocation system is cost. Generally, the cost of revocation gets increased with the number of certificates revoked per time slot. To evaluate this factor, the average number of certificates revoked varies from 10 to 30 as shown in Fig. 23.When a trust-based strategy is proposed, the cost of revocation gets decreased with the increasing number of regions. The revocation cost drops greatly when regions are proposed when

Table 2 Comparison of various MANET models

Parameters	Proposed trust scheme	CBTRP [59]	2ACK [39]
Rate of detection	High	Low	Low
Detection method	Hybrid trust-based clustering method	Trust based	Acknowledgement based
False alarm	Low	High	High
Attacks analysed	Flooding attack, wormhole attack, black hole attack, rushing attack, impersonation and Sybil attack	Routing attacks, packet dropping, packet spoofing	Routing attacks, packet dropping

Acknowledgements
The authors would like to thank the Computer Network Laboratory of Thiagarajar College of Engineering for the support in the development and simulation of the concept.

Funding
The authors declare that no funding sources support in the design of the study and collection, analysis and interpretation of data and in writing the manuscript should be declared.

Authors' contributions
All the authors contribute to the concept, the design and developments of the methodology, and the simulation results in this manuscript. Both authors read and approved the manuscript.

Competing interests
The authors declare that they have no competing interests.

References
1. V Cahill et al., Using trust for secure collaboration in uncertain environments. IEEE Pervasive Comput. **2**(3), 52–61 (2003)
2. C English, W Wagealla, P Nixon, S Terzis, H Lowe, A McGettrick, Trusting collaboration in global computing systems. Lect. Notes Comput. Sci. Springer-Verlag **2692**, 136–149 (2003)
3. M. Deutch, Cooperation and trust: some theoretical notes, Nebraska Symposium on Motivation, Nebraska University Press,1962, pp. 275–319.
4. C Zhu, H Nicanfar, VCM Leung, LT Yang, An authenticated trust and reputation calculation and management system for cloud and sensor networks integration. IEEE Trans. Inf. Forensics Secur. **10**(1), 118–131 (2015)
5. J Jiang, G Han, F Wang, L Shu, M Guizani, An efficient distributed trust model for wireless sensor networks. IEEE Trans. Parallel Distrib. Syst. **26**(5), 1228–1237 (2015)
6. Y Wu, Y Zhao, M Riguidel, G Wang, P Yi, Security and trust management in opportunistic networks: a survey. Secur. Commun. Netw. **8**(9), 1812–1827 (2015)
7. H Zhu, S Du, Z Gao, M Dong, Z Cao, A probabilistic misbehaviour detection scheme toward efficient trust establishment in delay-tolerant networks. IEEE Trans. Parallel Distrib. Syst. **25**(1), 22–32 (2014)
8. K S Cook (ed.), *Trust in society*, vol 2 (Russell Sage Foundation Series on Trust, New York, 2003)
9. M Blaze, J Feigenbaum, J Lacy, Decentralized trust management. Proc. IEEE Symp. Secur. Priv. **6-8**, 164–173 (1996)
10. A. Boukerch, L. Xu and K. EL-Khatib, Trust-based security for wireless ad hoc and sensor networks, Computer Communications, no. 30, 2007, pp. 2413–2427.
11. L Kagal, T Finin, A Joshi, Trust-based security in pervasive computing environments. IEEE Comput. **34**, 154–157 (2001)
12. H Sarvanko, M Hyhty, M Katz, F Fitzek, in *4th ERCIM eMobility Workshop in conjunction with WWIC'10*. Distributed resources in wireless networks: discovery and cooperative uses (2010)
13. MA Ayachi, C Bidan, T Abbes, A Bouhoula, in *International Symposium on Trusted Computing and Communications, Trustcom*. Misbehaviour detection using implicit trust relations in the AODV routing protocol (2009), pp. 802–808
14. Janani V. S and M.S.K. Manikandan, "Trust-based hexagonal clustering for efficient certificate management scheme in mobile ad hoc networks", Sadhana, Springer, Vol 41, Issue 10, October 2016, pp 1135-1154.
15. K Govindan, P Mohapatra, Trust computations and trust dynamics in mobile adhoc networks: a survey. IEEE Commun. Surv. Tutorials **14**(2), 279–298 (2012)
16. JH Cho, A Swami, IR Chen, A survey on trust management for mobile ad hoc networks. IEEE Commun. Surv. Tutorials **13**(4), 562–583 (2011)
17. A Ahmed, KA Bakar, MI Channa, K Haseeb, AW Khan, A survey on trust based detection and isolation of malicious nodes in ad hoc and sensor networks. Front. Comp. Sci. **9**(2), 280–296 (2015)
18. S. Marsh, "Formalising Trust as a Computational Concept", PhD Thesis, University of Stirling, 1994.
19. J Li, R Li, J Kato, Future trust management framework for mobile ad hoc networks: security in mobile ad hoc networks. IEEE Commun. Mag. **46**(4), 108–114 (2008)
20. M Blaze, J Feigenbaum, J Lacy, in *IEEE Symposium on Security and Privacy*. Decentralized trust management (1996), pp. 164–173
21. X Wang, W Cheng, P Mohapatra, T Abdelzaher, in *INFOCOM 2013*. Artsense: anonymous reputation and trust in participatory sensing (2013), pp. 2517–2525
22. Hui Xia, Zhiping Jia, Xin Li, Lei Ju, Edwin H.-M. Sha, "Trust prediction and trust-based source routing in mobile ad hoc networks", Ad Hoc Networks, Elsevier, Vol 11, Issue 7, September 2013, Pages 2096–2114.
23. A.M Shabut, K.P Dahal, S.K Bista, I.U Awan, Recommendation based trust model with an effective defence scheme for MANETs, IEEE Trans. Mob. Comput. 14(10), 2101–2115 2015
24. S. Marti, T.J. Giuli, K. Lai, and M. Baker, "Mitigating routing misbehavior in mobile ad hoc networks", Mobicom 2000 2000, pp. 255-265.
25. D Kukreja, SK Dhurandher, BVR Reddy, *Enhancing the security of dynamic source routing protocol using energy aware and distributed trust mechanism in MANETs* (Intelligent Distributed Computing, Springer, Switzerland, 2015)
26. SA Thorat, PJ Kulkarni, in *Computing, Communication and Networking Technologies (ICCCNT)*. Design issues in trust based routing for MANET (2014), pp. 1–7
27. S Buchegger, JY Le Boudec, in *Proceedings of the 10th Euromicro Workshop on Parallel, Distributed and Network-Based Processing PDP*. Nodes bearing grudges: Towards routing security, fairness, and robustness in mobile ad hoc networks (2002), pp. 403–410
28. S Buchegger, JY Le Boudec, in *Proceedings of the Second Workshop on the Economics of Peer-to-Peer Systems, P2PEcon 2004, Harvard University Press*. A robust reputation system for P2P and mobile ad hoc networks (2004)
29. S Buchegger, JY Le Boudec, Self-policing mobile ad hoc networks by reputation systems. IEEE Commun. Mag. **43**(7), 101107 (2005)
30. K Thirunarayan, P Anantharam, C Henson, A Sheth, Comparative Trust Management with Applications: Bayesian Approaches Emphasis. Future Generation Computer Systems, Elsevier, 2014.
31. ECH Ngai, MR Lyu, Trust and Clustering Based Authentication Services in Mobile ad hoc Networks, W4: MDC (ICDCSW'04), 2004, 323-324.
32. Z Hosseini, Z Movahedi, A Trust-Distortion Resistant Trust Management Scheme on Mobile Ad Hoc Networks", Wireless Personal Communications, Special Issue on Advances and Challenges in Convergent Communication Networks, Springer; 2016 [Online 1]. https://doi.org/10.1007/s11277-016-3734-6.
33. Z Movahedi, Z Hosseini, F Bayan, G Pujolle, Trust-distortion resistant trust management frameworks on mobile ad hoc networks: a survey. IEEE Commun. Surv. Tutorials **18**(2), 1287–1309 (2016)
34. G Nagaraja, C Pradeep Reddy, Mitigate lying and on-off attacks on trust based group key management frameworks in MANETs. Int. J. Intell. Eng. Syst. **9**(4), 215–222 (2016)
35. J-H Cho, I-R Chen, KS Chan, Trust threshold based public key management in mobile ad hoc networks. Ad hoc Netw. J. **44**(1), 58–75 (2016)
36. K. Gai, M. Qiu, M. Chen, and H. Zhao. SA-EAST: security-aware efficient data transmission for ITS in mobile heterogeneous cloud computing. ACM Transactions on Embedded Computing Systems, 2016.
37. Y Li, K Gai, Z Ming, H Zhao, M Qiu, Intercrossed access control for secure financial services on multimedia big data in cloud systems. ACM Trans. Multimed. Comput. Commun. Appl. **12**(4) (2016). https://doi.org/10.1145/2978575
38. K Gai, M Qiu, Z Ming, H Zhao, L Qiu, Spoofing-Jamming Attack Strategy Using Optimal Power Distributions in Wireless Smart Grid Networks. IEEE Trans. Smart Grid **8**(5), 2431-2439 (2017)
39. H Safa, H Artail, D Tabet, A cluster-based trust-aware routing protocol for mobile ad hoc networks. Wirel. Netw **16**(4), 969–984 (2010)
40. JH Cho, KS Chan, IR Chen, *Composite trust-based public key management in mobile ad hoc networks* (ACM 28th Symposium on Applied Computing, Coimbra, 2013)
41. J.H. Cho and I.-R. C. Kevin Chan, "A composite trust-based public key management in mobile ad-hoc networks," ACM 28th Symposium on Applied Computing, Trust, Reputation, Evidence and other Collaboration Know-how (TRECK), 2013.
42. RH Jhaveri, NM Patel, Attack-pattern discovery based enhanced trust model for secure routing in mobile ad-hoc networks. Int. J. Commun. Syst. (2016). https://doi.org/10.1002/dac.3148

43. Z Movahedi, Z Hosseini, F Bayan, G Pujolle, Trust-distortion resistant trust management frameworks on mobile ad hoc networks: a survey. IEEE Commun. Surv. Tutorials **18**(2), 1287–1309 (2016)

44. J.M.Nichols and J.V. Michalowicz, "Distance distribution between nodes in a 3D wireless network", J. Parallel Distrib. Comput., Vol 102, 2017, pp 71-79.

45. B Kao, SD Lee, F Lee, D Cheung, WS Ho, Clustering uncertain data using voronoi diagrams and R-tree index. IEEE Trans. Knowl. Data Eng. **22**(9), 1219–1233 (2010)

46. X Xie, R Cheng, M Yiu, L Sun, J Chen, Uv-diagram: a voronoi diagram for uncertain spatial databases. VLDB J. **22**(3), 319–344 (2013)

47. ML Elwin, RA Freeman, KM Lynch, Distributed Voronoi neighbor identification from inter-robot distances. IEEE Robot. Autom. Lett. **2**(3), 1320–1327 (2017)

48. Dingjiang Zhou, Zijian Wang, Saptarshi Bandyopadhyay, and Mac Schwager, " Fast, On-line Collision Avoidance for Dynamic Vehicles using Buffered Voronoi Cells", IEEE ROBOTICS AND AUTOMATION LETTERS, 2017.

49. P Fan, G Li, K Cai, KB Letaief, On the geometrical characteristic of wireless ad-hoc networks and its application in network performance analysis. IEEE Trans. Wireless Commun. **6**(4), 1256–1265 (2007)

50. Y Zhuang, TA Gulliver, Y Coady, On planar tessellations and interference estimation in wireless ad-hoc networks. IEEE Wireless Commun. Lett. **2**(3), 331-334 (2013)

51. Fei Tong ,Jianping Pan, Ruonan Zhang, "Distance Distributions in Finite Ad Hoc Networks: Approaches, Applications, and Directions", Ad Hoc Networks, 2016, pp 167-179.

52. W Li, H Song, ART: an attack-resistant trust management scheme for securing vehicular ad hoc networks. IEEE Trans. Intell. Transp. Syst. **17**(4), 960–969 (2016)

53. Wang, Yating, Ray Chen, Jin-Hee Cho, Ananthram Swami, Yen-Cheng Lu, Chang-Tien Lu, and Jeffrey Tsai. "Catrust: Context-Aware Trust Management for Service-Oriented Ad Hoc Networks", IEEE Transactions on Services Computing, 2016.

54. S Tan, X Li, Q Dong, A trust management system for securing data plane of ad-hoc networks. IEEE Trans. Veh. Technol. **65**(9), 7579–7592 (2016)

55. W. Li, H. Song and F. Zeng. Policy-based secure and trustworthy sensing for internet of things in smart cities.IEEE Internet of Things Journal 99,2017.

56. M Raya, P Papadimitratos, VD Gligor, J-P Hubaux, in *INFOCOM 2008. The 27th Conference on Computer Communications. IEEE.* On data-centric trust establishment in ephemeral ad hoc networks (2008), pp. 1238–1246

57. Li, Wenjia, and Anupam Joshi. Outlier detection in ad hoc networks using dempster-shafer theory. Mobile Data Management: Systems, Services and Middleware, 2009. MDM'09. Tenth International Conference on 112-121 2009.

58. H Zhao, X Yang, X Li, CTrust: trust management in cyclic mobile ad hoc networks. IEEE Trans. Veh. Technol. **62**(6), 2792–2806 (2013)

59. K Liu, J Deng, PK Varshney, K Balakrishnan, *An acknowledgment-based approach for the detection of routing misbehavior in MANETs,' IEEE Transactions on Mobile Computing* (2007), pp. 536–550

60. H Luo, J Kong, P Zerfos, S Lu, L Zhang, URSA: Ubiquitous and Robust Access Control for Mobile Ad Hoc Networks. IEEE/ACM Trans. Networking **12**(6), 1049–1063 (2004)

2

Efficient packet transmission in wireless ad hoc networks with partially informed nodes

Sara Berri[1,2]* (iD), Samson Lasaulce[2] and Mohammed Said Radjef[1]

Abstract

One formal way of studying cooperation and incentive mechanisms in wireless ad hoc networks is to use game theory. In this respect, simple interaction models such as the forwarder's dilemma have been proposed and used successfully. However, this type of models is not suited to account for possible fluctuations of the wireless links of the network. Additionally, it does not allow one to study the way a node transmits its own packets. At last, the repeated game models used in the related literature do not allow the important scenario of nodes with partial information (about the link state and nodes actions) to be studied. One of the contributions of the present work is precisely to provide a general approach to integrate all of these aspects. Second, the best performance the nodes can achieve under partial information is fully characterized for a general form of utilities. Third, we derive an equilibrium transmission strategy which allows a node to adapt its transmit power levels and packet forwarding rate to link fluctuations and other node actions. The derived results are illustrated through a detailed numerical analysis for a network model built from a generalized version of the forwarder's dilemma. The analysis shows in particular that the proposed strategy is able to operate in the presence of channel fluctuations and to perform significantly better than the existing transmission mechanisms (e.g., in terms of consumed network energy).

Keywords: Packet transmission, Power control, Game theory, Repeated games, Incentive mechanisms, Wireless ad hoc networks

1 Introduction

In wireless ad hoc networks, nodes are interdependent. One node needs the assistance of neighboring nodes to relay the packets or messages it wants to send to the receiver(s). Therefore, nodes are in the situation where they have to relay packets, but have at the same time to manage the energy they spend for helping other nodes, and therefore exhibiting selfish behavior. To stimulate cooperation, incentive mechanisms have to be implemented [1–7]. The vast majority of incentive mechanisms either rely on the idea of reputation [4, 5, 8] or the use of a credit system [9, 10]. Indeed, to capture the trade-off between a cooperative behavior (which is necessary to convey information through an ad hoc network) and a selfish behavior

(which is necessary to manage the node energy), the authors of [5] and [8], and many other papers, exploited a simple but efficient model, which consists in assuming, whatever the size of the network, that the local node interaction only involves two neighboring nodes having a decision-making role; one of the virtues of considering the interaction to be local is the possibility of designing distributed transmission strategies. In the original model, a node has two possible choices, namely, forward or drop the packets it receives from the neighboring node. As shown in [5] and [8], modeling the problem at hand as a game appears to be natural and relevant; in the corresponding game, the node utility function consists of the addition of a data rate term (which is maximized when the other node forwards its packets) and an energy term (which is maximized when the node does not forward the packets of the other node). At the Nash equilibrium of the strategic form static game (called the forwarder's dilemma in the corresponding literature), nodes do not transmit at

---

*Correspondence: berri.sara2012@gmail.com
[1]Research Unit LaMOS (Modeling and Optimization of Systems), Faculty of Exact Sciences, University of Bejaia, 06000 Bejaia , Algeria
[2]L2S (CNRS-CentraleSupelec-Univ. Paris-Saclay), 91192 Gif-sur-Yvette, France

all. To avoid this situation to happen in a real network, cooperation has to be stimulated by studying the repeated interaction between the nodes [1, 5, 8]. While providing an efficient solution, all the corresponding models still have some limitations, especially regarding the link quality fluctuations and partial knowledge at the nodes; indeed, they do not take into account the quality of the link between the transmitting and the receiving nodes, which may be an important issue since the link quality may strongly fluctuate if it is wireless. The solution in [3, 6], and [8] referred to as ICARUS (hybrId inCentive mechAnism for coopeRation stimUlation in ad hoc networkS) combines the two ideas, namely, reputation and credit system, but it is not suited to scenarios where the actions of the other nodes are not perfectly observed, which results, e.g., in inappropriate punishment (a node is declared selfish while it is cooperative) and therefore in a loss of efficiency. Additionally, in these works, when a node is out of credit, the transmission is blocked and the node cannot send any packet anymore; this might be not practical in some wireless networks where a certain quality of service has to be provided. Also, the authors propose a mechanism to regulate the credit when a node has an excessive number of credits, but the proposed mechanism may be too complex. At last but not least, no result is provided on the strategic stability property, which is important and even necessary to make the network robust against selfish deviations. The purpose of this paper is precisely to overcome the limitations of the aforementioned previous works. More precisely, the contributions of the present paper are as follows.

▶ The first key technical difference with the closely related works is that the proposed formulation accounts for the possible presence of quality fluctuations of the different links that are involved in the considered local interaction model. In particular, this leads us to a game model which generalizes the existing models since the forwarding game has now a state and the discrete action sets are arbitrary, not just binary; additionally, the node does not only choose the cooperation power but also the power used to send its own packets.

▶ An important and useful contribution of the paper is to characterize the feasible utility region of the considered problem, by exploiting implementability theorems provided by recent works [11–13]. This problem is known to be non-trivial in the presence of partial information and constitutes a determining element of folk theorems; this difficult problem turns out to be solvable in the proposed reasonable setting (the channel gains are i.i.d. and the observation structure is memoryless). The knowledge of the utility region is very useful since it allows one to measure the efficiency of any distributed algorithm relying on the assumed partial information.

▶ A third contribution of the paper is that we provide a new transmission strategy whose main features is to be able to deal with the presence of fluctuating link qualities and to be efficient. To design the proposed strategy, we show that the derived utility region can be used in a constructive manner to obtain efficient operating points and propose a new incentive mechanism to ensure that these points are equilibrium points. The proposed incentive mechanism combines the ideas of credit and reputation. To our knowledge, the closest existing incentive mechanism to the one proposed in the present paper is given by ICARUS in [3, 6], and [8]. Here, we go further by dealing with the problem of imperfect observation and that of credit outage or excess. Indeed, the credit evolution law we propose in this paper prevents, by construction, the number of credits from being too large; therefore, one does not need to resort to an additional credit regulation mechanism, which may be too complex.

▶ In addition to the above analytical contributions, we provide a numerical study which demonstrates the relevance of the proposed approach. Compared to the closest transmission strategies, significant gains are obtained both in terms of packet forwarding rate, network consumed power, and combined utilities. As a sample result, the network-consumed power is shown to be divided by more than two w.r.t. state-of-the art strategies [1, 3, 5, 8].

The remainder of the paper is organized as follows. In Section 2.1, we present the system model; the assumed local interaction model involves two neighboring nodes of an ad hoc network with arbitrary size and generalizes the model introduced in previous works [5, 8]. The associated static game model is also provided in Section 2.1. In Section 2.2, the repeated game formulation of the generalized packet forwarding problem is provided; one salient feature of the proposed model is that partial information is assumed both for the network state and the node actions. In Section 2.3, the feasible utility region of the studied repeated game with partial observation is fully characterized. We also provide an algorithm to determine power control policies that are shown to be globally efficient in Section 3.1. The proposed incentive mechanism and equilibrium transmission strategy are provided in Section 2.4; the proposed transmission strategy allows both the packet forwarding rate and the transmit power to be adapted. A detailed numerical performance analysis is conducted in Section 3.1. Section 4 concludes the paper.

2 Methods/experimental
2.1 System model
The present work concerns wireless ad hoc networks, namely, networks in which a source node needs the assistance of other nodes to communicate with the destination node(s). As well motivated in related papers such as [5]

and [8], we will assume the interaction among nodes to be local, i.e., it only involves neighboring nodes. This means that the network can have an arbitrary size and topology, but a node only considers local interactions to take its decision although it effectively interacts with more nodes. One of the virtues of such an interaction model is to be able *to design distributed transmission strategies* for every node. More specifically, we will assume the model in which local interactions take place in a pairwise manner, which not only allow us to design distributed strategies but also to easily compare the proposed transmission strategy with existing strategies. The key idea of this relevant model is to take advantage of the fact that the network is wireless to simplify the interaction model. For a given node, the dominant interaction will only involve its closest neighbors (see Fig. 1). If several neighboring nodes lie within the radio range of the considered node, then it is assumed to have several pairwise interactions in parallel, as explained in detail in the numerical part.

The nodes are assumed to be non-malicious, i.e., each of them does not aim at damaging the communication of the other. Additionally, they are assumed to operate in an imperfect promiscuous mode, which means that each node imperfectly overhears all packets forwarded by their neighbors. The proposed model generalizes the previous models for at least four reasons. First, the action of a node has two components instead of one: the transmit power used to help the other node, which is denoted by p_i', and

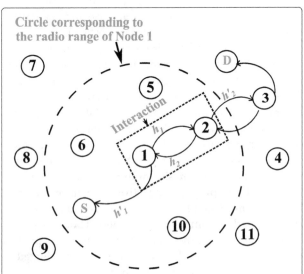

Fig. 1 In this example, the focus is on what node 1 does to allow node S (source) to route its packets to node D (destination). The dashed circle represents the radio range for node 1 and defines its neighbors. To ensure a distributed design, two key elements are exploited: (a) node 1 adopts its transmission behavior to each of its neighbors. Here, node 1 interacts with node 2 (indicated by the dotted box); (b) only the available knowledge of the quality of the most influential links is accounted for (denoted generically by h_1, h_1', h_2, h_2')

the transmit power used to send its own packets, which is denoted by p_i. Second, the transmit power levels are not assumed to be binary but to lie in a general discrete set[1] $\mathcal{P}_i = \mathcal{P}_i' = \mathcal{P} = \{P_1, P_2, ..., P_L\} = \{P_{\min}, ..., P_{\max}\}$, $|\mathcal{P}_i| = |\mathcal{P}_i'| = |\mathcal{P}| = L$. Assuming that the sets are discrete is of practical interest, as there exist wireless communication standards in which the power can only be decreased or increased by step and in which quantized wireless channel state information (CSI) is used (see, e.g., [14, 15]). Similarly, the channel may be quantized to define operating modes (e.g., modulation coding scheme (MCS)) used by the transmitter. Even when the effective channel is continuous, assuming it to be discrete in the model and algorithm part may be very relevant. At last, note that from the limiting performance characterization point of view, the analysis of the continuous case follows from the discrete case but the converse is not true [16]. As a third new feature compared to the related works, the considered model accounts for the possible fluctuations of the quality of each link. With each link, a non-negative scalar is associated, which is called the *channel gain* of the considered link. For a node, the channel gains of the links used to send its own packets and to help the other node are denoted by h_i and h_i', respectively. These channel gains are assumed to lie in discrete sets (of states): $\mathcal{H}_i = \mathcal{H}_i' = \mathcal{H} = \{h_{\min}, ..., h_{\max}\}$ with $|\mathcal{H}_i| = |\mathcal{H}_i'| = |\mathcal{H}| = H$; the realizations of each channel gain will be assumed to be i.i.d.. Technically, continuous channel gains might be assumed. But, as done in the information theory literature for establishing coding theorems, we address the discrete case in the first place, since the continuous case can be obtained by classical arguments (such as assuming standard probability spaces), whereas the converse is not true. Now, from the practical aspect, quantizing the channel gains typically induces a small performance loss compared to the continuous case; one figure assuming a typical scenario illustrates this. The corresponding channel gain model naturally applies to time-selective frequency flat fading single-input single-output channels. If the channel gain is interpreted as the combined effect of path loss and shadowing, our model can also be used to study more general channel models such as multiple-input multiple-output channels. Fourth, the utility function of a node has a more general form than in the forwarder's dilemma. The *instantaneous utility function* for node $i \in \{1, 2\}$ expresses as follows:

$$u_i(a_0, a_1, a_2) = \varphi(\text{SNR}_i) - \alpha(p_i + p_i'), \qquad (1)$$

where

- $a_0 = (h_1, h_1', h_2, h_2')$ is the *global channel or network state*. The corresponding set will be denoted by $\mathcal{A}_0 = \mathcal{H}_1 \times \mathcal{H}_1' \times \mathcal{H}_2 \times \mathcal{H}_2' = \mathcal{H}^4$;
- $a_i = (p_i, p_i')$ is the *action* of node $i \in \{1, 2\}$;

- The function φ is a communication efficiency function which represents the *packet success rate*. It is assumed to be increasing and lie in $[0, 1]$. A typical choice for φ is, for example, $\varphi(x) = (1 - e^{-x})^{\ell}$, ℓ being the number of symbols per packet (see, e.g., [17–19]) or $\varphi(x) = e^{-\frac{c}{x}}$ with $c = 2^r - 1$, r being the spectral efficiency in bit/s/Hz [20];
- For $i \in \{1, 2\}$, the quantity SNR_i is, for node i, the equivalent signal-to-noise ratio (SNR) at the next node after the neighbor. It is assumed to express as follows:

$$\text{SNR}_i = \frac{p_i h_i p'_{-i} h'_{-i}}{\sigma^2},\tag{2}$$

σ^2 being the noise variance and the *index notation* $-i$ standing for the index of the other node.

Remark. The results derived in Section 2.3 hold for any utility function under the form $u_i(a_0, a_1, a_2)$ (under some assumptions which only concern the observation structure) and not only for the specific choice made above. This choice is made to allow comparisons with existing results (and more specifically with the large set of contributions on the forwarder's dilemma) to be conducted and discussed. □

Remark. The assumed expression of the SNR is also one possible pragmatic choice, but all the analytical results derived in this paper hold for an arbitrary SNR expression of the form $\text{SNR}_i(a_0, a_1, a_2)$; this choice is sufficiently general to study the problem of channel fluctuations which is the main feature to be accounted for. The proposed expression is relevant, e.g., when nodes implement the amplify-and-forward protocol to relay the signals or packets [21]. This simple but reasonable model for the SNR may either be seen as an approximation where the single-hop links dominate the multi-hop links or the talk/listen phases are scheduled appropriately. If another relaying protocol is implemented such as decode-and-forward, other expressions for the equivalent SNR may be used (see, e.g., [22]) without questioning the validity of the analytical results provided in this paper. At last, the parameter $\alpha \geq 0$ in (1) allows one to assign more or less importance to the energy consumption of the node. Indeed, the first term of the utility function represents the benefit of transmitting (i.e., the goodput) while the second term represents the cost of transmitting (i.e., the spent energy). □

The pair of functions (u_1, u_2) defines a strategic-form *static game* (see, e.g., [23]) in which the *players* are nodes 1 and 2 and the *action sets* are respectively $\mathcal{A}_1 = \mathcal{P}^2$ and $\mathcal{A}_2 = \mathcal{P}^2$. This game generalizes the forwarder's dilemma. The latter can be retrieved by assuming that φ is a step

function, p'_i is binary, p_i is constant, and all the channel gains are constant. In the next section, we describe mathematically the problem under investigation. It is shown how the problem can be modeled by a repeated game, which is precisely built on the stage or static game:

$$\mathcal{G} = (\mathcal{N}, \{\mathcal{A}_i\}_{i \in \mathcal{N}}, \{u_i\}_{i \in \mathcal{N}}),\tag{3}$$

where $\mathcal{N} = \{1, 2\}$.

The unique Nash equilibrium of $\mathcal{G}$ is $p_{i,\text{NE}} = P_{\min}$ and $p_{i',\text{NE}} = P_{\min}$. If the minimum power $P_{\min}$ is taken to be zero, then the situation where the nodes do not transmit at all corresponds to the equilibrium (and thus $(u_1, u_2) = (0, 0)$), which clearly shows one of the interests in modeling the packet transmission problem as a repeated game.

2.2 Repeated game formulation of the problem

The problem we want to solve in this paper is as follows. It is assumed that the nodes interact over an infinite number of stages. Over stage $t \in \{1, 2, ..., T\}$, $T \to \infty$, the channel gains are assumed to be fixed while the realizations of each channel gain are assumed to be i.i.d. from stage to stage. During a stage, a node typically exchanges many packets with its neighbors. At each stage, a node has to make a decision based on the knowledge it has. In full generality, the decision of a node consists in choosing a probability distribution over its set of possible actions. The knowledge of a node is in terms of global channel states and actions chosen by the other node. More precisely, it is assumed that node $i \in \mathcal{N}$ has access to a signal which is associated with the state a_0 and is denoted by $s_i \in \mathcal{S}_i$, $|\mathcal{S}_i| < \infty$. At stage t, the observation $s_i(t) \in \mathcal{S}_i$ therefore corresponds to the image (i.e., the knowledge) that node i has about the global channel state $a_0(t)$. This signal is assumed to be the output of a memoryless observation structure [21][2] whose conditional probability is denoted by $\daleth_i$:

$$\daleth_i(s_i|a_0) = \Pr[S_i = s_i|A_0 = a_0],\tag{4}$$

where capital letters stand for random variables, whereas small letters stand for realizations. Simple examples for s_i are $s_i = h_i$, $s_i = \widehat{h}_i$, $\widehat{h}_i$ being an estimate of h_i, and $s_i = (h_i, h'_i)$; $s_i = a_0 = (h_1, h'_1, h_2, h'_2)$. Now, in terms of observed actions, it is assumed that node $i \in \mathcal{N}$ has imperfect monitoring. In general, node $i \in \mathcal{N}$ has access to a signal $y_i \in \mathcal{Y}_i$, $|\mathcal{Y}_i| < \infty$, which is assumed to be the output of a memoryless observation structure whose conditional probability is denoted by Γ_i:

$$\Gamma_i(y_i|a_0, a_1, a_2) = \Pr[Y_i = y_i|(A_0, A_1, A_2) = (a_0, a_1, a_2)].\tag{5}$$

The reason why we distinguish between the observations s_i and y_i comes from the assumptions made in terms of causality. Indeed, practically speaking, it is relevant to assume that a node has access to the past realizations of s_i in the wide sense, namely, to $s_i(1), ..., s_i(t)$ at stage t. However, only the past realizations in the strict sense $y_i(1), ..., y_i(t-1)$ are assumed to be known at stage t. Otherwise, it would mean that a node would have access to the image of its current action and that of the others before choosing the former.

At this point, it is possible to define completely the problem to be solved. The problem can be tackled by using a strategic-form game model, which is denoted by $\overline{\mathcal{G}}$. As for the static game $\mathcal{G}$ on which the repeated game model $\overline{\mathcal{G}}$ is built on, the *set of players* is the set of nodes $\mathcal{N} = \{1, 2\}$. The *transmission strategy* of the node i is denoted by σ_i and consists of a sequence of functions and is defined as follows:

$$\sigma_{i,t} : \begin{array}{l} \mathcal{S}_i^t \times \mathcal{Y}_i^{t-1} \rightarrow \Delta\left(\mathcal{P}^2\right) \\ \left(s_i^t, y_i^{t-1}\right) \mapsto \pi_i(t), \end{array} \tag{6}$$

where

- $s_i^t = (s_i(1), ..., s_i(t)), y_i^{t-1} = (y_i(1), ..., y_i(t-1))$.
- $\Delta\left(\mathcal{P}^2\right)$ represents the unit simplex, namely, the set of probability distributions over the set $\mathcal{P}^2$.
- $\pi_i(t)$ is the probability distribution used by the node i at stage t to generate its action $(p_i(t), p_i'(t))$.

The type of strategies we are considering is referred to as a behavior strategy in the game theory literature, which means that at every game stage, the strategy returns a probability distribution. The associated randomness not only allows one to consider strategies which are more general than pure strategies, but also to model effects such as node asynchronicity for packet transmissions. At last, the performance of a node is measured over the long run, and nodes are therefore assumed to implement transmission strategies which aim at maximizing their long-term utilities. The *long-term utility function* of node $i \in \mathcal{N}$ is defined as:

$$\begin{aligned} U_i(\sigma_1, \sigma_2) &= \lim_{T \to \infty} \sum_{t=1}^{T} \theta_t \mathbb{E}\left[u_i(A_0(t), A_1(t), A_2(t))\right] \\ &= \lim_{T \to \infty} \sum_{t=1}^{T} \theta_t \sum_{a_0, a_1, a_2} P_t(a_0, a_1, a_2) u_i(a_0, a_1, a_2), \end{aligned} \tag{7}$$

where

- σ_i stands for the transmission strategy of node $i \in \mathcal{N}$.
- It is assumed that the limit in (7) exists.
- θ_t is a sequence of weights which corresponds to a convex combination, that is $0 \leq \theta_t < 1$ and $\sum_{t=1}^{T} \theta_t = 1$. For a repeated game with discount $\theta_t = (1 - \delta)\delta^t$ and for a classical infinitely repeated game $\theta_t = \frac{1}{T}$.
- As already mentioned, capital letters stand for random variables, whereas, small letters stand for realizations. Here, $A_0(t)$, $A_1(t)$, and $A_2(t)$ stand for

the random processes corresponding to the network state and the node actions.
- The notation P_t stands for the joint probability distribution induced by the strategy profile (σ_1, σ_2) at stage t.

This general model thus encompasses the two well-known models for the sequence of weights which are given by the model with discount and the infinite Cesaro mean. In the model with discount, note that the discount factor may model different phenomena, but in a wireless ad hoc network, the most relevant effect to be modeled seems to be the uncertainty that there will be a subsequent iteration of the stage game, for example, connectivity to an access point can be lost. With this interpretation in mind, the discounting factor represents, for example, the probability that the current round is not the last one , or in terms of mobility, it may also represent the probability that the nodes do not move for the current stage. Therefore, it may model the departure or the death of a node (e.g., due to connectivity loss) for a given routing path. More details about this interpretation can be found in [23] while [24] provides a convincing technical analysis to sustain this probabilistic interpretation.

At this point, we have completely defined the strategic form of the *repeated game* that is the triplet

$$\overline{\mathcal{G}} = \left(\mathcal{N}, \{\Sigma_i\}_{i \in \mathcal{N}}, \{U_i\}_{i \in \mathcal{N}}\right), \tag{8}$$

where Σ_i is the set of all possible transmission strategies for node $i \in \mathcal{N}$.

One of the main objectives of this paper is to exploit the above formulation to find a globally efficient transmission scheme for the nodes. For this purpose, we will characterize long-term utility region for the problem under consideration. It is important to mention that the characterization of the feasible utility region of a dynamic game (which includes repeated games as a special case) with an arbitrary observation structure is still an open problem [25]. Remarkably, as shown recently in [11] and [12], the problem can be solved for an interesting class of problems. It turns out that the problem under investigation belongs to this class provided that the channel gains evolve according to the classical model of block i.i.d. realizations.

In the next section, we show how to exploit [11, 12] to characterize the long-term utility region and construct a practical transmission strategy. In Section 2.4, we will show how to integrate the strategic stability[3] property into this strategy, this property being important to ensure that selfish nodes effectively implement the efficient strategies.

2.3 Long-term utility region characterization
When the number of stages is assumed to be large, the random process associated with the *network state*

$A_0(1), A_0(2), ..., A_0(T)$ is i.i.d, and the observation structure given by $(\daleth_1, \daleth_2, \Gamma_1, \Gamma_2)$ is memoryless, some recent results can be exploited to characterize the feasible utility region of the considered repeated game and to derive efficient transmission strategies. The main difficulty to determine the feasible utility region of $\overline{\mathcal{G}}$ is to find the set of possible average correlations between a_0, a_1, and a_2. Formally, the correlation averaged over T stages is defined by:

$$P^T(a_0, a_1, a_2) = \frac{1}{T} \sum_{t=1}^{T} P_t(a_0, a_1, a_2), \qquad (9)$$

where P_t is the joint probability at stage t. More precisely, a key notion to characterize the attainable long-term utilities is the notion of implementability, which is given as follows.

Definition 1 *An average correlation Q is said to be implementable if there exists a pair of transmission strategies (σ_1, σ_2) such that the average correlation induced by these transmission strategies verifies:*

$$\begin{aligned} &\forall (a_0, a_1, a_2) \in \mathcal{A}_0 \times \mathcal{A}_1 \times \mathcal{A}_2, \\ &\lim_{T \to \infty} \tfrac{1}{T} \sum_{t=1}^{T} P_t(a_0, a_1, a_2) = Q(a_0, a_1, a_2). \end{aligned} \qquad (10)$$

Using the above definition, the following key result can be proved.

Proposition 1 *The Pareto frontier of the achievable utility region of $\overline{\mathcal{G}}$ is given by all the points under the form $\left(\mathbb{E}_{Q_\lambda}(u_1), \mathbb{E}_{Q_\lambda}(u_2) \right)$, $\lambda \in [0, 1]$, where Q_λ is a maximum point of*

$$W_\lambda = \lambda \mathbb{E}_Q(u_1) + (1 - \lambda) \mathbb{E}_Q(u_2), \qquad (11)$$

and each maximum point is taken in the set of probability distributions which factorize as follows:

$$\begin{aligned} Q(a_0, a_1, a_2) = \sum_{v, s_1, s_2} \rho(a_0) P_V(v) \times \daleth(s_1, s_2 | a_0) \\ \times P_{A_1 | S_1, V}(a_1 | s_1, v) P_{A_2 | S_2, V}(a_2 | s_2, v), \end{aligned} \qquad (12)$$

where

- *λ denotes the relative weight assigned to the utility of the first player and can be chosen arbitrarily depending on some prescribed choice, e.g., in terms of fairness or global efficiency.*
- *ρ is the probability distribution of the network state a_0.*
- *$\daleth$ is the joint conditional probability which defines the assumed observation structure, i.e., a probability which is written as:[4]*

$$\daleth(s_1, s_2 | a_0) = Pr[(S_1, S_2) = (s_1, s_2) | A_0 = a_0]. \qquad (13)$$

- *$V \in \mathcal{V}$ is an auxiliary random variable or lottery.*

(See the proof in the Appendix). One interesting comment to be made concerns the presence of the "parameter" or auxiliary variable V. The presence of the auxiliary variable is quite common in information-theoretic performance analyses and in game-theoretic analyses through the notion of external correlation devices (such as those assumed to implement correlated equilibria). Indeed, $V \in \mathcal{V}$ is an auxiliary random variable or lottery which can be proved to improve the performance in general (see [11] for more details). Such a lottery may be implemented by sampling a signal which is available to all the transmitters, e.g., an FM or a GPS signal.

In (22), ρ and $\daleth$ are given. Thus, W_λ has to be maximized with respect to the triplet $(P_{A_1 | S_1, V}, P_{A_2 | S_2, V}, P_V)$. In this paper, we restrict our attention to the optimization of $(P_{A_1 | S_1, V}, P_{A_2 | S_2})$ for a fixed lottery P_V and leave the general case as an extension.

The maximization problem of the functional $W_\lambda(P_{A_1 | S_1, V}, P_{A_2 | S_2, V})$ with respect to $P_{A_1 | S_1, V}$ and $P_{A_2 | S_2, V}$ amounts to solving a bilinear program. The corresponding bilinear program can be tackled numerically by using iterative techniques such as the one proposed in [26], but global convergence is not guaranteed, and therefore, some optimality loss may be observed. Two other relevant numerical techniques have also been proposed in [27]. The first technique is based on a cutting plane approach while the second one consists of an enlarging polytope approach. For both techniques, convergence may also be an issue since for the first technique, no convergence result is provided and for the second technique, cycles may appear [28]. To guarantee convergence and manage the computational complexity issue, we propose here another numerical iterative technique, namely, to exploit the sequential best-response dynamics (see, e.g., [29] for a reference in the game theory literature, [23] for application examples in the wireless area, [30] for a specific application to power control over interference channels). Here also, some efficiency loss may be observed, but it will be shown to be relatively small for the quite large set of scenarios we have considered in the numerical performance analysis. The sequential best-response dynamics applied to the considered problem translates into the following algorithm.

Although suboptimal in general (as the available state-of-the art techniques), the proposed technique is of particular interest for at last three reasons. First, convergence is unconditional. It can be proved to be guaranteed, e.g., by induction or by identifying the proposed procedure as an instance of the sequential best-response dynamics for an exact potential game (any game with a common utility is an exact potential game). Second, convergence points are local maximum points, but in all the simulations performed, those maximums had the virtue of not being too far from the global maximum. At last but not least, it allows us to build a practical transmission strategy which outperforms all the state-of-the art transmission strategies, as explained next. Note that this is necessarily the case when Algorithm 1 is initialized with the state-of-the art transmission strategy under consideration. Although we will not tackle the classical issue of the influence of initialization on the convergence point, it is worth mentioning that many simulations have shown that the impact of the initial point on the performance at convergence is typically small, at least for the utilities under consideration. Therefore, initializing Algorithm 1 with naive strategies such as transmitting at full power $\forall t \in \{1, ..., T\}$, $(a_1(t), a_2(t)) = (P_{\max}, P_{\max}, P_{\max}, P_{\max})$ is well suited.

Remark. Algorithm 1 would typically be implemented offline in practice. The purpose of Algorithm 1 is to generate decision functions which are exploited by the proposed transmission strategy. To implement Algorithm 1, only statistics need to be estimated in practice (namely, the channel distribution ρ and the observation structure conditional distribution $\daleth$); estimating statistics such as the channel distribution information is known to be a classical issue in the communications literature.   □

2.4   Proposed equilibrium transmission strategy

The main purpose of this section is to obtain globally efficient transmission strategies. Here, global efficiency is measured in terms of social welfare, namely, in terms of the sum $U_1 + U_2$. This corresponds to choosing $\lambda = \frac{1}{2}$. This choice is pragmatic and follows to what is often done in the literature; it implicitly means that the network nodes have the same importance. Otherwise, this parameter can always be chosen to operate at the desired point of the utility region. Indeed, social welfare is a

---

Algorithm 1

1. **Initialization**. *The arguments of the functional* W_λ *are fixed to an initial value:* $(P_{A_1|S_1,V}, P_{A_2|S_2,V}) = \left(P^{(0)}_{A_1|S_1,V}, P^{(0)}_{A_2|S_2,V} \right).$

2. **Iteration**. *At iteration* $n \geq 1$, $P^{(n)}_{A_i|S_i,V}$ *is updated by being chosen in the argmax of* $W_\lambda \left(P_{A_i|S_i,V}, P^{(n-1)}_{A_{-i}|S_{-i},V} \right).$ *If there are several maximum points, choose one of them randomly and according to a uniform law.*

3. **Stopping criterion**.
$\left| W_\lambda \left(P^{(n)}_{A_i|S_i,V}, P^{(n)}_{A_{-i}|S_{-i},V} \right) - W_\lambda \left(P^{(n-1)}_{A_i|S_i,V}, P^{(n-1)}_{A_{-i}|S_{-i},V} \right) \right| <$
η *for some* $\eta \geq 0$.

---

well-known measure of efficiency, and it also allows one to build other famous efficiency measure of a distributed network such as the price of anarchy [31]. The proposed approach holds for any other feasible point of the utility region which is characterized in the preceding section.

The transmission strategy we propose comprises three ingredients: (1) a well-chosen operating point of the utility region, (2) the use of reputation [5, 8], and (3) the use of virtual credit [9, 10].

1. The proposed operating point is obtained by applying the sequential best-response dynamics procedure described in Section 2.3 and choosing $\lambda = 0.5$, $|\mathcal{V}| = 1$. Each individual maximization operation provides a probability distribution which is denoted by $\pi_i^\star$. Since W_λ is linear in $P_{A_i|S_i,V}$, the maximization operation returns a point which is one of the vertices of the unit simplex. The corresponding probability distribution has thus a particular form, namely, that of a decision function under the form $f_i^\star(s_i)$. Therefore, when operating at this point, at stage t, node i chooses its action to be $a_i(t) = f_i^\star(s_i(t))$. This defines for node i a particular choice of a lottery over its possible actions; this lottery is denoted by $\pi_i^\star(t)$ and is the unit simplex of dimension $2L$, i.e., $\Delta(\mathcal{P}^2)$. By convention, the possible actions for node i are ordered according to a *lexicographic ordering*. Having $\pi_i^\star(t) = (1, 0, 0, ..., 0) \in \Delta(\mathcal{P}^2)$ means that action (P_1, P_1) is used with probability 1 (wp1); having $\pi_i^\star(t) = (0, 1, 0, ..., 0) \in \Delta(\mathcal{P}^2)$ means that action (P_1, P_2) is used wp1; ...; having $\pi_i^\star(t) = (0, 0, 0, ..., 0, 1) \in \Delta(\mathcal{P}^2)$ means that action (P_L, P_L) is used wp1.

2. The reputation (see, e.g., [5, 8]) of a neighboring node is evaluated as follows. Over each game stage duration, the nodes exchange a certain number of packets which is denoted by K. This number is typically large. Since each node has access to the realizations of the signal y_i for each packet, it can exploit it to evaluate the reputation of the other node at stage t. In this section, we assume a particular observation structure Γ_1, Γ_2, which is tailored to the considered problem of packet forwarding in ad hoc networks. We assume that the signal y_i is binary: $y_i \in \{D, F\}$. Let $\epsilon \in [0, 1]$ be the parameter which represents the *probability of misdetection*. If node i chooses the action $a^{\min} = (P_{\min}, P_{\min})$ (resp. any other action of $\mathcal{P}_i^2$), then with probability $1 - \epsilon$, node $-i$ receives the signal D (resp. F). With probability ϵ, node $-i$ perceives what we define as the action Drop D (resp. Forward F) while the action Forward F (resp. Drop D) has been chosen by node i. Thus, Γ_i takes the following form:

$$\Gamma_i(y_i|x_0, a_1, a_2) = \begin{vmatrix} 1 - \epsilon & \text{if } y_i = \text{F and } a_{-i} \in \mathcal{A}_i^{\text{F}}, \\ \epsilon & \text{otherwise,} \end{vmatrix}$$
(14)

where $\mathcal{A}_i^{\text{F}} = \mathcal{P}_i \times \mathcal{P}_i \setminus \{0\}$.

Using these notations, node i can compute the reputation of node $-i$ as follows:

$$R_{-i}(t) = \frac{(1 - \epsilon)|\{y_i = \text{F}\}| + \epsilon|\{y_i = \text{D}\}|}{K},$$
(15)

where $|\{y_i = \text{F}\}|$ and $|\{y_i = \text{D}\}|$ are respectively the numbers of occurrences of the action Forward and Drop among the K packets node i has been needing the assistance of node $-i$ to forward its packets. The reputation $R_{-i}(t)$ is one of the tools we use to implement the transmission strategy which is described further. Note that one of the interesting features of the proposed mechanism is that reputation (15) of node $-i$ only exploits local observations (first-hand reputation information); node i does not need any information about the behavior of its neighboring nodes. This contrasts with the closest existing reputation mechanisms such as [5] and [8], for which the reputation estimation procedure exploits information obtained from other nodes (second-hand information). The corresponding information exchange induces additional signaling and additional energy consumption. At last, by using these techniques, selfish nodes may collude and disseminate false reputation values.

3. The idea of virtual credit is assumed to be implemented with a similar approach to previous works [9, 10], namely, we assume that the nodes have an initial amount of credits, impose a cost in terms of spent credits for a node that wants to transmit through a neighbor at a certain frequency or probability, and that are rewarded when they forward their neighbors' packets. The reward and cost assumed in this paper are defined next.

The proposed transmission strategy is as follows. While the node has not enough credit, it adopts a cooperative decision rule, which corresponds to operating at the point we have just described. Otherwise, it adopts a signal-based tit-for-tat decision rule, which has been found to be very useful to implement mutual cooperation [1, 32]. Existing tit-for-tat decision rules such as Generous-Tit-For-Tat (GTFT) or Mend-Tolerance Tit-For-Tat (MTTFT) [1] do not take into account the possible existence of a state for the game and therefore the existence of a signal associated with the realization of the state. Additionally, the proposed tit-for-tat decision rule also takes into account the fact that action monitoring is not perfect.

In contrast with the conventional setup assumed to implement tit-for-tat or its variants [1], the action set of a node is not binary. Therefore, we have to give a meaning to tit-for-tat in the considered setup. The proposed meaning is as follows. When node i receives the signal D and node $-i$ has effectively chosen the action Drop, it means that node $-i$ has chosen $a_{-i} = a^{\text{min}} = (P_{\text{min}}, P_{\text{min}})$. When node i receives the signal Forward and node $-i$ has effectively chosen the action Forward, it means that node $-i$ has chosen $a_{-i} = a^{\star}_{-i} = f^{\star}_{-i}(s_{-i})$. Implementing tit-for-tat for node i means choosing $a_i = a^{\text{min}} = (P_{\text{min}}, P_{\text{min}})$ (which represents the counterpart of the action Drop) if the node $-i$ is believed to have chosen the action $a_{-i} = a^{\text{min}} = (P_{\text{min}}, P_{\text{min}})$. On the other hand, node i chooses the best action $a^{\star}_i = f^{\star}_i(s_i)$ when it perceives that node $-i$ has chosen the action $a^{\star}_{-i} = f^{\star}_{-i}(s_{-i})$. Note that, contrarily to the conventional tit-for-tat decision rule, the actions $a^{\star}_i$ and $a^{\star}_{-i}$ differ in general. Denoting by $m_i(t) \geq 0$ the credit of node i at stage t, the proposed strategy expresses formally as follows. We will refer to this transmission strategy as SARA (for State Aware tRAnsmission strategy).

Proposed transmission strategy (SARA).

$$\sigma^{\star}_{i,t}(s_i^t, y_i^{t-1}) = \begin{vmatrix} \pi^{\star}_i(t) & \text{if } t = 0 \text{ or } m_i(t) < \mu, \\ \widehat{\pi}_{-i}(t-1) & \text{otherwise,} \end{vmatrix}$$
(16)

where

- *The virtual credit $m_i(t)$ obeys the following evolution law:*

$$m_i(t) = m_i(t-1) + \beta < \pi_i(t-1), e_k > -\beta v_i(t-1),$$
(17)

  $\beta v_i(t-1)$ represents the virtual monetary cost for node i when its packet arrival rate is $v_i(t-1)$, with $\beta \geq 0$; $<; >$ stands for the scalar product; e_k is the kth vector of the canonical basis of $\mathbb{R}^{2L}$, namely, all components equal 0 except the kth component which equals 1. The index k is given by the index of action $a^{\star}_i(t) = f^{\star}_i(s_i(t))$.
- *$\mu \geq 0$ is a fixed parameter which represents the cooperation level of the nodes. A sufficient condition on μ and β to guarantee that the nodes have always enough credits is that $\mu \geq 2\beta$.*
- *The distribution $\widehat{\pi}_{-i}(t-1)$ is constructed as follows:*

$$\widehat{\pi}_{-i}(t-1) = R_{-i}(t-1)\pi^{\star}_i(t) + [1 - R_{-i}(t-1)]\pi^{\text{min}},$$
(18)

  with $\pi^{\text{min}} = (1, 0, 0, \ldots, 0) \in \mathbb{R}^{2L}$ representing the pure action $a^{\text{min}} = (P_{\text{min}}, P_{\text{min}})$.

Comment 1 *The second term of the dynamical equation which defines the credit evolution corresponds to the reward a node obtains when it forwards the packets of the other node. On the other hand, the third term is the cost paid by the node for asking the assistance of the other node to forward. The same weight (namely, β) is applied on both terms to incite the node to cooperate. Additionally, such a choice allows one from preventing cooperative nodes to have an excessive number of credits, thus to avoid using a mechanism such as in [6] and [8] to regulate the number of credits. In [6] and [8], the credit excess occurs because the reward in terms of credits only depends on the node action and the cost only depends on the relaying node action. Thus, when the node is cooperative and the relaying node is selfish, there will be a reward but not a cost. As for the credit system, in practice, it might be implemented either by assuming the existence of an external central trusted entity [10] that stores and manages the node credits, or through a credit counter located in the node and maintained by a tamper-resistant security module. Thus, in practice, the operation of this module would not be altered, because it would be designed so that information be accessible only by specific software containing appropriate security measures.*

Comment 2 *Depending on whether packet forwarding rate maximization or energy minimization is sought, it is possible to tune the triplet of parameters $(m_i(0), \beta, \mu)$ according to what is desired. In this respect, it can be checked that the best packet forwarding rate is obtained by choosing any triplet under the form $(2\beta, \beta, 2\beta)$ for any $\beta > 0$. But the power (which is defined by the quantity average network power (ANP) defined through (21)) is then at its maximum. On the other hand, if the triplet of parameters takes the form $(m_i(0), 0, 0)$, with $m_i(0) \geq 0$, the consumed network power will be at its minimum and the packet forwarding rate will be minimized as well. Other choices for the triplet $(m_i(0), \beta, \mu)$ therefore lead to various tradeoffs in terms of transmission rate and consumed power.*

Comment 3 *The proposed strategy can always be used in practice whether or not it corresponds to an equilibrium point of $\overline{\mathcal{G}}$. However, if the strategic stability property is desired, some conditions have to be added to ensure that it corresponds to an equilibrium. Indeed, effectively operating at an efficient point in the presence of self-interested and autonomous nodes is possible if the latter have no interest in changing their transmission strategy. More formally, a point which possesses the property of strategic stability or Nash equilibrium is defined as follows.*

Definition 2 *A strategy profile $\left(\sigma_1^{NE}, \sigma_2^{NE}\right)$ is a Nash equilibrium point for $\overline{\mathcal{G}}$ if*

$$\forall i \in \mathcal{N}, \forall \sigma_i \in \Sigma_i, \ U_i\left(\sigma_1^{NE}, \sigma_2^{NE}\right) \geq U_i\left(\sigma_i, \sigma_{-i}^{NE}\right).$$
(19)

In order to obtain an explicit condition for the proposed strategy to be an equilibrium, we consider, as the closest related works [1, 8], a repeated game with discount. This also allows some effects such as the loss of network connectivity to be captured. Remarkably, for the repeated game model with discount, the subgame[5] perfection property is also available. This is useful in practice since it offers some robustness in terms of node behavior. Indeed, this property makes the equilibrium strategy robust against changes in terms of node behavior which might occur during the transmission process; even if some deviations from equilibrium occurred in the past, players have an interest in coming back to equilibrium. A necessary and sufficient condition for a strategy profile to be subgame perfect equilibrium is given by the following result.

Proposition 2 *Assume that $\forall t \geq 1, \theta_t = (1 - \delta)\delta^t, 0 < \delta < 1$. The strategy profile $(\sigma_1^\star, \sigma_2^\star)$ defined by (16) is a subgame perfect equilibrium of $\overline{\mathcal{G}}$ if and only if:*

$$\delta \geq \max\left\{\frac{c_i}{(1 - 2\epsilon)r_i}, \frac{c_{-i}}{(1 - 2\epsilon)r_{-i}}, 0\right\},$$
(20)

where: $c_i = \sum_{a_0} \rho(a_0)\left(u_i^1 - u_i^k\right),$ $r_i = \sum_{a_0} \rho(a_0)$ $\left(u_i^k - u_i^2\right).$ $u_i^1 = u_i\left(a_0, a_1^\star, a^{\min}\right),$ $u_i^k = u_i\left(a_0, a_1^\star, a_2^\star\right),$ $u_i^2 = u_i\left(a_0, a^{\min}, a_2^\star\right),$ *and* $a_i^\star = f_i^\star(s_i).$

(See the proof in the Appendix)

Comment 4 *The proposed transmission strategy is compatible with a packet delivery mechanism such as an ACK/NACK mechanism. Indeed, in the definition of the transmission strategies (6), the observed signal y_i may correspond to a binary feedback such as an ACK/NACK feedback. Indeed, y_i corresponds to an image of (a_0, a_1, a_2). Such image might then be a binary version of the receive SNR or SINR (e.g., if the receive SNR is greater than a threshold then the packet is well received and the corresponding feedback signal y_i will be ACK). More generally, a binary feedback of the form Forward/Drop is completely compatible with the presence of ACK/NACK feedback-type mechanisms. Simply, the signal Drop may combine the effects of a selfish behavior and bad channel conditions.*

Comment 5 *This section shows that the proposed transmission strategy has five salient features.*

1. First of all, in contrast with the related works on the forwarder's dilemma, it is able to deal with the problem of time-varying link qualities.
2. Second, it not only deals with the adaptation of the cooperation power p_i' of node i (which is the power to forward the packets of the other node) but also the power to transmit its own packets p_i.
3. Third, the proposed strategy is built in a way to exploit the available arbitrary knowledge about the global channel state (a_0) as well as possible. The key observation for this is to exploit the provided utility region characterization. Ideally, the nodes should operate on the Pareto frontier. This is possible if a suited optimal algorithm is used.
4. Fourth, the proposed transmission strategy is shown to possess the strategic stability property in games with discount under an explicit sufficient condition on the discount factor. Note that, here again, each node has only imperfect monitoring of the actions chosen by the other node. Additionally, the equilibrium strategy is subgame perfect.
5. Fifth, the proposed strategy does not induce any problem of credit outage or excess. Credit outage is avoided only if the conditions $\mu \geq 2\beta$ and (20) are satisfied. Therefore, if there is no credit outage problem, there is no need for assisting distant nodes as required in [6] and [8].

3 Results and discussion

3.1 Numerical performance analysis

All simulations provided in this section have been obtained by an ad hoc simulator developed under *Matlab*. The simulation setup we consider in this paper is very close to those assumed in the closest works and [3] and [8] in particular. The setup we assume by default is provided in Section 3.1.1. When other values for some parameters are considered, this will be explicitly mentioned. In addition to the simulation setup subsection, the simulation section comprises three subsections. The first subsection (Section 3.1.2) aims at conducting a performance analysis in terms of utility function (1), which captures the trade-off between the transmission rate and the energy spent for transmitting. Section 3.2 focuses on the transmission rate aspect while Section 3.2.1 is dedicated to a performance analysis in terms of consumed network energy.

3.1.1 Simulation setup assumed by default

We consider a network of N nodes. When N is considered to be fixed, it will be taken to be equal to 50. The N nodes are randomly placed (according to a uniform probability distribution) over an area of 1000×1000 m^2; only network topology draws which guarantee every node to have a neighbor (in the sense of its radio range) are kept. The assumed topology corresponds to a random topology since the node locations are drawn from a given spatial distribution law (which is uniform for the simulations). Each node only considers the behavior of its neighbors to choose its own behavior. As assumed in the related literature, if a node has several neighbors, it is assumed to play a given game with each of its neighbors. In fact, averaging the results over the network topology realizations has the advantage of making the conclusions less topology dependent. Provided simulations are averaged over 1200 draws for the network topology. Routes are supposed to be fixed and known. Indeed, the proposed transmission strategy is compatible with any routing algorithm. One node can communicate with another node only if the inter-node distance is less than the radio range, which is taken to be 150 m. When a node has several neighbors, it may be involved in several routing paths; then, it is assumed to play several independent forwarding games in parallel and have a given initial credit $m_i(0)$ for each neighbor. The credits are updated separately based on the corresponding forwarding game. This means that the credits a node receives by cooperating with one of its neighbors can only be used for forwarding via the considered neighbor. As a node without neighbors does not need credits and the nodes do not obtain an initial credit, the problem of credit excess is avoided. By default, 50% of the nodes are assumed to be selfish but the network does not comprise any malicious node. The initial packet forwarding rate for cooperative nodes and selfish nodes are respectively set to 1 and 0.1, respectively. Each source node transmits at a constant bit rate of 2 packets/s. For each draw for the network topology, the simulation is run for 1000 s. This period of time is made of 20 frames of 50 s. A frame corresponds to a game stage and to a given draw for the channel vector h. The fact that channel gains are assumed to fluctuate over time is a way of accounting for mobility; in the simulations, they are assumed to evolve according to a (discrete version of the) Rayleigh fading law. Averaging over network topologies allows one to average the results over the path losses. Each channel gain is thus drawn according to an exponential law, which corresponds to a Rayleigh law for the amplitude; if one denotes by h_i the considered channel gain, we have that $h_i \sim \frac{1}{\bar{h}_i} e^{-\frac{h_i}{\bar{h}_i}}$, where $\bar{h}_i = \mathbb{E}(h_i)$ represents the path loss effects. As mentioned above, the channel gain is discrete and the discrete realizations are obtained by quantizing the realizations given by a Rayleigh distribution. The effect of quantization on the performance is typically small. Simulations, which are provided here, show that the loss induced by implementing Algorithm 1 by using quantized channel gains in the presence of actual channel gains which are continuous is about a few percents for the size of channel gain sets used for the simulations. If d denotes the inter-node distance of the considered pair of nodes, then the path loss is assumed to depend on the distance according to $\bar{h}_i = \frac{\text{const}}{d^2+\kappa^2}$; $\kappa > 0$ is a distance which is used to avoid numerical divergence in $d = 0$. In practice, κ may typically represent the antenna height. During each frame, 100 packets are exchanged. Table 1 recaps the values chosen for the main network parameters.

Concerning the game parameters, the following choices are made by default. The parameter α is set to 10^{-2}. The receive variance σ^2 is always set to 0.1. The sets of possible power levels are defined by: $\forall i \in \{1,2\}, \mathcal{P}_i = \mathcal{P}'_i, L = 10$, $P_{\min} = 0$, and $P_{\max} = 10$ W. The power increment is uniform over a dB scale, starting from the minimal positive power which is taken to be equal to 10 mW. The sets of possible channel gains are defined by: $\forall i \in \{1,2\}, \mathcal{H}_i = \mathcal{H}'_i$: $H = 10$, $h_{\min} = 0.04$, and $h_{\max} = 10$, and the channel gain increment equals $\frac{10-0.04}{10}$. The different means of the channel gains are given by $\left(\bar{h}_i, \bar{h}'_i, \bar{h}_{-i}, \bar{h}'_{-i}\right) = (1,1,1,1)$. The communication efficiency function is chosen as in [20]: $\varphi(x) = e^{-\frac{c}{x}}$ with $c = 2^r - 1$, r being the spectral efficiency in bit/s/Hz [20]. In the simulations provided we always have $r = 1$ bit/s/Hz; one simulation will assume a higher spectral efficiency, namely, $r = 3$ bit/s/Hz.

3.1.2 Utility analysis

Here, to be able to easily represent the utility region for the considered problem and to be able to compare our approach with previous models, we consider two neighboring nodes.

The first question we want to answer is to what extent the ability for a node to properly adapt to the link qualities which have an impact on the weighted utility $w_\lambda = \lambda u_1 + (1 - \lambda)u_2$; it is related to its knowledge about these qualities, i.e., the global channel state $a_0 = \left(h_1, h'_1, h_2, h'_2\right)$. To this end, we have represented in Fig. 2 the achievable utility region under various information assumptions. The top curve in solid line represents the Pareto frontier which is obtained when implementing the transmission strategy given by Algorithm 1 when $\forall i \in \mathcal{N}, s_i = a_0 = \left(h_1, h'_1, h_2, h'_2\right)$ that is each node has *global CSI*. The disks correspond to the performance of the *centralized transmission strategy*, namely the best performance possible. It is seen that for a typical scenario the proposed algorithm does not involve any optimality loss. The curve with squares is obtained with Algorithm 1 under *local CSI*, i.e., $s_i = (h_i, h'_i)$. Interestingly, the loss for moving from global CSI to local CSI is relatively small. This shows that it is possible to implement a distributed transmission strategy without sacrificing too much the global performance. This result is not obvious since the weighted utility w_λ depends on the whole vector a_0. When *no CSI* is available (i.e., $s_i =$ constant), the incurred loss is more significant. Indeed, the curve with diamonds (which is obtained by choosing for each $\lambda \in [0,1]$ the best action profile in terms of the expected weighted utility[6] (11)) shows that the gain in terms of sum-utility or social welfare when moving from no CSI to global CSI is about 10%. The point marked by a star indicates the operating point for which transmitting at full power $a_i = (P_{\max}, P_{\max}, P_{\max}, P_{\max})$ is optimal under no CSI.

As a second step, we compare the performance of SARA, ICARUS [3, 6, 8] and GTFT [1], which do not take into account the channel fluctuations. The three corresponding equilibrium points are particular points of the achievable or feasible utility region represented by Fig. 3. The outer curve is the achieved utility region of Fig. 2 when Algorithm 1 is implemented under local CSI (it is the same as the curve with squares of Fig. 2). The social optimum corresponds to the point indicated by the small disk. The point marked by a square corresponds to the performance of SARA, whereas the points marked by a star and a diamond respectively represent the equilibrium points obtained when using ICARUS [3, 6, 8] and GTFT [1]. Note that the way the strategies ICARUS and GTFT have been designed is such that they are able to adapt the packet forwarding rates but not the transmit power level. As a consequence, they cannot exploit any available knowledge in terms of CSI, which induces a quite significant performance loss; it is assumed that GTFT and ICARUS use a pair of actions (a_1, a_2) which maximizes the expected sum-utility. The gain obtained by the proposed

Table 1 Simulation settings

Network parameters	Value
Space	1000 m × 1000 m
Number of nodes	$N = 50$
Radio range	150 m
Const	10^3
κ	5 m
Initial credit	$m_i(0) = 35$
Initial credit of ICARUS [8]	$m_i(0) = 220$
Parameter cost	$\beta = 10$
Cooperation degree	$\mu = 20$
Probability of misdetection	$\epsilon = 0$
Packet arrival rate	$\nu = 1$
Simulation time	1000 s
Frame/stage duration	50 s
Generosity parameter of GTFT	0.1
IFN of ICARUS [6, 8]	5
edp$_{\text{th}}$ of ICARUS [6, 8]	0.85
a of ICARUS [6, 8]	0.5
b of ICARUS [6, 8]	2.3
Number of topology draws	1200

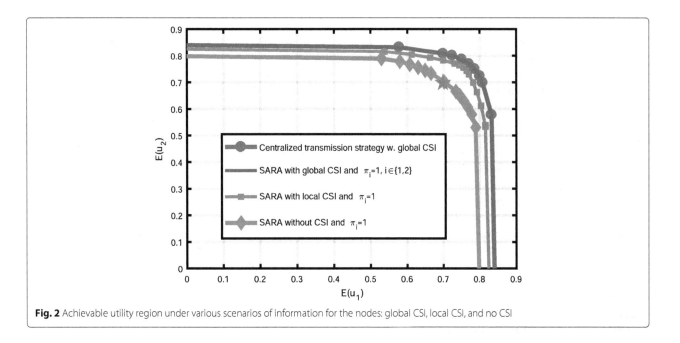

Fig. 2 Achievable utility region under various scenarios of information for the nodes: global CSI, local CSI, and no CSI

transmission strategy comes not only from the fact that the transmit power level can adapt to the wireless link quality fluctuations, but also from the proposed cooperation mechanism. The latter both exploits the idea of virtual credit and reputation, which allows one to obtain a better packet forwarding rate than ICARUS and GTFT. We elaborate more on this aspect in the next subsection. At last, when implementing a transmission strategy built from the one-shot game model given in Section 2.1, the

NE of would be obtained, i.e., the operating point would be $(0, 0)$, which is very inefficient.

3.2 Packet forwarding rate analysis

In the previous subsection, we have been assessing the benefits from implementing the proposed transmission strategy in terms of utility. The utility implements a trade-off between the transmission rate and the consumed energy. Here, we want to know how good is the proposed

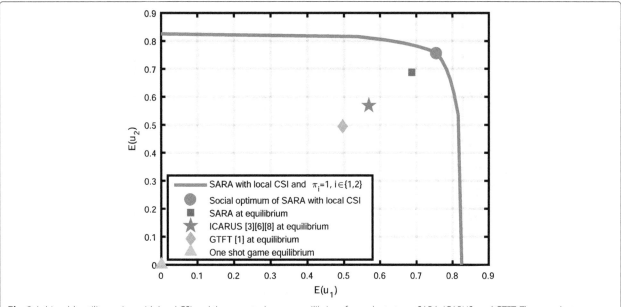

Fig. 3 Achievable utility region with local CSI and the repeated game equilibrium for each strategy: SARA, ICARUS, and GTFT. The one-shot game Nash equilibrium is also represented. The strategies ICARUS and GTFT do not take into account the channel fluctuations

strategy in terms of packet forwarding rate, that is the packet transmission probability.

Figure 4 depicts the evolution of the packet forwarding rate for SARA, ICARUS, and GTFT for a network of 50 nodes. We look at the influence of the fraction of selfish nodes. SARA is very robust to selfishness. Whatever the fraction of selfish nodes, SARA provides a high performance in terms of packet forwarding rate. We see that ICARUS is less efficient than SARA in terms of stimulating cooperation in the presence of selfish nodes, which shows that the proposed punishment mechanism is effectively relevant. The GTFT strategy performance decreases in a significant manner with the number of selfish nodes. For the latter transmission strategy, it is seen that when the network is purely selfish, the operating packet forwarding rate is about 50%; this shows the significant loss induced by using a cooperation scheme which is not very robust to selfishness.

The robustness to observation errors is assessed. More precisely, we want to evaluate the impact of not observing the action Forward or Drop perfectly on the packet forwarding rate. Figure 5 depicts the packet forwarding rate as a function of the probability of misdetection ϵ (see (15)). When $\epsilon > 10\%$, the performance of ICARUS sharply decreases. This is because the retaliation aspect becomes a dominant effect. Nodes punish each other, whereas they should not; this is due to the fact that the estimates of the forwarding probabilities become poor and the ICARUS mechanism is sensitive to estimation errors; illegitimate punishments are implemented, leading

to a very inefficient network. On the other hand, observation errors have little influence on SARA because under the equilibrium condition, provided that the credit is less than μ, nodes keep on cooperating. Estimating the forwarding rate does not intervene in the decision process of a node. Note that we have only considered $\epsilon \leq 50\%$. The reason for this is as follows. When $\epsilon > 50\%$, it is always possible, by symmetry, to decrease the effective probability of misdetection to $\epsilon' = 1 - \epsilon$. For this, it suffices to declare the used action to be Forward, whereas the action Drop was observed and vice-versa.

3.2.1 Consumed network energy analysis

Based on the preceding two sections, we know that SARA provides improvements in terms of utility and packet forwarding rate. But the most significant improvements are in fact obtained in terms of consumed energy. Indeed, ICARUS and GTFT have not been designed to account for link quality fluctuations, whereas SARA adapts both the packet forwarding rate and the transmit power level using the parameters assumed by default, except for the path loss $\bar{h}_i = \frac{const}{d^2 + \kappa^2}$, where const=$10^3$ and $\kappa = 5$. In the current formulation of ICARUS and GTFT, the transmit power is fixed (as in Section 3.1.2) according to the best pair of actions in terms of expected sum-utility. In this subsection, the advantage of adapting the power to the quality of the wireless link is clearly observed. Since the consumed network energy is proportional to the network sum-power averaged over time, we will work with the *average network power* (ANP). Here, we consider the

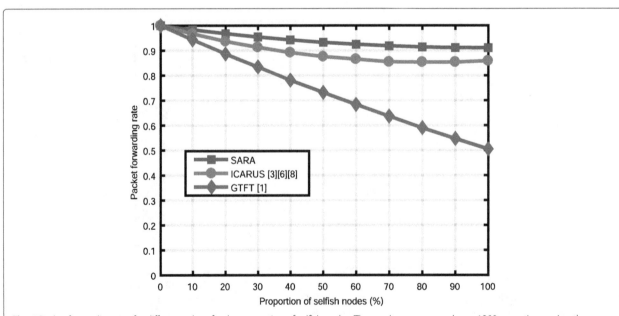

Fig. 4 Packet forwarding rate for different values for the proportion of selfish nodes. The results are averaged over 1200 executions, using the simulation setup assumed by default

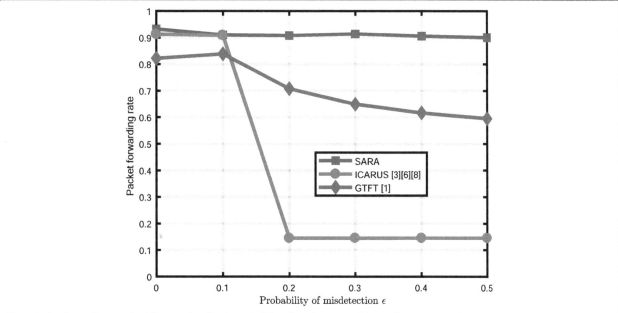

Fig. 5 Packet forwarding rate for different values for the probability of misdetection ϵ. The results are averaged over 1200 executions, using the simulation setup assumed by default

total power which is effectively consumed by the node and not the radiated powers p_i and p_i' (the consumed power therefore includes the circuit power in particular). As explained, for example, in [33, 34], and [35], a reasonable and simple model for relating the radiated power and the consumed power is the affine model: $p_{i,\text{total}} = a(p_i + p_i') + b$. The parameter b is very important since it corresponds to the power consumed by the node when no packet is transmitted; in [34] and [35], it represents the circuit power, whereas in [33], it represents the node computation power. Here, we assume as in [35] that b is comparable to the $P_{\max}$ and choose the same typical values as in [35], namely, $b = P_{\max} = 1$ W. Eventually, the ANP is obtained by averaging the following quantity $\sum_{i=1}^{N} \{a[p_i(t) + p_i'(t)] \pi_i(t) + b\}$ over all channel and network topology realizations, where N is always the number of nodes in the network and $\pi_i(t)$ the forwarding probability for stage or frame t:

$$\text{ANP} = \frac{1}{T'} \sum_{i=1}^{N} \left\{ a \left[p_i(t) + p_i'(t) \right] \pi_i(t) + b \right\} \qquad (21)$$

where T' corresponds to the number of realizations used for averaging. Here, this quantity is averaged over 1200 × 20 stages, the number of network realizations being 1200, and the number of channel realizations being 20. Figure 6 shows how the ANP in dBm scales with the number of nodes for SARA, ICARUS, and GTFT. It is seen that the ANP and therefore the total energy consumed by the network can be divided by more than 2 (the gain is about 4 dB

to be more precise) showing the importance of addressing the problem of packet forwarding and power control jointly.

3.2.2 Impact of quantizing channel gains
As motivated in Section 2.1, one strong argument for assuming discrete channel gains is that, technically, it corresponds to the most general case; the continuous case follows by using standard information-theoretic arguments. But, from a practical point of view, it matters to assess the loss induced by using an algorithm which exploits quantized channel gains instead of continuous ones. Figure 7 represents the performance in terms of the average utility as a function of cardinality of the set of channel gains. It is seen that the performance obtained by using discrete channel gains in the proposed algorithm, whereas the actual channel gains are continuous is typically small. Here, the simulation setup assumed by default is used.

4 Conclusion
One of the contributions of this work is to generalize the famous and insightful model of the forwarder's dilemma [5, 8] by accounting for channel gain fluctuations. Therefore, the problem of knowledge about global channel state appears, in addition to the problem of imperfect action monitoring when the interaction is repeated. In this paper, we have seen that it is possible to characterize the best performance of the studied system even in the presence of partial information; the corresponding observation structure is arbitrarily provided. The observations

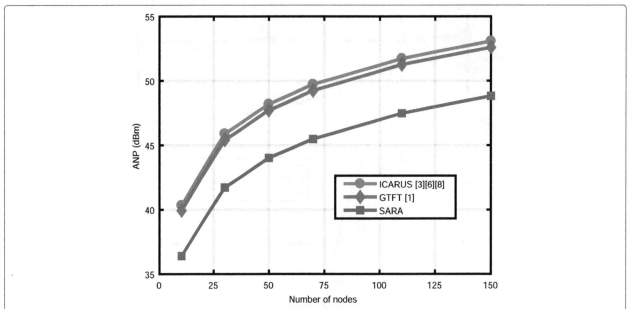

Fig. 6 The figure depicts the average total power (in dBm) or equivalently the total energy consumed by the network against the total number of nodes in the network. It is seen that SARA allows the energy consumed by the network to be divided by more than 2 when compared to the state-of-the art strategies ICARUS and GTFT that do not take into account the channel fluctuations

are generated by discrete observation structures denoted by $\daleth$ and Γ. In terms of performance, designing power control policies which exploit the available knowledge as well as possible is shown to lead to significant gains. Since, we are in the presence of selfish nodes, we propose a mechanism to stimulate cooperation among nodes. The proposed mechanism is both reputation-based and credit-based. For the reputation aspect, one of the novel features of the proposed strategy is that it generalizes the concept

of tit-for-tat to a context where actions are not necessarily binary. For the credit aspect, we propose an evolution law for the credit which is shown to be efficient and robust to selfishness and especially imperfect action monitoring.

From the quantitative aspect, the proposed transmission strategy (referred to as SARA) Pareto dominates ICARUS and GTFT for the utility, the packet forwarding rate, and the energy consumed by the network. Significant gains have been observed; one very convincing result is

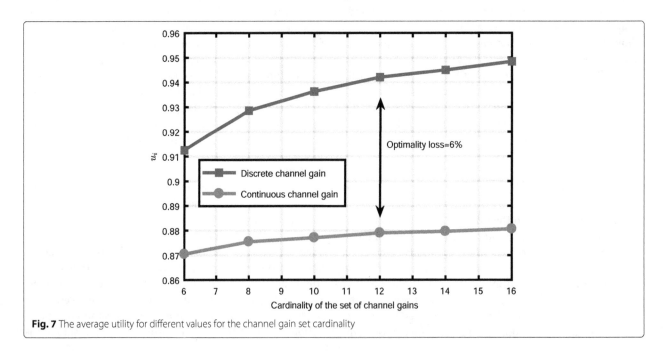

Fig. 7 The average utility for different values for the channel gain set cardinality

that the energy consumed by the network can be divided by 2 when the packet forwarding problem and the power control problem are addressed jointly.

This paper provides the characterization of the best performance in terms of transmission strategy under partial information. Although all performed simulations show that the optimality loss appears to be small, there is no guarantee that the proposed algorithm provides an optimal solution of the optimization problem to be solved to operate on the Pareto frontier or the utility region. Providing such a guarantee would constitute a valuable extension of the present work. Another significant extension would be to relax the i.i.d. assumption on the network state. In this work, the network state corresponds to the global channel state and the i.i.d. assumption is known to be reasonable, but in other setups, where the state represents, e.g., a queue length, a buffer size, or a battery level, the used framework would need to be extended since Markov decision processes would be involved.

Endnotes

[1] The notation $|.|$ stands for the cardinality of the considered set.

[2] The memoryless assumption means that for sequences of realizations of size t (t being arbitrary), $\Pr\left(y_i^t | a_0^t, a_1^t, a_2^t\right) = \Pi_{t'=1}^{t} \Pr(y_i(t') | a_0(t'), a_1(t'), a_2(t'))$.

[3] We will refer to the stability of a point to single deviations as strategic stability.

[4] Note that $\daleth_1$ and $\daleth_2$ are directly obtained from $\daleth$ by a simple marginalization operation.

[5] A subgame of the repeated game is a game that starts at a stage t with a given history.

[6] We therefore assume that the corresponding statistical knowledge is available and exploited.

Appendix 1: Proof of Proposition 1

First, it has to be noticed that long-term utilities are linear images of the implementable distribution. Therefore, characterizing the achievable utility region amounts to characterizing the set of implementable distributions. Note that the set of implementable characterization does not depend on the assumed choice for the infinite sequence of weights $(\theta_t)_{t \geq 1}$, making the result valid for both considered models of repeated games (namely, the classical infinitely repeated game and the model with discount).

Second, to obtain the set of implementable distributions, we exploit the implementability theorem derived in [11]. Therein, it is proved that under the main assumptions of the present paper (namely, the network state is i.i.d. and the observation structure is memoryless), a joint distribution is implementable if and only if it factorizes

as in (22). That is, a joint probability distribution or correlation $Q(a_0, a_1, a_2)$ is implementable if and only if it factorizes as:

$$Q(a_0, a_1, a_2) = \sum_{v, s_1, s_2} \rho(a_0) P_V(v) \times \daleth(s_1, s_2 | a_0) \times$$
$$P_{A_1 | S_1, V}(a_1 | s_1, v) P_{A_2 | S_2, V}(a_2 | s_2, v). \quad (22)$$

Third, a key observation to be made now is that if two probability distributions Q_1 and Q_2 are implementable, then the convex combination $\mu Q_1 + (1 - \mu) Q_2$ is implementable. Indeed, if there is a transmission strategy to implement Q_1 and another to implement Q_2 then by using the first one $\frac{T_1}{T}$ of the time and the second one $\frac{T - T_1}{T}$ of the time, and making T_1 large such that $\frac{T_1}{T} \to \mu$, $\mu Q_1 + (1 - \mu) Q_2$ becomes implementable. It follows that the long-term utility region is convex. Therefore, the Pareto frontier of the utility region, which characterizes the utility region, can be obtained by maximizing the weighted utility W_λ. This concludes the proof.

Appendix 2: Proof of Proposition 2

We want to prove the following result.

The strategy profile $(\sigma_i^\star, \sigma_{-i}^\star)$ is a subgame perfect equilibrium of $\bar{\mathcal{G}}$ if and only if:

$$\delta \geq \max\left\{\frac{c_i}{(1 - 2\epsilon)r_i}, \frac{c_{-i}}{(1 - 2\epsilon)r_{-i}}, 0\right\}, \quad (23)$$

where $c_i = \sum_{a_0} \rho(a_0)\left(u_i^1 - u_i^k\right)$, and $r_i = \sum_{a_0} \rho(a_0)\left(u_i^k - u_i^2\right)$.

$u_i^1 = u_i\left(a_0, a_1^\star, a^{\min}\right)$, $u_i^k = u_i\left(a_0, a_1^\star, a_2^\star\right)$, $u_i^2 = u_i\left(a_0, a^{\min}, a_2^\star\right)$, and $a_i^\star = f_i^\star(s_i)$.

As a preliminary, we first review the one-shot deviation "principle" in the context of interest. This "principle" is one of the elements used to prove the desired result.

One-shot deviation principle: For node i, the one-shot deviation principle from strategy σ_i is a strategy $\tilde{\sigma}_i$ writes as:

$$\exists! \ \tau, \ \forall \ t \neq \tau \sigma_{i,t}\left(s_i^t, y_i^{t-1}\right) = \tilde{\sigma}_{i,t}\left(s_i^t, y_i^{t-1}\right). \quad (24)$$

The two strategies $\tilde{\sigma}_i$ and σ_i therefore produce identical actions except at stage τ.

Definition 3 *For node i, the one-shot deviation $\tilde{\sigma}_i$ from strategy σ_i is not profitable if:*

$$U_i(\sigma_i, \sigma_{-i}) \geq U_i(\tilde{\sigma}_i, \sigma_{-i}), \quad (25)$$

with $\tilde{\sigma}_i \neq \sigma_i$.

Let us exploit the one-shot deviation principle to prove the result, since it is well known that a strategy profile σ is a subgame perfect equilibrium if and only if there are no profitable one-shot deviations. Assume that for a given game history, the distributions used by nodes i and $-i$ are respectively π_i and π_{-i}. Following the proposed strategy $\sigma_{i,t}^\star$ defined by (16), and by using (15) and (18), one can obtain the distribution of a node i, $\pi_i(t)$, from π_{-i} for each stage t. As defined by relation (18), at each stage t, if $m_i(t) \geq \mu$, node i chooses a distribution $\pi_i(t) = \hat{\pi}_{-i}(t-1)$ stipulating that $a^{min} = (P_{min}, P_{min})$ and $a_i^\star(t) = f_i^\star(s_i(t))$ are the only actions that could be chosen with a positive probability. Thus, it would be sufficient to provide only the kth component of $\pi_i(t)$, which is denoted by $\pi_i^k(t)$. Note that k is the index of action $a_i^\star(t) = f_i^\star(s_i(t))$. Thus, for $t \geq 1$ we have that:

$$\pi_i^k(t) = \begin{cases} 1, & \text{if } m_i(t) < \mu; \\ (1-2\epsilon)^t \pi_i + \epsilon \sum_{k=0}^{t-1}(1-2\epsilon)^k, & \text{if } \mathrm{mod}(t,2) = 1; \\ (1-2\epsilon)^t \pi_i + \epsilon \sum_{k=0}^{t-1}(1-2\epsilon)^k, & \text{if } \mathrm{mod}(t,2) = 0. \end{cases}$$

Now, we define a one-shot deviation. We consider that node i deviates unilaterally at one stage from the proposed strategy $\sigma_{i,t}^\star$, by choosing $\tilde{\sigma}_{i,t}$. If node i deviates, it will be in order to save energy; thus, it chooses a^{min} with a higher probability than the one provided by the proposed strategy $\sigma_{i,t}^\star$. Therefore, we consider that $\tilde{\sigma}_{i,t}$ defines a distribution over the action set as follows (26):

$$\tilde{\pi}_i(t) = \pi_i(t) - d.(\underbrace{-1}_{a^{min}}, 0, \ldots, \underbrace{1}_{a^\star}, 0, \ldots), \tag{26}$$

Using the one-shot deviation $\tilde{\pi}_{-i}(t)$, we have for $t \geq 1$:

$$\tilde{\pi}_i^k(t) = \begin{cases} 1, & \text{if } m_i(t) < \mu; \\ \pi_i^k(t), & \text{if } \mathrm{mod}(t,2) = 0; \\ \pi_i^k(t) - d(1-2\epsilon)^{t-1}. & \text{if } \mathrm{mod}(t,2) = 1. \end{cases}$$

$$\tilde{\pi}_{-i}^k(t) = \begin{cases} 1, & \text{if } m_i(t) < \mu; \\ \pi_{-i}^k(t) - d(1-2\epsilon)^{t-1}, & \text{if } \mathrm{mod}(t,2) = 0; \\ \pi_{-i}^k(t). & \text{if } \mathrm{mod}(t,2) = 1. \end{cases}$$

Now, to accomplish the proof, we need to define the associated expected utilities for each stage provided by $\pi_i(t)$ and $\tilde{\pi}_i(t)$, which are denoted by $u_{i,t}^\star$ and $\tilde{u}_{i,t}$, respectively.

$$u_{i,t}^\star = \sum_{a_0,a_1,a_2} P_t(a_0,a_1,a_2)u_i(a_0,a_1,a_2), \tag{27}$$

where P_t is the joint probability distribution, and u_i the instantaneous utility (1). Denote by $u_i^k = u_i(a_0, a_1^\star, a_2^\star)$, $u_i^1 = u_i(a_0, a_1^\star, a^{min})$, $u_i^2 = u_i(a_0, a^{min}, a_2^\star)$, and $u_i^{min} = u_i(a_0, a^{min}, a^{min})$. $a_i^\star = f_i^\star(s_i)$, and $a^{min} = (P_{min}, P_{min})$. By means of these notations, we obtain:

$$u_{i,t}^\star = \sum_{a_0} \rho(a_0) \Big[\pi_i^k(t)\pi_{-i}^k(t)u_i^k + \pi_i^k(t)\left(1 - \pi_{-i}^k(t)\right)u_i^1 +$$
$$\left(1 - \pi_i^k(t)\right)\pi_{-i}^k(t)u_i^2 + \left(1 - \pi_i^k(t)\right)\left(1 - \pi_{-i}^k(t)\right)u_i^{min} \Big]. \tag{28}$$

We now define the expected utility of the deviation for each stage, denoted $\tilde{u}_{i,t}$.

$$\tilde{u}_{i,t} = \sum_{a_0} \rho(a_0) \Big[\tilde{\pi}_i^k(t)\tilde{\pi}_{-i}^k(t)u_i^k + \tilde{\pi}_i^k(t)\left(1 - \tilde{\pi}_{-i}^k(t)\right)u_i^1 +$$
$$\left(1 - \tilde{\pi}_i^k(t)\right)\tilde{\pi}_{-i}^k(t)u_i^2 + \left(1 - \tilde{\pi}_i^k(t)\right)\left(1 - \tilde{\pi}_{-i}^k(t)\right)u_i^{min} \Big]. \tag{29}$$

As the deviation distribution $\tilde{\pi}_i$ depends on the distribution provided by the proposed strategy $\sigma_i^\star$, π_i, one can also define $\tilde{u}_{i,t}$ as a function of $u_{i,t}^\star$, by using the definitions of $\pi_i(t)$ and $\tilde{\pi}_i(t)$. Hence, we have the following result:

$$\tilde{u}_{i,t} = \begin{cases} u_{i,t}^\star, & \text{if } m_i(t) < \mu; \\ u_{i,t}^\star - (1-2\epsilon)^{t-1}d\sum_{a_0}\rho(a_0)\breve{U}^1(t) & \text{if } \mathrm{mod}(t,2) = 0; \\ u_{i,t}^\star - (1-2\epsilon)^{t-1}d\sum_{a_0}\rho(a_0)\breve{U}^2(t) & \text{if } \mathrm{mod}(t,2) = 1, \end{cases}$$

where $\breve{U}^1(t) = \pi_i^k(t)\left(u_i^k - u_i^1 + u_i^{min} - u_i^2\right) + u_i^2 - u_i^{min}$, and $\breve{U}^2(t) = \pi_{-i}^k(t)\left(u_i^k - u_i^1 + u_i^{min} - u_i^2\right) + u_i^1 - u_i^{min}$. Thus, the deviation utility of node i in the repeated game $\bar{\mathcal{G}}$ is:

$$U_i(\tilde{\sigma}_i, \sigma_{-i}^\star) = u_{i,0}^\star + \sum_{t=1}^{z}\delta^t\tilde{u}_{i,t} + \sum_{t=z+1}^{\infty}\delta^t u_{i,t}^\star,$$

where z is the number of stages until the condition $m_i < \mu$ is satisfied. The unilateral deviation from the proposed strategy $\sigma_i^\star$ is not profitable if:

$$U_i\left(\tilde{\sigma}_i, \sigma_{-i}^\star\right) \leq U_i\left(\sigma_i^\star, \sigma_{-i}^\star\right). \tag{30}$$

The equilibrium condition could be determined using relation (30). It is defined as follows:

$$u_{i,0}^\star + \sum_{t=1}^{z}\delta^t\tilde{u}_{i,t} + \sum_{t=z+1}^{\infty}\delta^t u_{i,t}^\star \leq \sum_{t=0}^{\infty}\delta^t u_{i,t}^\star.$$

By substituting $\tilde{u}_{i,t}$ by its value, the equilibrium condition writes as:

$$\sum_{t=0}^{t=\frac{z}{2}-1} \delta^{2t}\left((1-2\epsilon)^{2t}d\sum_{a_0}\rho(a_0)\breve{U}^1(2t+1)\right)$$
$$+ \delta\sum_{t=0}^{t=\frac{z}{2}-1}\delta^{2t}\left((1-2\epsilon)^{2t+1}d\sum_{a_0}\rho(a_0)\breve{U}^2(2t+2)\right) \geq 0. \tag{31}$$

We have $\breve{U}^1(2t+1) = \pi_i^k(2t+1)\left(u_i^k - u_i^1 + u_i^{min} - u_i^2\right) + u_i^2 - u_i^{min}$ and $\breve{U}^2(2t+2) = \pi_{-i}^k(2t+2)\left(u_i^k - u_i^1 + u_i^{min} - u_i^2\right) + u_i^1 - u_i^{min}$. We provide results for $\pi_i^k(2t+1) = \pi_{-i}^k(2t+2) = 1$, which implies that relation (31) is satisfied for each $\pi_i^k(2t+1)$ and $\pi_{-i}^k(2t+2)$. With this assumption, the relation (31) becomes:

$$\sum_{a_0} \rho(a_0) \left(u_i^k - u_i^1 \right) \sum_{t=0}^{t=\frac{z}{2}-1} \delta^{2t} (1 - 2\epsilon)^{2t}$$

$$+ \sum_{a_0} \rho(a_0) \left(u_i^k - u_i^2 \right) \delta \sum_{t=0}^{t=\frac{z}{2}-1} \delta^{2t} (1 - 2\epsilon)^{2t+1} \geq 0.$$

This is satisfied if and only if:

$$\delta \geq \frac{\sum_{a_0} \rho(a_0) \left(u_i^1 - u_i^k \right) \sum_{t=0}^{t=\frac{z}{2}-1} \delta^{2t} (1 - 2\epsilon)^{2t}}{\sum_{a_0} \rho(a_0) \left(u_i^k - u_i^2 \right) (1 - 2\epsilon) \sum_{t=0}^{t=\frac{z}{2}-1} \delta^{2t} (1 - 2\epsilon)^{2t}}. \tag{32}$$

The equilibrium condition is thus:

$$\delta \geq \max \left\{ \frac{c_i}{(1 - 2\epsilon)r_i}, 0 \right\}, \tag{33}$$

where $c_i = \sum_{a_0} \rho(a_0) \left(u_i^1 - u_i^k \right)$, $r_i = \sum_{a_0} \rho(a_0) \left(u_i^k - u_i^2 \right)$, $u_i^1 = u_i \left(a_0, a_1^\star, a^{\min} \right)$, $u_i^k = u_i \left(a_0, a_1^\star, a_2^\star \right)$, $u_i^2 = u_i \left(a_0, a^{\min}, a_2^\star \right)$, and $a_i^\star = f_i^\star(s_i)$.

Thus, the strategy profile $(\sigma_i^\star, \sigma_{-i}^\star)$ is a subgame perfect equilibrium if and only if:

$$\delta \geq \max \left\{ \frac{c_i}{(1 - 2\epsilon)r_i}, \frac{c_{-i}}{(1 - 2\epsilon)r_{-i}}, 0 \right\}. \tag{34}$$

Abbreviations

ANP: Average network power; CSI: Channel state information; GTFT: Generous-Tit-For-Tat; ICARUS: hybrId inCentive mechAnism for coopeRation stimUlation in ad hoc networkS; MTTFT: Mend-Tolerance Tit-For-Tat; NE: Nash equilibrium; SARA: State Aware tRAnsmission strategy; SNR: Signal-to-noise ratio

Acknowledgements
Not applicable.

Funding
Not applicable.

Authors' contributions
All authors read and approved the final manuscript.

Competing interests
The authors declare that they have no competing interests.

References
1. M. Tan, T. Yang, X. Chen, G. Yang, G. Zhu, P. Holme, J. Zhao, A game-theoretic approach to optimize ad-hoc networks inspired by small-world network topology. Physica A. **494**, 129–139 (2018)
2. L. Feng, Q. Yang, K. Kim, K. Kwak, Dynamic rate allocation and forwarding strategy adaption for wireless networks. IEEE Signal Proc. Lett. **25**(7), 1034–1038 (2018)
3. N. Samian, W. K. G. Seah, in *Proceedings of the 14th EAI International Conference on Mobile and Ubiquitous Systems*. Trust-based scheme for cheating and collusion detection in wireless multihop networks (Computing, Networking and Services, Melbourne, VIC, Australia, 2017)
4. S. Berri, V. Varma, S. Lasaulce, M. S. Radjef, J. Daafouz, in *Proceedings of the 4th International Symposium on Ubiquitous Networking, Lecture Notes in Computer Science*. Studying node cooperation in reputation based packet forwarding within mobile ad hoc networks (Springer LNCS, Casablanca, Maroc, 2017)
5. C. Tang, A. Li, X. Li, When reputation enforces evolutionary cooperation in unreliable MANETs. IEEE Trans. Cybern. **45**(10), 2091–2201 (2015)
6. N. Samian, Z. Ahmad Zukarnain, W. K. G. Seah, A. Abdullah, Z. Mohd Hanapi, Cooperation stimulation mechanisms for wireless multihop networks: a survey. J. Netw. Comput. Appl. **54**, 88–106 (2015)
7. J. M. S. P. J. Kumar, A. Kathirvel, N. Kirubakaran, P. Sivaraman, M. Subramaniam, A unified approach for detecting and eliminating selfish nodes in MANETs using TBUT. EURASIP J. Wirel. Commun. Netw. **2015**(143), 1–11 (2015)
8. D. E. Charilas, K. D. Georgilakis, A. D. Panagopoulos, ICARUS: hybrId inCentive mechAnism for coopeRation stimUlation in ad hoc networkS. Ad Hoc Netw. **10**(6), 976–989 (2012)
9. A. Krzesinski, Promoting cooperation in mobile ad hoc networks. Int. J. Inf. Commun. Technol. Appl. **2**(1), 24–46 (2016)
10. Q. Xu, Z. Su, S. Guo, A game theoretical incentive scheme for relay selection services in mobile social networks. IEEE Trans. Veh. Technol. **65**(8), 6692–6702 (2016)
11. B. Larrousse, S. Lasaulce, M. Wigger, in *Proceedings of the IEEE Information Theory Workshop (ITW)*. Coordinating partially-informed agents over state-dependent networks (IEEE xplore, Jerusalem, 2015)
12. B. Larrousse, S. Lasaulce, in *Proceedings of the IEEE International Symposium on Information Theory*. Coded power control: performance analysis, (2013)
13. B. Larrousse, S. Lasaulce, M. Bloch, Coordination in distributed networks via coded actions with application to power control. IEEE Trans. Inf. Theory. **64**(5), 3633–3654 (2018)
14. A. Gjendemsj, D. Gesbert, G. E. Oien, S. G. Kiani, Binary power control for sum rate maximization over multiple interfering links. IEEE Trans. Wirel. Commun. **7**(8), 3164–3173 (2008)
15. S. Sesia, I. Toufik, M. Baker, *LTE-the UMTS long term evolution: from theory to practice*. (Wiley Publishing, Hoboken, 2009)
16. M. R. Gray, *Entropy and Information Theory*. (Springer, New York, 2013)
17. D. J. Goodman, N. Mandayam, A power control for wireless data. IEEE Pers. Commun. **7**(2), 48–54 (2000)
18. F. Meshkati, M. Chiang, H. V. Poor, S. C. Schwartz, A game theoretic approach to energy-efficient power control in multicarrier CDMA systems. J. Sel. Areas Commun. **24**(6), 1115–1129 (2006)
19. S. Lasaulce, Y. Hayel, R. El Azouzi, M. Debbah, Introducing hierarchy in energy games. IEEE Trans. Wirel. Commun. **8**(7), 3833–3843 (2009)
20. E. V. Belmega, S. Lasaulce, Energy-efficient precoding for multiple-antenna terminals. IEEE Trans. Signal Process. **59**(1), 329–340 (2011)
21. A. El Gamal, Y. Kim, *Network information theory*. (Cambridge University Press, Cambridge, 2011)
22. B. Djeumou, S. Lasaulce, A. G. Klein, in *Proceedings of the IEEE International Sympsium on Signal Processing and Information Technology (ISSPIT)*. Combining decoded-and-forwarded signals in Gaussian cooperative channels (IEEE xplore, Vancouver, 2006)
23. S. Lasaulce, H. Tembine, *Game theory and learning for wireless networks: fundamentals and applications*. (Academic Press, Elsevier, Amsterdam, 2011), pp. 1–336
24. A. Neyman, S. Sorin, Repeated games with public uncertain duration process. Int. J. Game Theory. **39**(1), 29–52 (2010)
25. M. Maschler, E. Solan, S. Zamir, *Game theory*. (Cambridge University Press, Cambridge, 2013)
26. H. Konno, A cutting plane algorithm for solving bilinear programs. Math. Program. **11**(1), 14–27 (1976)
27. G. Gallo, A. Olkticti, Bilinear programming: an exact algorithm. Math. Program. **12**(1), 173–194 (1977)
28. H. Vaish, C. M. Shetty, The bilinear programming problem. Nav. Res. Logist. Q. **23**(2), 303–309 (1976)
29. D. Fudenberg, J. Tirole, *Game theory*. (MIT Press, Cambridge, 1991)
30. A. Agrawal, S. Lasaulce, O. Beaude, R. Visoz, in *Proceedings of the IEEE Fifth International Conference on Communications and Networking (ComNet)*. A framework for decentralized power control with partial channel state information (IEEE xplore, Hammamet, Tunisia, 2015)
31. C. H. Papadimitriou, in *Proceedings of the 33rd Annual ACM Symposium on Theory of Computing (STOC)*. Algorithms, games, and the internet (ACM Digital Library, New York, 2001)
32. S. D. Yi, S. K. Baek, J. K. Choi, Combination with anti-tit-for-tat remedies problems of tit-for-tat. J. Theor. Biol. **412**, 1–7 (2017)

33. S. M. Betz, H. V. Poor, Energy rfficient communications in CDMA networks: a game theoretic analysis considering operating costs. IEEE Trans. Signal Process. **56**(10), 518–5190 (2008)
34. F. Richter, A. J. Fehske, G. Fettweis, in *Proceedings of the IEEE 70th Vehicular Technology Conference Fall*. Energy efficiency aspects of base station deployment strategies for cellular networks (IEEE xplore, Anchorage, 2009)
35. V. Varma, S. Lasaulce, Y. Hayel, S. E. El Ayoubi, A cross-layer approach for energy-efficient distributed interference management. IEEE Trans. Veh. Technol. **64**(7), 3218–3232 (2015)

3

Identity attack detection system for 802.11-based ad hoc networks

Mohammad Faisal[1*], Sohail Abbas[2] and Haseeb Ur Rahman[1]

Abstract

Due to the lack of centralized identity management and the broadcast nature of wireless ad hoc networks, identity attacks are always tempting. The attackers can create multiple illegitimate (arbitrary or spoofed) identities on their physical devices for various malicious reasons, such as to launch Denial of Service attacks and to evade detection and accountability. In one scenario, the attacker creates more than one identity on a single physical device, which is called a Sybil attack. In the other one, the attacker creates cloned/replicated nodes. We refer collectively to these attacks as identity attacks. Using these malicious techniques, the attacker would perform activities in the network for which the attacker may not be authorized. In the existing literature, these attacks are often counteracted separately. However, in this paper, we propose a solution to counteract both attacks jointly. Our proposed scheme uses the received signal strength for the detection without using extra hardware (such as GPS, antennae or air monitors) and centralized entities (such as trusted third party or certification authority). Upon the detection of malicious identities, they will be quarantined and will be blacklisted for future data communication by the mobile nodes. Our proposed attack detector detects the presence of Sybil attacks and replication attacks locally by analysing the received signal strength captured by each node. Moreover, we propose a technique that will identify these attacks in the overall network. In both local and global cases, we evaluate our solutions theoretically and via simulation in NS-2. The obtained results demonstrate that it is possible to detect identity attacks with considerable accuracy without causing extra overhead in the form of extra hardware, periodic beacons or expensive localization operations in the wireless ad hoc networks.

Keywords: Impersonation, Sybil attack, Replication attack, Intrusion detection, Mobile ad hoc networks

1 Introduction

The IEEE 802.11-based wireless ad hoc architecture represents networks that consist of mobile nodes that randomly construct ad hoc topologies in a self-organized manner. The mobile nodes may be laptops, tablets, or smartphones irrespective of the operating systems that they use, such as Windows, Linux/Unix, Apple, Android, Blackberry, Symbian and iOS. These networks facilitate users in infrastructure-less environments where intentional or unintentional catastrophic violent situations may occur, such as earthquakes, floods, battlegrounds and search and rescue operations. In such situations, infrastructure-based networks are difficult to be installed because of their ad hoc or ephemeral nature [1–3].

Traditionally, the networks were formed using homogenous nodes (computers). However, due to the emergence of 5G and the Internet of Things (IoT) paradigms, heterogeneous networks comprising heterogeneous nodes such as sensors, phones, computers and satellites form, thereby providing various ubiquitous services. These wireless ad hoc networks are fully applicable in the field of the IoT. In the IoT, virtual objects, services, processes and devices are considered as nodes that are interconnected through the Internet. Soon, the IoT will amalgamate different technologies wirelessly in which ad hoc networks will play an integral part. Examples of such systems are smart cities, the Internet of connected vehicles (intelligent transportation system), etc. [4, 5].

Similar to all other networks, communications in 802.11-based ad hoc networks are commonly conducted based on a unique identifier that represents a network entity called a node. These identifiers are used for inter-nodal

* Correspondence: mfaisal@uom.edu.pk; mfaisal_1981@yahoo.com
[1]Department of Computer Science and IT, University of Malakand, Chakdara, KPK, Pakistan
Full list of author information is available at the end of the article

communications. This forms a one-to-one relationship between an entity (i.e. a node) and an identity used by a single user. This implication is commonly followed by many protocols directly or indirectly [6]. The ad hoc networks require each node to have a unique and distinct identifier to ensure the correct operations of the networked system and secure transactions. However, this one-to-one relationship may not be followed, thus resulting in serious security threats. There are two major types of identity attacks that violate this one-to-one entity-identity philosophy, the Sybil attack and the replication attack, which are discussed below.

Sybil attack: In this attack, one-to-one becomes one-to-many, which means that the attacker creates and manages more than one identity on a single physical device [7]. These newly forged identities will all be used simultaneously or in sequential order. That is, they may be used one-after-the-other, but only one identity will be up and running at a time. In the former case, the attacker uses the identity group to launch different types of attacks where the number of identities plays an important role, in some ways, to counteract a working system, such as Denial of Service (DoS) or Distributed Denial of Service (DDoS) attacks or altering the outcome of voting-based protocol(s). In the latter case, the identities may be used to escape accountability and/or traceability, such as evading a detection system where after the detection of one identity another forged identity emerges [8], thereby whitewashing all the malicious actions committed. The Sybil attacker may either adopt arbitrary identifiers for his Sybil, virtual identities or spoof the already existing nodes' identities. In the latter case, the attacker can gather the identity information of network nodes by snooping or sniffing in which an attacker sniffs the identity of privileged nodes and adopts them for malicious activities. Sybil attacks are detrimental to the correct network functioning and can disrupt the entire functionality of such systems in multiple ways. For example, a Sybil attacker can disrupt the routing of packets by giving a false impression of being distinct nodes on different locations or disjointed nodal paths. In trust or reputation-based schemes, a Sybil node can deteriorate the system by increasing or decreasing the reputation or trust by exploiting its virtual identities. In wireless sensor networks and smart grids, a Sybil attacker can change the whole aggregated reading outcome by contributing many times as a different node. In voting-based protocols, a Sybil attacker can manipulate the resulting outcome by rigging the polling process using Sybil identities. In vehicular ad hoc networks (VANETs), Sybil attackers can forge virtual non-existent vehicles and communicate false information in the network for malicious intents, such as to give false impressions of

traffic congestion to divert traffic. Similarly, in a distributed system environment, a Sybil attacker can access and gain more resources using its forged identities [9–12].

Replication: In this attack, a many-to-one (entity-to-identity, i.e. multiple nodes have the same identity) strategy is used. In it, an attacker captures a node (most probably a sensor node in a wireless sensor network) and creates multiple clones of that physical node. Then, the malicious node or authority deploys these cloned or replicated nodes at various important locations in the network for malicious purposes, such as data analysis or launching a DoS or DDoS attack in the network. Moreover, these replicated nodes at various locations may hardly be detected by an intrusion detection system in place. The distinction between cloned identities and original ones is considered a challenging task [2, 13–17]. It is worth mentioning here that the Sybil attack with spoofed identities and replication attacks are logically identical because in both cases multiple nodes exist with the same identifier.

There are mainly three techniques used in the literature for Sybil and replication attack detection and/or protection, which are described below.

Trusted certification: The traditional approaches to detect or prevent these attacks use cryptographic-based authentication or trusted certification [16]. However, these approaches do not suit the infrastructure-less domain of the IoT since these schemes are costly in terms of their initial setup and the overhead incurred for maintaining and distributing cryptographic keys [8, 18, 19]. Moreover, they are also not scalable.

Resource testing: Some of the schemes have been proposed to counteract Sybil attacks based on the physical resource testing, such as radio, storage and computational resource testing [8, 17]. The goal of these schemes is to check (by employing some tests) whether an identity possesses resources greater than the resources normally possessed by a single node. These schemes are not effective due to the unrealistic bounds imposed on the attackers.

Position verification: Some authors, such as [10, 18], proposed position verification-based techniques to counteract Sybil attackers. These schemes work based on the assumption that each identity is bound by a single distinct location at any particular time. These approaches use the received signal strength indicator (RSSI) for position verification and hence are more promising than the others because of their lightweight and distributed nature [11, 20, 21]. However, since

RSSI varies with time, they rely mostly on extra hardware, such as directional antennae or GPS (Global Positioning System) or periodic beacon messages [11, 20, 21].

In this paper, we propose a received signal strength (RSS)-based scheme for identity-based attacks' counteraction. Since the RSS is a rough indicator of distance, we use the distance parameter to detect identity attacks in two ways: in a single radio range and, locally and globally, in the network. Each node will use the proposed algorithm in a distributed manner using the distance parameter to detect malicious identities in its own radio range. In addition to that, each node will also cooperate and collaborate with its neighbours to detect these attacks using our proposed global map algorithm. As soon as a malevolent identity is detected, it will be quarantined and will be blacklisted for future communication that are either detected in same radio range or different radio ranges within the network. It is worth mentioning that our scheme neither takes the services of any other third party nor does it use any sort of extra hardware, such as directional antennae or GPS. Furthermore, our proposed scheme is lightweight since it does not create any extra overhead on the overall architecture of the network.

We articulate our model through statistical and analytical analyses. Through these analyses, we are able to prove our detection rationale. Our detection works in two phases. First, attackers are detected locally (i.e. in a single radio range) and then globally (i.e. in the network). In the local detection, we use statistical analysis to generalize our scheme, create empirical cumulative distributive functions of three different radio ranges and compare the RSSI fluctuation in each range by employing a greater than 90% confidence interval for improved accuracy. In the global detection, we analytically analyse a criterion by which replicated nodes can be detected. Finally, to evaluate our proposed scheme, we use the NS-2 simulator where we plug in the thresholds that were analysed empirically into the simulated scenarios. The obtained results indicate a greater than 90% detection accuracy with less than 10% false positives.

The remainder of the paper is organized as follows. In Section 2, we discuss the literature review in our proposed classification manner. In Section 3, we discuss the feasibility of the identity attacks and the detection of such attacks in two scenarios, which are the detection in a single radio range and the detection across the network. We also develop a theoretical threshold for the detection using statistical significance testing. Section 4 is about the simulation-based evaluation of our proposed scheme using the NS-2 simulator and the result analysis. In Section 5, we discuss the main pros and cons of our

proposed work. The paper is concluded in Section 6 in which we also highlight future work.

2 Literature review

We classify and briefly discuss the traditional countermeasures proposed in the literature for the detection or prevention of identity attacks in the following subsections.

2.1 Trusted third party or certification authority

In these schemes, a centralized or semi-centralized trusted third party (TTP) is used to create, maintain and revoke the identity certificate for each node. However, the certification authority (CA) suffering from an expensive initial installation setup is deficient in its scalability and is vulnerable to single point intentional (attack) and unintentional (failure) shutdowns [1, 22, 23].

In securing the wireless ad hoc networks, Hoeper and Gong [24] listed the various schemes based on the TTP or CA. In all these schemes, the centralized architecture is responsible for countering identity attacks in 802.11-based ad hoc networks. In practical security for disconnected nodes, Hoeper and Gong [24] proposed a concrete cryptosystem for ad hoc networks by focusing on hierarchical identity-based cryptography (HIBC). The authors introduced the concept of anonymity in the HIBC because of its roaming capability in different regions since the scheme is distributed in different hierarchies and can cover multiple regions.

Xing and Cheng [14] proposed two node replication schemes based on the time reference and space tradeoff called the Time Domain Detection (TDD) and the Space Domain Detection (SDD), respectively. The schemes generate a cryptographic one-way hash function with high accuracy and resilience to collusion, which is stored with a TTP for validation. However, the schemes did not tackle the problem of the overhead incurred, which affects the resource constraint nodes of mobile ad hoc networks (MANETs). In addition, the authors focused on node replication only.

Hoeper and Gong [24] proposed bootstrapping security in MANETs using identity-based schemes with key revocation. The authors combined the identity-based authentication and the key exchange (IDAKE) mechanism in one scheme, thereby using the symmetric key cryptography for reduced computational overhead. The TTP initialized all the nodes of MANETs with unique identities for communications. The proposed scheme relied heavily on the TTP.

Chen et al. [25] proposed a cluster-based certificate revocation with proof capability for MANETs. In this scheme, the authors focus on the revocation to isolate the problem of reuse by the intruders, which increases the accuracy threshold that is imposed on the mobile nodes for certificate activation. However, the scheme assumed that all nodes already have certificates before joining the network and the nodes must be uniformly distributed among the radio range of the ad hoc network, which is not practicable.

In preventing impersonation attacks in MANETs with multi-factor authentication, Glynos et al. [26] proposed a framework via a cryptographic association to secure the physical device with the logical address. The framework achieved this target by using the certification of keys and nodes with the use of additional hardware and firmware installed on each node of the ad hoc network.

2.2 Software-based approaches

These schemes work as standalone solutions that use cryptographic-based authentication that imposes greater computation and communication overhead, which is usually caused by key distribution, maintenance and revocation operations. Hence, these issues make these approaches unsuitable for infrastructure-less environments, such as 802.11-based ad hoc networks, due to the resource constraint devices and distributed nature of the network [22, 25, 26].

Bouassida and Shawky [27] proposed anonymous multipath routing protocol based on secret sharing in mobile ad hoc networks. The protocol delivered the location, identity, data and traffic anonymity using cryptography. The protocol was also capable of countering interceptions and tampering attacks. However, the protocol assumed that every legal node in the ad hoc network must possess the same session and symmetric key, which could be easily impersonated by the intruders.

Hall et al. [28] proposed a novel secure identity-based cryptographic-based scheme for the hybrid wireless mesh protocol for IEEE 802.11s. The authors, in their proposed scheme, used a software-based approach to counter identity attacks. The scheme focused on securing the route discovery mechanism in which the route request and route reply control messages are communicated during the routing. The authors concentrated on securing the data exchange in both the route request and route reply mechanisms. However, the scheme increased the overhead, which does not suit IEEE 802.11-based ad hoc networks.

Bouassida and Shawky [27] proposed a fuzzy logic system that predicted the behaviours of nodes, such as how much a node in question can be trusted, while considering the partial history of the nodes' actions. The logic delivered the shortest possible route to ensure security against identity attacks and to identify the intruders. However, the scheme assumed that the nodes can predict the neighbouring node behaviours and broadcasting packets are reached to all nodes appropriately, which is a challenge in MANETs due to the presence of intruders.

2.3 Transceiver fingerprinting

In this category of schemes, the main logic of attack detection assumes that each radio transceiver generates a distinct radio frequency signal that reflects some physical characteristics that make it distinguishable from other transceivers, which is called its frequency fingerprint. These frequency fingerprints are used to counter identity attacks. However, the schemes may not be able to detect the attack if multiple devices are installed close to each other, thus creating the same fingerprints. Hence, DoS attacks can be launched easily. Moreover, this transceiver requires higher costs in terms of fingerprinting measurements and extra hardware implementation with enough precision, which restricts its use [27, 28]. Some of the schemes that use frequency fingerprinting are given below.

He and Wang [29] proposed a robust biometrics-based authentication scheme for the multi-server environment using transceiver fingerprinting. In the traditional system, for user authentication, authorization passwords were used. However, passwords may be stolen, shared or lost, and thus, here, the authors introduced the fingerprinting concept using elliptic curve cryptography in a multi-server environment. It is difficult to lose or forget copies or share, distribute, forge, guess or break a biometric generated key. However, in addition to the costs incurred for equipping each node with a biometric device, the computational costs of the scheme were also non-affordable for the 802.11-based ad hoc networks.

Faria and Cheriton [2] proposed the signal print-based solution to detect identity-based attacks in wireless networks. The author identified the transmitting devices by their distinct signal prints, which were also correlated with their physical location. By keeping the signal print information of all participating nodes, the network can identify the signal print of each node, which can help categorize the packets used for identity attacks. The authors assumed that the intruder would use an omnidirectional antenna equipped with standardized transmitters. However, the scheme may not be able to detect the intruders' signal prints if the devices were installed close to each other with below standard specifications, thus generating odd single prints.

Debdutta and Rituparna [30] proposed a three-pronged method based upon the transceiver fingerprinting. The scheme measured (normalization phase), attuned (database for patterns) and approximated (via data mining techniques) the RSSI values irrespective of the hardware devices to counter the effects of doors and walls on RSSI. The scheme used artificial neural networks to make clusters of the mobile nodes. Since the scheme used artificial neural networks, it is expensive for the low energy networks of MANETs.

Debdutta and Rituparna [30] proposed a scheme for detecting masquerading attacks in 802.11-based wireless networks. The authors identified the intruders by assigning unique fingerprints to each host of the wireless ad hoc network. The fingerprints were calculated by the Bayesian classifier from the networking activities of the wireless host. The scheme was precise and accurate in detection and was the pioneer in the category of transceiver fingerprinting. The scheme did not require any specialized extra hardware or any change in the existing hardware architecture; it may even be able to be implemented in the firmware of the already installed hardware. However, the scheme did not propose the solution of the new architecture hardware's fingerprints and address the limitations and computational overhead of the Bayesian classifier.

2.4 Received signal strength

The received signal strength (RSS) is used to localize nodes and hence is used to detect identity attacks since these schemes assume that each location must be bound by a unique and distinct identity. Messages emanating from the same location bearing two or more than two identities will be detected as Sybil attacks. Similarly, messages received from different locations with the same identities will be classed as replication attacks. The RSS is also hard to be forged. Existing RSS-based countermeasures use diverse antennas and Global Positioning System (GPS) technology. Some authors [31, 32] use air monitors (AM) as an additional hardware device that sniffs the traffic passively to detect spoofing attacks on the MAC layer. However, this approach creates additional hardware overhead for the infrastructure-less 802.11 [27]-based ad hoc networks [31–33].

Bouassida and Shawky [27] proposed an intrusion detection technique based on the degree of distinguishability analysis. In this technique, the nodes can verify each other's location and can authenticate the incoming nodes to the network based on the physical characteristics of the signals. However, the scheme was designed for static scenarios only. That is, the authors assumed that the verifier node's location and distance were fixed.

Yang et al. [22] proposed a detection mechanism for spoofing attacks in mobile environments. It used RSSI values when the attacker node was in motion. In this scheme, the authors developed a DEMOTE (DEtecting MObile spoofing aTtacks in wireless Environments) system. The scheme can predict the best RSS alignment over the radio range spread in the 802.11-based ad hoc networks without any supervision.

Dhamodharan and Vayanaperumal [34] proposed a Sybil attack detection technique in wireless sensor networks by using the message authentication and passing method. The message is passed to the new nodes joined

to the sensor network to check its trustworthiness. If the message is verified and is not found to be a duplicate communication, the base station will allow it to start communications. Otherwise, it will be declared as a Sybil identity. The proposed work differs from ours in that it relies on the centralized entity, which is the base station in sensor networks without using mobility.

Zdonik et al. [35] proposed a scheme to detect Sybil attacks using the spatial variance of the physical layer in the WSN. The scheme focused on accurate detection without creating any extra overhead. The scheme used the iterative least squares (ILS) channel estimation method to determine the channel identification (CI). The detection idea is very interesting. However, the proposed scheme differs from ours in that the scheme strongly relied on a "powerful" base station in the WSNs, which may not work in 802.11-based ad hoc networks.

Qabulio et al. [36] proposed a scheme to counter the clone node attack in the WSN, irrespective of the node's location. The scheme does not use extra processing time and memory to calculate the distances between the nodes similar to its predecessor schemes, which is why it is a lightweight solution. Again, this paper implements its solution in wireless sensor networks while we implement our work in 802.11-based ad hoc networks that were mentioned in the topic of the paper. Qabulio et al. [37] also surveyed and analysed approximately 25 schemes that detect replication attacks in WSNs.

3 The proposed detection methodology

In wireless networks, the RSS is considered as a rough estimator of the distance between any two nodes. RSS-based schemes are usually based on a simple radio model where the received power approximately decays with the mth power of the distance. That is,

$$P_r \propto \frac{P_t}{d^m}$$

where P_r is the received power at the receiver, P_t is the transmission power at the transmitter node (which is considered constant at each transmitter) and d is the distance between the transmitter and the receiver. The value of m is called the path loss exponent, and its value depends on the environment being used. For outdoor line-of-sight (LoS) conditions, its value is 2, and for indoor environments, its value is 4. For a known transmitted power, the receiver node can compute the distance between itself and the transmitter, and by using simple geometric triangulation, the receiver can locate the transmitter.

In this section, we will use the RSS to detect Sybil attacks and replication attacks. Due to the lack of centralized identity management and administration in ad hoc networks where nodes have no information about their remote neighbours at the nth hop, in the following subsection, we will devise local and non-local detection strategies. In the former, each node will detect the attackers in its local radio range, whereas in the latter, nodes will construct partial global maps of neighbours for the detection by collecting topological information.

3.1 Local detection

In identity attacks, multiple nodes can illegitimately acquire the same identity. In 802.11-based ad hoc networks, mobile nodes frequently find new neighbours by periodically broadcasting beacon packets in which these nodes ascertain their identities. Due to the non-predictable nature of these ad hoc networks, a malevolent node can claim the same identities without being detected. Our goal is to detect the identity attacks by ensuring that each physical node is bound with only one legal identity.

In the local detection approach, each node must store the RSS that is captured along with its capture time in a table from all the neighbouring nodes. Furthermore, each node monitors and records the RSS that is received from every 1-hop neighbour, as shown in Fig. 1. For any two successive RSS messages that originated from the same identity, the receiver will need to determine whether the messages that are received are from the same legitimate node or from two distinct nodes with the same identity. However, there are some intricacies involved here, which make the task of detection slightly more complicated. For example, in Fig. 1, node A receives messages $f1$ and $f2$ at times $t1$ and $t2$,

respectively, from same identity m. Now, node A needs to determine whether the messages received are from the same node moved from location $l1$ to $l2$ with the induced change in distance from $d1$ to $d2$ or from two distinct nodes. In this case, one is legitimate and the other would be the attacker node.

Before we answer the above question, it is important to build some terminology. Let $R_i^j(t_k)$ be the RSS value of node i received at node j at time t_k. Similarly, the change in RSS from successive messages from the same sender at the receiver will be

$$\Delta R = R_i^j(t_k) - R_i^j(t_{k-1})$$

We put an upper bound on the speed V that nodes can have by assuming that the maximum speed a node can have in the network is V_{max} ms^{-1}. In addition, R_{max} is the change induced in the RSS when a node's covering distance is d_{max}, which is the distance covered by a node moving from any arbitrary location $l1$ to any $l2$ with V_{max} in Δt time. Hence, a node can never induce more change in the RSS than that of the R_{max} at the receiver since no node can cover more than d_{max} distance in Δt time.

In the above example, for the detection of spoofing attacks and replication attacks where attackers take on duplicate identities, node A will test the following condition.

$$\Delta R = R_A^m(t_k) - R_A^m(t_{k-1})$$

$$\text{Detection} = \begin{cases} \dfrac{\Delta R}{\Delta t} > R_{max} & \text{Attack} \\ \dfrac{\Delta R}{\Delta t} \leq R_{max} & \text{Normal} \end{cases}$$

3.1.1 Attack formulation

Due to its reusability feature, the RSS is an attractive choice for us to adopt it for attack detection. In addition, it is almost appropriate to meet the accuracy constraints of many applications. For that reason, we develop an attack detector using the RSS properties for identity-based attack detections [38, 39].

Here, we formulate the detection of an identity attack as a statistical significance testing problem in which the null hypothesis is

H_0 normal (no attack).

In this kind of testing, we will consider a test statistic T and will keep it under observation to establish whether the data under consideration belong to the hypothesis or not.

For a specific significance level α (defined as the probability of rejecting the hypothesis if it is true), there is a

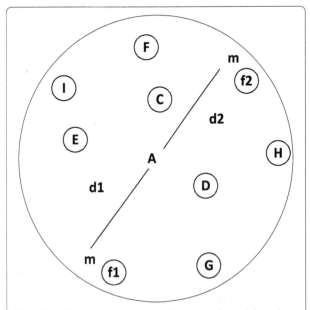

Fig. 1 Spoofing in same radio range. Represents the mobile nodes A, C, D, F, G, H, I and M that lie in the local radio range for which experiment is conducted

corresponding *acceptance region* Ω such that we declare the null hypothesis to be valid if an observed value of the test statistic $T_{obs} \in \Omega$ and reject the null hypothesis if T_{obs} Ω. In other words, we declare that an attack is present if $T_{obs} \in \Omega_c$, where Ω_c is the *critical region* of the test.

Here, in our identity attack detection problem, we use the distance in the signal space and make the decision in comparison with the calculated threshold. Then, the acceptance region Ω and the detection rate are based on the specified T. If the attack is present, then our proposed null hypothesis will be rejected.

3.1.2 Statistical analysis of the RSS

Given that the RSS is usually affected by noise, environmental factors and multipath fading, we measure the RSS with a reference node and determine the distance to that reference node. Here, we consider three different scenarios: scenario I, scenario II and scenario III. In scenario I, the reference node is separated from the receiving node by 1, 6 and 11 ft. In scenario II, the reference node is separated from the receiving node by 20, 25 and 30 ft. In scenario III, the reference node is separated from the receiving node by 50, 55 and 60 ft. Thus, the RSS readings show a strong statistical correlation. We will assume that the reference node has the same transmission power.

According to the propagation model, the RSS at a receiving node from the reference node is given by [18].

$$P(d_i)[\text{dBm}] = P_i(d_0)[\text{dBm}] - 10\gamma \, \log\left(\frac{d_i}{d_i}\right) \qquad (1)$$

where i is the ith receiving wireless node, $P_i(d_0)$ represents the transmission power of the reference node i at the reference distance d_0, d_i is the distance between the receiving wireless node and the reference node and γ is the path loss exponent whose value for free space communications is 2 and is greater than 4 for indoor communications [25, 40, 41].

Hence, the RSS distance between two nodes (reference node and receiving node) in signal space is given by

$$\Delta P = 10\gamma \, \log\left(\frac{d_i}{d_0}\right) \qquad (2)$$

Here, we show the empirical cumulative distribution function (CDF) of the RSS distance between the reference and the receiving nodes in the signal spaces of scenario I, scenario II and scenario III, which are shown by Figs. 2, 3 and 4 respectively.

In all the abovementioned three scenarios, we find that the observed changes in variance and standard deviation increase with the increase in the distance. A minimum change in the skewness (a measure of the asymmetry of the probability distribution of a real-valued random variable about its mean) is also observed when the nodes are near each other.

It is important to analyse how well we can derive a threshold under which the distance in the signal space can effectively be exploited to perform attack detection. We refer to Eq. (2), where the two wireless nodes are at two different positions, such as the reference and the receiving node, with their respective means (μ_0 and μ_i) and standard deviations (δ_0 and δ_i).

The probability density functions (*pdfs*) of the distance under these two different conditions can be represented as follows

$$f_{\Delta P} \, (p \mid \text{Reference point}) = \frac{1}{\sqrt{\pi\delta}} e^{\frac{-(x-\mu_0)^2}{\delta^2}} \qquad (3)$$

$$f_{\Delta P} \, (p \mid \text{consideration point}) = \frac{1}{\sqrt{\pi\delta}} e^{\frac{-(x-\mu_i)^2}{\delta^2}} \qquad (4)$$

$$\text{DR} = \text{Prob} \, (\Delta P > t \mid \text{Reference point}) = 1 - \phi \, (t - \mu_0)/\delta_0 \qquad (5)$$

$$\text{FPR} = \text{Prob} \, (\Delta P > t \mid \text{consideration point}) = 1 - \phi \, (t - \mu_i)/\delta_0, \qquad (6)$$

where t is the detection threshold.

If we calculate the detection rate taken from the real test bed data, calculate it based on the detection rate by using Eq. (5) and then plot the result with the normal expected data without any attack; the following graph can be constructed.

We take the detection threshold $t = d_{\max}$, which is the maximum distance that can be covered by the mobile node, where t can be given as

$$t = \frac{\mu_0 + \mu_1}{2}$$

$$t = \frac{0 + 10\alpha \, \log\left(\frac{d_i}{d_0}\right)}{2}$$

$$t = 5\alpha \, \log\left(\frac{d_i}{d_0}\right)$$

To tune our detection accuracy by increasing true positives and reducing false positives using a 95% confidence interval, we add two standard deviations to our calculated threshold t, which can be seen in Figs. 5 and 6.

3.1.3 Threshold tuning

To theoretically obtain approximate estimates for the nodes moving with the certain speeds of 1, 2, 3 and 4 m/s in the signal space, we obtain the following:

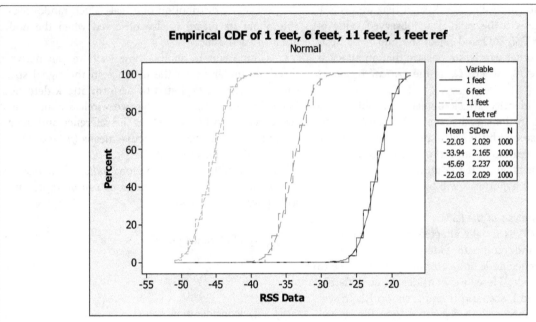

Fig. 2 Empirical CDF of scenario I. Calculates the empirical cumulative distribution functions for 1, 5 and 11 ft. The *x*-axis represents the RSS data while the *y*-axis represents the percentage value of each graph. It also shows the mean and standard deviation values of the three instances taken in scenario I

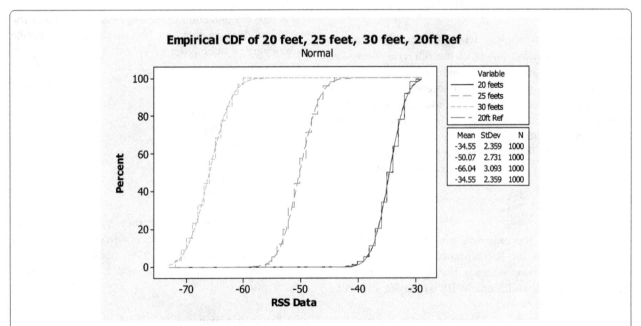

Fig. 3 Empirical CDF of scenario II. Calculates the empirical cumulative distribution function for 20, 25, and 30 ft. The *x*-axis represents the RSS data, while the *y*-axis represents the percentage value of each graph. It also shows the mean and standard deviation values of the three instances taken in the scenario II

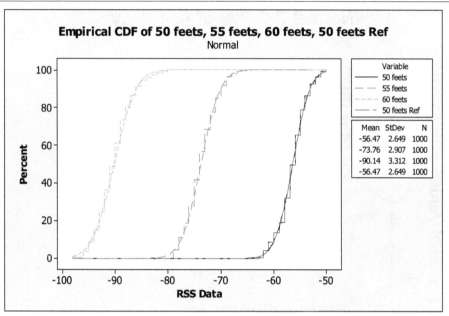

Fig. 4 Empirical CDF of scenario III. Calculates the empirical cumulative distribution function for 50, 55 and 60 ft. The x-axis represents the RSS data, while the y-axis represents the percentage value of each graph. It also shows the mean and standard deviation values of the three instances taken in the scenario III

For 1 m/s

$$\Delta P = |\ P0{-}P1m\ |$$

$$= |\ {-}18{-}({-}31)\ | = 13\ dbm$$

For 2 m/s

$$\Delta P = |\ P0{-}P2m\ |$$

$$= |\ {-}18{-}({-}38.6)\ | = 20\ dbm$$

For 3 m/s

$$\Delta P = |\ P0{-}P3m\ |$$

$$= |\ {-}18{-}({-}43)\ | = 25\ dbm$$

For 4 m/s

$$\Delta P = |\ P0{-}P3m\ |$$

$$= |\ {-}18{-}({-}46)\ | = 28\ dbm$$

Hence, it can be deduced that each RSS must not induce a change greater than 13 dbm for 1 m/s, 20 dbm for 2 m/s, 25 dbm for 3 m/s and 28 dbm for 4 m/s. This is also depicted in Fig. 7.

3.2 Non-local detection

Here, in this scenario, we consider multiple radio ranges named as radio range I, II, III, IV and V. In each region, a reference node is considered and shown as an underlined alphabet, as shown in Fig. 8. Please note that these reference nodes are not different than the other nodes

and they are just for the explanation. We assume that each node constructs its 1-hop neighbours using the captured RSS directly or via overhearing. This 1-hop list will be shared periodically in order to enable the nodes to construct partial or complete network topology maps, as shown in Table 1. Table 1 shows the maps of the reference nodes only. We will use these maps to detect the replicated identities in the network.

In this section, we will try to solve a problem: how do we distinguish a replicated node from a legitimate node? The replicated identity may either be a distinct malicious node or it may be a spoofed identity created and adopted by a Sybil attacker. For example, in Fig. 8, node *m* shows its presence in three reference nodes' lists of *A*, *D* and *J*, as shown in Table 1. Our aim here is to detect the replicated nodes or identities in the network.

Please note that if the Sybil attacker spawned an identity that does not previously exist in the network, it can be detected by our local detection scheme discussed in Section 3.

3.2.1 System model

To develop the criteria for the replicated identity detection, we will introduce some terminology first, which may be given as follows.

Let N nodes be uniformly distributed in an area A. Let $n(p)$ be the immediate or 1-hop neighbours of node p, which are in p's radio range and share a bidirectional link with p. Two nodes, p and q, can communicate directly with each other if they are 1-hop neighbours of

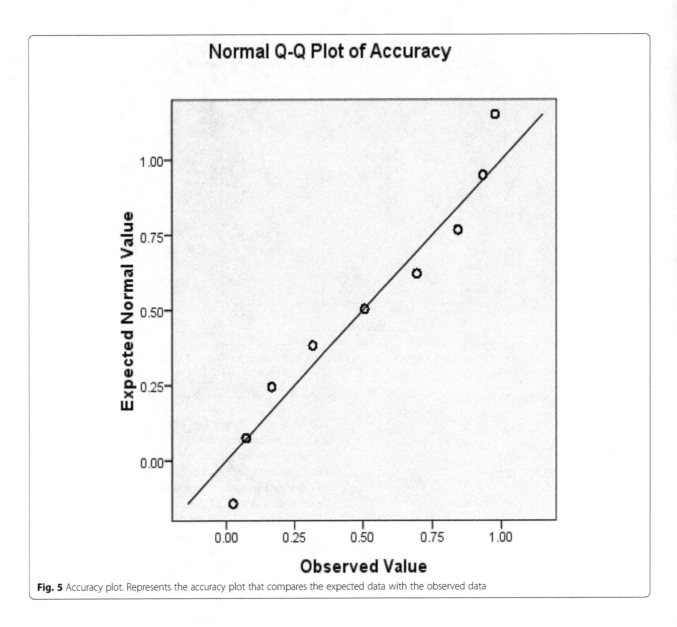

Fig. 5 Accuracy plot. Represents the accuracy plot that compares the expected data with the observed data

each other. That is, if $p \in n(q)$, then $q \in n(p)$. Let $n_2(p)$ denote the 2-hop neighbours of p that are the set of nodes that are neighbours of at least one node of $n(p)$. However, they do not belong to $n(p)$, and $n_2(p) = \{t | \exists z \in n(p) | t \in n(z) \backslash \{p\} \cup \{n(p)\}\}$. For a node $q \in n(p)$, let $\Delta_p^+(q)$ be the number of nodes belonging to $n_2(p)$ that also belong to $n(q)$ such that $\Delta_p^+(q) = | n_2(p) \cap n(q) |$. In other words, these are the number of nodes in $n_2(p)$ that node p can reach via node q. Similarly, for a node $q \in n_2(p)$, let $\Delta_p^-(q)$ represent the $n(p)$ nodes that also belong to $n(q)$ such that $\Delta_p^-(q) = | n(p) \cap n(q) |$. In other words, this quantity denotes the number of overlapping nodes in $n(p)$ that acts as bridge nodes and connects p and q in 2-hops.

PROPOSITION: *Let $\mathcal{G}(V, E)$ be an undirected graph with $V = \{v_1, v_2, v_3, \ldots \ldots v_n\}$ vertices connected together* *using $E = \{e_1, e_2, e_3, \ldots \ldots e_m\}$ edges. Let $\mathcal{G}$ be k-connected, where $k \geq 2$. Let node m be a node under observation. Let $m \in n(p)$ and $m \in n(q)$. Then, for a normal situation, there must exist a bridge node b that connects p to q. Otherwise, m will be deemed as a spoofed identity.*

PROOF: $\mathcal{G}$ is k-connected, and node m happens to be $m \in n(p)$ and $m \in n(q)$. Since $k \geq 2$, then there must be at least one other node x (other than m) such that $x \in \Delta_p^-(q)$, which also implies that $q \in \Delta_p^+(x)$. However, if $k = 1$, then m will be the only node that will belong to the set $\Delta_p^-(q)$.

The average number of 1-hop, 2-hop, Δ^- and Δ^+ nodes of a perspective arbitrary node in the network remain to be shown. We use the unit disk graph in order to model the network, such as that modeled by [42]. Since we are interested in the above mean values only, it is sufficient for us to capture the

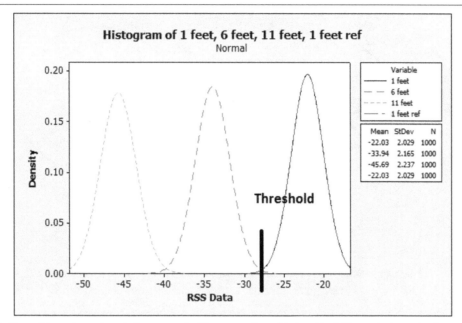

Fig. 6 Threshold with confidence interval graph. Represents the histogram values of 1, 6 and 11 ft. The *x*-axis shows the density value of the data, while the *y*-axis shows the RSS data. The figure also displays the mean and standard deviation of the three separate instances taken into consideration. The confidence interval with the threshold value has also been calculated statistically

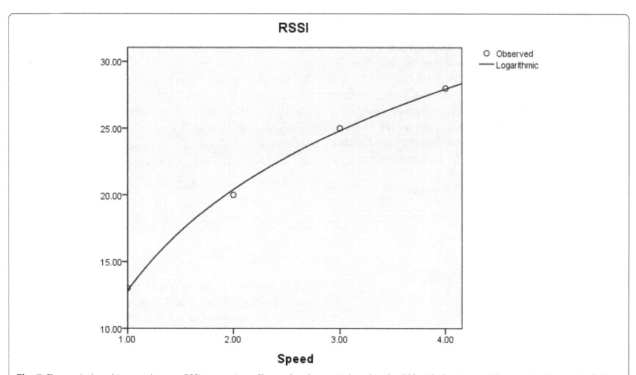

Fig. 7 Change induced in speed versus RSSI comparison. Shows the change induced in the RSS with the increase/decrease in the speed of the mobile nodes. The *x*-axis represents the speed of the mobile nodes in metres per second, while the *y*-axis represents the density of the RSS data

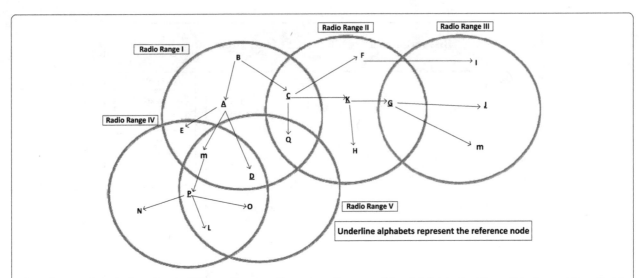

Fig. 8 Connected graph of spoofing in different radio ranges. Shows the radio ranges of the global detection. The participating mobile nodes are A, B, C, D, E, F, G, H, I, J, K, L, M, N, O, P and Q, which are divided in the five different radio ranges of I, II, III, IV and V. The underlined letters represent the reference nodes

portion of the network rather than modelling the complete network.

Let $\mathcal{D}(p, R)$ represent a disk with radius R to imitate the radio range R of node p. Let p lie at the origin of the disk. Then, by using the Poisson Point Process, the average number of points (nodes) of the process by the surface unit on $\mathcal{D}(0, R)$ is λ, which is called the intensity of the process, where $\lambda > 0$. The same is true for 2-hop nodes where $\mathcal{D}(0, 2R)$. It is worth mentioning that the points are uniformly and independently distributed in each disk. In other words, the points in $\mathcal{D}(0, R)$ are independent of the points distributed in $\mathcal{D}(0, 2R)$. As discussed above, we assume bidirectional links between each pair of nodes. These links would exist if and only if $d(p, q) \leq R$, where $d(p, q)$ is called the Euclidean distance between the pair p and q. As shown in Fig 9a, $A(r)$ is the area of the intersection of two disks with radii of R that have r as the distance between their centres, which can be computed as follows:

$$A(r) = 2R^2 \arccos\left(\frac{r}{2R}\right) - r\sqrt{\left(R^2 - r^2/4\right)} \tag{7}$$

Table 1 The 1-hop list

A	K	J	P	D
B	C	G	E	Q
C	F	I	D	M
D	G	M	M	P
E	H		N	L
M	Q		L	O
Q			O	

Let p be a point that is uniformly distributed in $\mathcal{D}(0, R)$. Then, the average number of points in $\mathcal{D}(0, R)$ is given as

$$\mathbb{E}[n(p)] = \lambda \pi R^2 \tag{8}$$

To calculate the average number of process points belonging to n_2 (2-hop neighbours) that are accessible to p nodes via q, which is denoted by $\Delta_p^+(q)$, we assume that p and q are the process points uniformly distributed in $\mathcal{D}(0, R)$ and $\mathcal{D}(0, 2R)$, respectively (consult Fig. 9a). The quantity $\Delta_p^+(q)$ is basically the q number of nodes that do not belong to $n(p)$ and reside on the $\pi R^2 - A(r)$ surface. By definition of the Poisson Point Process, on average, we have $\lambda \pi R^2$ nodes lying on a surface, and by proportionality, we have $\frac{\lambda}{\pi R^2}\left(\pi R^2 - A(r)\right)$ nodes in $\mathcal{D}(0, 2R) \setminus \mathcal{D}(0, R)$. Therefore, by integrating all such points, we obtain the average number of nodes lying in $\mathcal{D}(0, 2R) \setminus \mathcal{D}(0, R)$ as

$$\mathbb{E}\left[\Delta_p^+(q)\right] = \frac{\lambda}{\pi R^2} \int_0^{2\pi} \int_0^R \left(\pi R^2 - A(r)\right) r \, dr \, d\theta$$
$$= \lambda R^2 \frac{3\sqrt{3}}{4} \tag{9}$$

To compute the average number of Δ^- nodes, assuming Fig. 9b, let p and q be the process points that are uniformly distributed in $\mathcal{D}(0, R)$ and $\mathcal{D}(0, 2R)$, respectively. Please note here that some of the q nodes will belong to $\mathcal{D}(0, 2R) \setminus \mathcal{D}(0, R)$ without being 2-hop neighbours of p since there must be a node on $\mathcal{D}(0, R)$'s surface to connect p to q. We assume that $A(r)$ is an overlapping area where common neighbours exist for

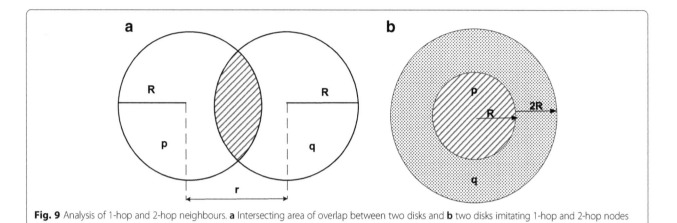

Fig. 9 Analysis of 1-hop and 2-hop neighbours. **a** Intersecting area of overlap between two disks and **b** two disks imitating 1-hop and 2-hop nodes

every r between R and $2R$. There are $\frac{2r}{3R^2}$ nodes at distance r. By integrating these points between R and $2R$, we get

$$\mathbb{E}\left[\Delta_p^-(q)\right] = \lambda \frac{2}{3R^2} \int_R^{2R} A(r) r dr = \lambda R^2 \frac{\sqrt{3}}{4} \qquad (10)$$

To calculate the average number of 2-hop neighbours, there must be at least one common neighbour that connects 1-hop to 2-hop nodes, such as

$$\mathbb{E}\left[\Delta_p^-(q)\mid q \in n_2\right] = \frac{\mathbb{E}\left[\Delta_p^-(q)\right]}{\mathrm{P}\left(\Delta_p^-(q) > 0\right)}.$$

The probability that a node in $\mathcal{D}(0, 2R)\backslash\mathcal{D}(0, R)$ has at least one common neighbour in $\mathcal{D}(0, R)$ that makes q a 2-hop neighbour of p is

$$P\left(\Delta_p^-(q) > 0\right) = 1 - \frac{2}{3R^2} \int_R^{2R} \exp\{-\lambda A(r)\} r dr.$$

Therefore, the average number of 2-hop nodes can be written as

$$\mathbb{E}[n_2(p)] = 3\lambda\pi R^2$$
$$\times \left(1 - \frac{2}{3R^2} \int_R^{2R} \exp\{-\lambda A(r)\} r dr\right) \qquad (11)$$

4 Simulation-based evaluation

4.1 Simulation setup

After evaluating our scheme with the help of statistical testing and analytical modelling, we also evaluate our scheme with the help of the NS2 simulator. We use the simulation parameters listed in Table 2. Here, we conduct our experiment for density versus accuracy in two scenarios, the true positive rate (TPR) versus speed and the false positive rate (FPR) versus speed. In each scenario, we again took three instances with different numbers of nodes.

To evaluate our attack detector, we investigate the effect of the node density on the detection accuracy. In the accuracy, we consider TPR and FPR. Meanwhile, in node density, we take three instances of 20, 30 and 40 nodes in each case of TPR and FPR separately with respect to speed.

4.2 Results

In the first experiment, as shown in Fig. 10, it is observed that the detection accuracy (TPR) of our scheme increases with the increase in node density. In the first instance where the number of nodes is 20, the TPR is almost 93.73%. In the second instance where the number of nodes is 30, the TPR is almost 95.21%. In the last instance where the number of nodes is the maximum of all, the TPR is 99.01%.

In the second experiment, as shown in Fig. 11, again there is no significant effect of the number of nodes on the FPR of our scheme. In the first instance, we take 20 nodes, which produce a low FPR of almost 7.13%. In the second instance, we take 30 nodes, and the resulting FPR is almost 5.51%. In the third instance, we take 40 nodes, and the ensued FPR is almost 0.98%. Hence, we can conclude that high node densities improve our

Table 2 Simulation parameters

Parameter	Value
Used area	1000 m × 1000 m
Maximum speed	10 m/s
Pause time used	60 s
Radio range	250 m
Nodes used	50–60
Connection established	5–15
The MAC	802.11
The application	CBR
Simulation time	900 s
Mobility model	Random wave point model

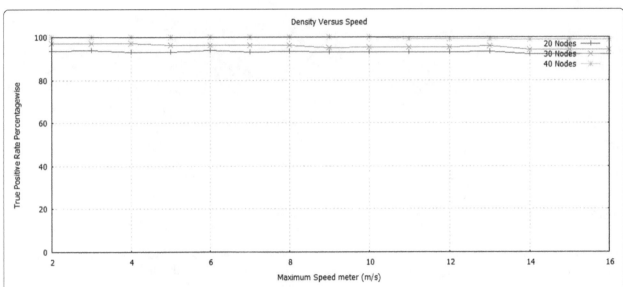

Fig. 10 Density versus speed true positive rate. Shows the simulation results of the density versus speed experiment. The *x*-axis represents the maximum speed per metre second, while the *y*-axis represents the TPR. The experiment was conducted in three different instances of 20, 30 and 40 mobile nodes

scheme's detection accuracy, such as high TPR and low FPR. This is since with high node density, the attackers fall into more radio ranges, which improves the detection accuracy.

5 Discussion

In the above sections, we discussed our proposed detection system and its design rationale. We analysed the RSS using statistical significance testing and theoretically calculated the PDR and FPR. Finally, we evaluated the scheme using the NS-2 simulator in order to analyse the overall detection accuracy of the system in different mobile environments. However, some important points of our proposed work remain to be discussed, which are given below.

First, one of the main issues in our proposed scheme that we discovered during the simulations is that although it performed well in dense environments, the detection accuracy decreased in sparse environments. One of the main reasons for this is that due to the mobility and

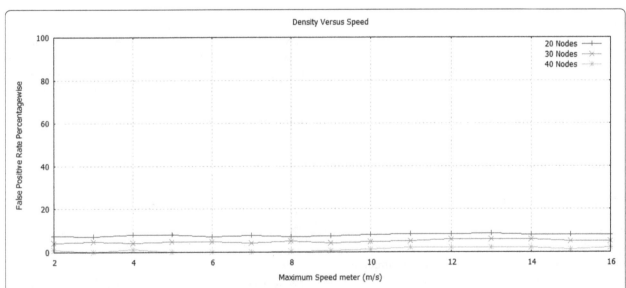

Fig. 11 Density versus speed false positive rate. Shows the simulation results of the density versus speed experiment. The *x*-axis represents the maximum speed per metre second, while the *y*-axis represents the FPR. The experiment was conducted in three different instances of 20, 30 and 40 mobile nodes

fluctuating RSS, some of the neighbouring nodes may not be able to detect the required change in the attackers' RSSs. This induced false negatives in some nodes. However, if the number of nodes was greater in the vicinity, the false negatives decreased.

Second, throughout our detection system, we established nodes to capture and store the RSS during the four-way handshake of the MAC 802.11 protocol. They were the RTS, CTS, DATA and ACK. This sometimes makes the RSS insufficient for the detection. For instance, an attacker node during its lifetime may move to a region in the network where it does not happen to be involved in the active routing path(s) (not forwarding any routing packets). Therefore, no RSS can be collected from this attacker and it will be undetected. This type of false negative can be mitigated by using periodic beacons in which each node broadcasts periodic control frames that are called beacons. Periodic beacons can also improve the topological construction in our non-local detection process. For example, each node will maintain a fresh map of its neighbours. However, periodic beacons are still a problem for a few reasons. First, they cause much overhead in the network. Second, attackers may not be forced to broadcast beacons in the prescribed manner. Third, attackers may broadcast spoofed and fabricated beacons, thus disrupting the overall network operations and deteriorating the detection process.

We assume a maximum speed for the network. By using it, we effectively distinguished between normal nodes and attackers. We believe that this assumption is valid and this will not make our scheme impractical for real environments. For instance, our scheme can still be used in vehicular ad hoc networks (VANETs) where vehicles cannot move faster than a limit. The limit might be the vehicle's maximum speed (from a speed metre) or it might be the permissible speed limit on a particular road or on a highway.

We have observed considerable fluctuations in the RSS, as discussed in Section 3. This fluctuation may produce a few metres of inaccuracy, which normally creates a weak degree of distinguishability of the closely lying nodes. However, our scheme can still be used in various application domains. For example, in VANETs where nodes are vehicles occupying few metres of space, they can easily be distinguished from other nearby vehicles in the signal space. Similarly, in the e-healthcare scenario where each patient's mobile phone can be considered in a single room, it can easily be distinguished from another patient's mobile phone in another room in the hospital.

6 Conclusions

In this paper, we proposed an RSS-based scheme to counter the identity attacks on IEEE 802.11-based ad hoc networks without using any additional hardware or a third-party guarantor. Unlike other schemes, our scheme did not incur any overhead in the form of periodic beacons and expansive localization computations. Similarly, we have validated our scheme theoretically and by using the NS-2 simulator. We have used our empirically collected data in our theoretical analysis. The result obtained from our analysis and the simulation showed that our proposed scheme produced good detection accuracy with negligible false positives.

Throughout our problem formulation, we have assumed homogenous transmission power in all nodes. In our future work, we will aim at improving our work for heterogeneous transmission powers at each node.

Abbreviations
CA: Certification authorityGPSGlobal Positioning System; DDoS: Distributed Denial of Service attack; DoS: Denial of Service attack; FPR: False positive rate; IEEE: Institute of Electrical and Electronics Engineering; IoT: Internet of Things; LoS: Line-of-sight; MANETs: Mobile ad hoc networks; NS-2: Network simulator; RSS: Received signal strength; RSSI: Received signal strength indicator; TPR: True positive rate; TTP: Trusted third party; VANETs: Vehicular ad hoc networks

Acknowledgements
The authors are grateful to Dr. Nathalie Mitton at INRIA labs France for the help with understanding some concepts related to analytical modelling. The authors also thank the anonymous reviewers for their valuable comments to improve this paper.

Authors' contributions
MF is the PhD scholar that performed all the work in this paper. HUR is the main research supervisor of MF who helped him in fine-tuning the proposed scheme. SA is the co-supervisor of MF who helped him in the modelling and simulation of the detection rationale. All authors read and approved the final manuscript.

Competing interests
The authors declare that they have no competing interests.

Author details
[1]Department of Computer Science and IT, University of Malakand, Chakdara, KPK, Pakistan. [2]Department of Computer Science, College of Science, University of Sharjah, Sharjah, UAE.

References
1. S Abbas et al., Lightweight Sybil attack detection in MANETs. Systems Journal, IEEE **7**(2), 236–248 (2013)
2. DB Faria, DR Cheriton, in *Proceedings of the 5th ACM workshop on wireless security*. Detecting identity-based attacks in wireless networks using signalprints (ACM, 2006)
3. A Cheng, E Friedman, *Sybilproof reputation mechanisms, in Proceedings of the 2005 ACM SIGCOMM workshop on economics of peer-to-peer systems* (ACM, Philadelphia, 2005)
4. M Presser et al., The SENSEI project: integrating the physical world with the digital world of the network of the future. IEEE Commun. Mag. **47**(4), 1–4 (2009)
5. DG Reina et al., *The role of ad hoc networks in the internet of things: a case scenario for smart environments, in Internet of Things and Inter-Cooperative Computational Technologies for Collective Intelligence* (Springer, 2013), pp. 89–113
6. Dorri, A., S.R. Kamel, and E. Kheirkhah, Security challenges in mobile ad hoc networks: a survey. arXiv preprint arXiv:1503.03233, 2015.

7. MA Jan et al., A Sybil attack detection scheme for a forest wildfire monitoring application. Futur. Gener. Comput. Syst. **80**, 613-626 (2016)

8. X Feng et al., A method for defensing against multi-source Sybil attacks in VANET. Peer-to-Peer Netw Appl, 1–10 (2016)

9. SR Jan et al., *An innovative approach to investigate various software testing techniques and strategies*, International Journal of Scientific Research in Science, Engineering and Technology (IJSRSET), Print ISSN (2016), pp. 2395–1990

10. AK Pal et al., A discriminatory rewarding mechanism for Sybil detection with applications to Tor. Electron. Mark. **208**, 12786 (2010)

11. B Yu, C-Z Xu, B Xiao, Detecting Sybil attacks in VANETs. J Parallel Distrib Comput **73**(6), 746–756 (2013)

12. S Raza, L Wallgren, T Voigt, SVELTE: real-time intrusion detection in the Internet of Things. Ad Hoc Netw. **11**(8), 2661–2674 (2013)

13. L Maccari, RL Cigno, A week in the life of three large wireless community networks. Ad Hoc Netw. **24**, 175–190 (2015)

14. K Xing, X Cheng, *From time domain to space domain: detecting replica attacks in mobile ad hoc networks, INFOCOM, 2010 Proceedings IEEE* (IEEE, 2010)

15. M Conti, S Giordano, Mobile ad hoc networking: milestones, challenges, and new research directions. IEEE Commun. Mag. **52**(1), 85–96 (2014)

16. R Di Pietro et al., Security in wireless ad-hoc networks: a survey. Comput. Commun. **51**, 1–20 (2013)

17. VM Agrawal, H Chauhan, An overview of security issues in mobile ad hoc networks. Int J Comput Eng Sci **1**(1), 9–17 (2015)

18. S Marti, H Garcia-Molina, Taxonomy of trust: categorizing P2P reputation systems. Comput. Netw. **50**(4), 472–484 (2006)

19. MN Mejri, J Ben-Othman, M Hamdi, Survey on VANET security challenges and possible cryptographic solutions. Vehicular Communications **1**(2), 53–66 (2014)

20. G Yan, S Olariu, MC Weigle, Providing VANET security through active position detection. Comput. Commun. **31**(12), 2883–2897 (2008)

21. Jan, M., et al., PAWN: a payload-based mutual authentication scheme for wireless sensor networks. Concurrency and Computation: Practice and Experience, 2016.

22. J Yang, Y Chen, W Trappe, *Detecting spoofing attacks in mobile wireless environments. In Sensor, Mesh and Ad Hoc Communications and Networks, 2009. SECON'09. 6th Annual IEEE Communications Society Conference on* (IEEE, 2009)

23. NB Margolin, BN Levine, *Quantifying resistance to the Sybil attack, in Financial Cryptography and Data Security* (Springer, 2008), pp. 1–15

24. Hoeper, K. and G. Gong, Bootstrapping security in mobile ad hoc networks using identity-based schemes with key revocation. Centre for Applied Cryptographic Research (CACR) at the University of Waterloo, Canada, Tech. Rep. CACR, 2006. 4: p. 2006.

25. Y Chen et al., Detecting and localizing identity-based attacks in wireless and sensor networks. IEEE Trans. Veh. Technol. **59**(5), 2418–2434 (2010)

26. D Glynos, P Kotzanikolaou, C Douligeris, in *Modeling and Optimization in Mobile, Ad Hoc, and Wireless Networks, 2005. WIOPT 2005. Third International Symposium on.* Preventing impersonation attacks in MANET with multi-factor authentication (IEEE, 2005)

27. MS Bouassida, M Shawky, in *Communications and Information Technologies, 2007. ISCIT'07. International Symposium on.* Localization verification and distinguishability degree in wireless networks using received signal strength variations (IEEE, 2007)

28. J Hall, M Barbeau, E Kranakis, *Using transceiverprints for anomaly based intrusion detection*, Proceedings of 3rd IASTED, CIIT (2004), pp. 22–24

29. D He, D Wang, Robust biometrics-based authentication scheme for multiserver environment. IEEE Syst. J. **9**(3), 816–823 (2014)

30. BR Debdutta, C Rituparna, MADSN: mobile agent based detection of selfish node in MANET. International Journal of Wireless & Mobile Networks (IJWMN) **3**, No. 4 (2011)

31. B Parno, A Perrig, in *Workshop on Hot Topics in Networks (HotNets-IV)*. Challenges in securing vehicular networks (2005)

32. Hoeper, K. and G. Gong, Bootstrapping security in mobile ad hoc networks using identity-based schemes. Security in Distributed and Networking Systems, 2007.

33. Y Sheng et al., in *INFOCOM 2008. The 27th Conference on Computer Communications. IEEE.* Detecting 802.11 MAC layer spoofing using received signal strength (IEEE, 2008)

34. USRK Dhamodharan, R Vayanaperumal, Detecting and preventing Sybil attacks in wireless sensor networks using message authentication and passing method. Sci. World J. **2015**, 841267;7 (2015) https://doi.org/10.1155/2015/841267

35. Zdonik, S., et al., SpringerBriefs in computer science. 2012.

36. M Qabulio, YA Malkani, A Keerio, in *Information Assurance and Cyber Security (CIACS), 2015 Conference on.* Securing mobile wireless sensor networks (WSNs) against clone node attack (IEEE, 2015)

37. M Qabulio, YA Malkani, AA Keerio, On node replication attack in wireless sensor networks. Mehran Univ Res J Eng Technol **34**(4), 413–424 (2015)

38. P Bahl, VN Padmanabhan, in *INFOCOM 2000. Nineteenth Annual Joint Conference of the IEEE Computer and Communications Societies. Proceedings.* RADAR: an in-building RF-based user location and tracking system (IEEE, 2000)

39. MA Youssef, A Agrawala, AU Shankar, in *Pervasive Computing and Communications, 2003.(PerCom 2003). Proceedings of the First IEEE International Conference on.* WLAN location determination via clustering and probability distributions (IEEE, 2003)

40. A Goldsmith, *Wireless communications* (Cambridge university press, 2005)

41. TK Sarkar et al., A survey of various propagation models for mobile communication. IEEE Antennas Propag Mag **45**(3), 51–82 (2003)

42. Busson, A., N. Mitton, and E. Fleury. An analysis of the MPR selection in OLSR and consequences. In Mediterranean Ad Hoc Networking Workshop (MedHocNet'05). 2005.

4

Traffic-predictive QoS on-demand routing for multi-channel mobile ad hoc networks

Jipeng Zhou[*] (ID), Liangwen Liu and Haisheng Tan

Abstract

Mobile multimedia applications have recently attracted numerous interests in mobile ad hoc networks (MANETs) supporting quality-of-service (QoS) communications. Multiple non-interfering channels are available in 802.11- and 802.15-based wireless networks. Channel assignment depends on the available bandwidth at involved nodes and the bandwidth consumption required by a new flow. Predicting available bandwidth of a node in wireless networks is challenging due to the shared and open nature of the wireless channel. This paper proposes a traffic-predictive QoS on-demand routing(TPQOR) protocol to support QoS bandwidth and delay requirements. A distributed channel assignment scheme and routing discovery process are presented to support multimedia communication and to satisfy QoS bandwidth requirement. The proposed channel assignment and reuse schemes can reduce the channel interference and enhance channel reuse rate. The proposed bandwidth prediction scheme can estimate the bandwidth requirement of each node for future traffic by the history information of its channel usage. Unlike many existing routing protocols, we take the traffic prediction as an important factor in route selection. The simulation results show that TPQOR protocol can effectively increase throughput, reduce loss ratio as well as delay, and avoid the influences of future interference flows, as compared to AODV protocol for a different number of channels.

Keywords: Mobile ad hoc network, QoS routing, Bandwidth prediction, Channel assignment

1 Introduction

In mobile ad hoc networks (MANETs), the topology of the network can change frequently and the network routing becomes a crucial task [1]; while routing with mobility prediction is well-studied, routing with traffic prediction is still considered as an open, but meaningful problem. This is particularly important for routing with quality-of-service (QoS) constraints, since QoS resource-reservations affect future network traffic. There have been significant progress in using mobility prediction to build a more stable route in MANETs, and some predictive and reliable routing schemes [2, 3] have been reported. These accomplishments inspire us to study further on the design of a routing scheme with traffic prediction. The prediction of future traffic can be a powerful tool in QoS routing. For example, if there are two candidate paths and it is predicted that a new traffic will be produced along one of them right after the QoS route has been builded, thus, with the help of traffic prediction, we can choose the

more "peaceful" route for the QoS flow so that resources reserved by QoS route and resources used by that new traffic will not affect with each other. Node mobility is modeled as a time-homogeneous semi-Markov process for disruption-tolerant networks in [2], which predicts the future contacts of two specified nodes at a specified time. It can predict not only whether two nodes would have a contact, but also the time of contact. With this model, a node estimates the future contacts of its neighbors and the destination and then selects a proper neighbor as the next hop to forward the message. In [3], an interference aware metric with a prediction algorithm is proposed to reduce the interference between nodes at the MAC layer and works in an on-demand routing scheme in multichannel vehicular ad hoc networks.

Traffic routing and channel assignment jointly play a critical role in determining the performance of MANETs. The scare wireless channel resource, high dynamic link quality, and the uncertain traffic demands are big challenges for routing in MANETS. There are several approaches for traffic routing. One approach is based on traffic-predictive models, such as the method proposed

---
*Correspondence: tjpzhou@jnu.edu.cn
Department of Computer Science, Jinan University, Guangzhou 510632, People's Republic of China

in [4], which has a competitive performance when traffic can be predicated accurately, but may result in unbounded worst-case performance when forecasts go wrong. Traffic prediction by employing multi-layer feeding forward neural network model is proposed in [4], which learns the effect of spatio-temporal-spectral parameters on traffic patterns and predicts future traffic load on each of the channels. It is also hard to choose the parameters of neural network model. In another approach, routing can be made with the focus towards maximally unbalanced demand, such that the worst case is contained (known as oblivious routing) such as [5]. It is an open question how these two approaches would compare with each other in real networks. A weighted average predictive algorithm is presented for mesh networks in [5], which gives detailed simulation studies of predictive and oblivious routing. Their results show that the proposed algorithm can accommodate the changing conditions in the predictability of the traffic and has good performance, but it is difficult to give the accurate weights. One natural approach to address the traffic uncertainty in network routing is predictive routing [6], which infers the traffic demand with maximum probability based on history and optimizes the routing strategy for the predicted traffic demand. Underlying predictive routing is the assumption that past behavior is a good indicator of the future. Many researchers have studied predictive routings in MANETs; for example, in [7], authors proposed an algorithm that utilizes the prediction of vehicle position and navigation information to improve the routing protocol in vehicular ad hoc networks with a cross-layer approach. Paper [8] focuses on link quality prediction and estimation. A routing algorithm based on traffic prediction is proposed for DTN in [9]. Paper [10] investigates the cross-layer optimization problem of congestion and power control in cognitive radio ad hoc networks (CRANETs) under predictable contact constraint; a predictable contact model is presented by deriving the probability distribution of contact via a mathematical statistics theory; they do not adapt to MANETs. A method of the prediction of link residual lifetime using Kalman filter is proposed for vehicular ad hoc networks in [11]; it focuses on route selection; an analytical model for predictable contact between two cognitive users is proposed in the intermittently connected cognitive radio ad hoc networks [12]; it is based on mobility model, which do not consider the bandwidth usage and channel assignment. The scheme of opportunistic routing with autonomic forwarding angle adjustment (FAOR) is proposed for cognitive radio ad hoc networks in [13], which is a different network model from this paper.

QoS applications are usually sensitive to available bandwidth. The available bandwidth of each node is directly affected by the existing traffic, which makes traffic prediction very meaningful in QoS routing. The suitability of linear predictors for traffic prediction is discussed in packet and burst switching networks [14], where both the prediction method and the prediction interval are considered in traffic prediction. However, the performance of the network is limited by the packet arrival distribution. A priority aware dynamic source routing protocol is proposed in order to enhance QoS for MANETs in [15]; it assigns the priority for different flows in accordance with their data rates in dynamic source routing. As far as we have known, there is no QoS routing scheme based on the traffic prediction in ad hoc networks.

For channel assignment, a fully distributed channel assignment algorithm is proposed in [16], which can adapt to traffic loads dynamically for wireless mesh networks. The mentioned scheme can improve the utilization of network resource, but does not supply any solution of QoS routing. In [17], a QoS-aware routing mechanism to support real-time multimedia communication is proposed for ad hoc networks. A node estimates the usage of its wireless channels and disseminates the information about its available bandwidth to other nodes in the whole network. Thus, each node obtains a view of topology and bandwidth information of the whole network. Based on the obtained information, a source node determines a logical path with the maximum available bandwidth to satisfy the QoS requirements of applications. However, the above two papers do not take channel reuse into consideration.

The main goal of this paper is to design a routing scheme that can meet QoS bandwidth and delay requirements by taking traffic variations into account. We propose a traffic predictive QoS on-demand routing(TPQOR) protocol to pursue a better network performance. Particularly, TPQOR protocol uses channel reuse mechanism described in our previous work [18], in which we present a cross-layer protocol that solves channel assignment, reuse, and routing problem jointly. In TPQOR protocol, future traffic is predicted according to nodes' history traffic patterns, based on an assumption that the traffic of network nodes has a certain pattern. Such assumption is reasonable in reality considering the device utilization and routing characteristics. Different devices produce different traffic for their use, such as a video monitor usually produces higher traffic than those "light-weight" counterparts such as humidity or temperature sensors, since a node acts as a terminal as well as a router in ad hoc networks, which makes traffic of a node include not only the traffic it produces, but also the traffic it forwards. The shortest path routing has been simulated in a distributed 9×9 grid [19] , where the routing characteristic of the network is that nodes in the center area have much higher traffic load than nodes in other areas.

The remainder of this paper is structured as follows: we firstly describe network model as research basement in Sections 2, 3, and 4, depicting our proposed channel reuse scheme and traffic prediction mechanism respectively; Section 5 describes TPQOR routing procedure; Section 6 shows simulation results; at last, we make conclusions for this paper in Section 7.

2 Network model

We first present the network model used throughout this paper. An ad hoc network can be modelled as a graph $G = (V, E)$, where V is the set of nodes and E is the set of edges that represent wireless links. A link is assumed to exist between two nodes if and only if the two nodes are within each other's transmission range. For each link $e = (u, v) \in E$, u is the transmitter and v is the receiver. Each node n has a transmission range $R_t(n)$, which allows only those nodes within distance $R_t(n)$ to receive the signal from node n correctly. We assume that each terminal n also has an interference range $R_i(n)$ such that every unintended receiver would be interfered by the signal from node n when it is using the same channel as node n does simultaneously. For simplicity, it is assumed that all nodes have the same transmission range R_t and the same interference range R_i, and the interference range of nodes is two times of their transmission range, that is, $R_i = 2R_t$. We define that two distinct links (u_1, u_2) and (v_1, v_2) are interference links if one of two pairs (u_1, v_2) and (v_1, u_2) is less than R_i apart. In order to transmit simultaneously, two interference links need to be assigned with different channels.

In our network model, each node can operate on one common control channel for control information and other several data channels for data packets, each channel can be switched among the network interface cards (NIC) of each node. Let $CT = \{ch1, ch2, ..., chk\}$ denote the set of K orthogonal data channels that can be used by all nodes. We assume that every channel in CT has its unique serial number and has the same bandwidth Bw_{ch}. The set of available channels of node n is $A(n)$, which are free channels and can be assigned for future use at node n. $C_t(n)$ denotes the set of active transmitting channels of node n, and $C_r(n)$ denotes the set of active receiving channels of node n.

Interference neighbors of each node n are those nodes locating in the interference range of node n. Let $N_i(n)$ denote the set of interference neighbors of node n. To reach an interference-free channel assignment for link $e = (u, v)$, both the active transmitting channels of $N_i(v)$ and the active receiving channels of $N_i(u)$ should be excluded. The active transmitting channels of $N_i(v)$ is $C_t(N_i(v)) = \cup_{\forall x \in N_i(v)} C_t(x)$, and the active receiving channels of $N_i(u)$ is $C_r(N_i(u)) = \cup_{\forall x \in N_i(u)} C_r(x)$. That is, in order to avoid conflicts, link $e = (u, v)$ should not be assigned any channel in the set of $C_t(N_i(v)) \cup C_r(N_i(u))$. The notations used in this paper are as shown in Table 1.

Table 1 Notation used in the paper

Symbol	Comments
V	The set of nodes
E	The set of edges that represents wireless links
$R_t(n)$	The transmission range of node n
$R_i(n)$	The interfernce range of node n
CT	K orthogonal data channel set $\{ch1, ch2, ..., chk\}$
$N_i(v)$	Interference node set of node v
$C_t(v)$	Active transmitting channel set of node v
$C_r(v)$	Active receiving channel set of node v
$C_r(X(v))$	Active receiving channel set of node set $X(v)$
$C_t(X(v))$	Active transmitting channel set of node set $X(v)$
Bw_{req}	QoS request bandwidth for the flow
Bw_{ch}	Bandwidth of a channel
$A(u)$	Available channel set of node u
$AL(l)$	Available channel set of link l

3 Channel assignment scheme

In multichannel ad hoc networks, channel assignment is a real hot topic. A good channel assignment scheme should decrease collisions and enhance the network throughput as much as possible. Channel reuse scheme is to economize the number of assigned channels on the precondition of avoiding conflicts as much as possible. In other words, a distinct advantage of channel reuse scheme is that it can leave more available channels for other use. An example for channel assignment with the channel reuse scheme is shown in Fig. 1, where we assume that data channel set $CT = \{ch1, ch2, ch3, ch4, ch5\}$ and $R_i = 2R_t$. The channels are assigned from source node $u1$ to destination node $u6$, where the first four adjacent links from $u1$ to $u5$ must be assigned with different channels to avoid conflicts. We assume that links $(u1, u2)$, $(u2, u3)$, $(u3, u4)$, $(u4, u5)$ are assigned channels $ch1, ch2, ch3, ch4$ respectively. If the channel $ch5$ is assigned to link $(u5, u6)$ for interference-free assignment, it will leave no available

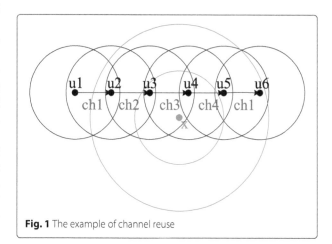

Fig. 1 The example of channel reuse

channel for other use. For example, there is no available channels for node x to transmit signals. If $ch1$ is reused by links $(u1, u2)$ and $(u5, u6)$ as shown in Fig. 1, $ch5$ will be available for the transmission channel of node x. The channel reuse can leave more free channels for other use, and this can enhance the throughput of the network.

In order to implement the channel reuse scheme, we should obtain the information of every node's channel assignment. Obtaining the information from the whole network produces too much traffic and acts against to network scalability. We design a distributed scheme that solves the problem of channel reuse elegantly. In our proposed mechanism, each channel is assigned to an unique serial number. The interference-free channel with the minimum number is firstly assigned. The lower a channel's number is, the higher probability it is reused.

The available channel set $AL(l)$ for a link $l = (u, v)$ is calculated as follows: let $C_t(N_i(v))$ the active transmitting channel set of v's interference neighbors and $C_r(N_i(u))$ the active receiving channel set of u's interference neighbors; $AL(l)$ can be represented as $AL(l) = CT - (C_t(N_i(v)) \cup C_r(N_i(u)))$. The proposed channel assignment scheme selects channels by sequence from $AL(l)$ for channel reuse. Let Bw_{req} represent the minimum bandwidth requirement of a QoS flow, and Bw_{ch} represent the physical maximum bandwidth of each channel. The ceiling integer of quotient of Bw_{req} divided by Bw_{ch} is the number of channels that need to be assigned for the QoS flow. The channel assignment algorithm is given in Algorithm 1.

Algorithm 1 Channel Assignment Algorithm(CA) for a link $l = (u, v)$

1: **initial:** for each link $l = (u, v)$, $AL(l)$ is set to empty
2: $AL(l) = CT - C_t(N_i(v)) \cup C_r(N_i(u))$
3: **if** the flow is not a QoS flow **then**
4:     **if** $AL(l)$ is empty **then**
5:         return a FAIL
6:     **else**
7:         return the channel with the minimum number in $AL(l)$ for link l
8:     **end if**
9: **else**
10:     $Req_{ch} = \lceil Bw_{req}/Bw_{ch} \rceil$
        /*Req_{ch} is the number of the QoS flow required channels */
11:     **if** $|AL(l)| < Req_{ch}$ **then**
12:         return a FAIL
13:     **else**
14:         return Req_{ch} channels with the minimum number in $AL(l)$ for link l
15:     **end if**
16: **end if**

---

4 Traffic prediction scheme

Traffic is an abstract concept referring to data flows which are shuttling to and from in the network, and can be described as various ways, such as consumed bandwidth. Since every new traffic would be assigned one or more interference-free channels in this paper and each channel denotes a certain value of bandwidth, we can use the number of active channels to represent the amount of traffic. When no enough channels in $A(n)$ are available for node n to communicate, it is called traffic overflowing. If a node occurs traffic overflowing, it will not be able to assign enough channels for new QoS flows. The traffic prediction scheme is designed to reduce such a situation in this paper. After receiving a QoS routing request, TPQOR protocol would predict the probability of the traffic overflowing in the networks. When a node receives a QoS routing request, the future probability of its traffic overflowing is estimated by the history of its bandwidth use.

Traffic history of a node is periodically recorded in an FIFO (first in first out) queue T_{his}. Let his_length be the length of the queue. Each element in T_{his} denotes the number of channels in $A(n)$ at one past time. The minimum request channel number Req_{ch} is calculated for a QoS flow in one channel assignment. Let $ovflw\_time$ be the number of times of overflowing. The value of $ovflw\_time$ will be increased by one each time when the number of available channels is not enough for QoS flows. The probability $P_{ovflw}(n)$ for node n to occur traffic overflowing can be represented as:

$$P_{ovflw}(n) = ovflw\_time/his\_length \qquad (1)$$

We define a route with one or more traffic overflowing nodes as traffic overflowing route. According to the definition, the probability $PR_{ovflw}(R)$ of traffic overflowing for a route can be deduced as Eq. 2, where R is a route and v represents each node along the route, and $\bar{R}$ denotes the set of all nodes on the route R.

$$PR_{ovflw}(R) = 1 - \prod_{v \in \bar{R}}(1 - P_{ovflw}(v)) \qquad (2)$$

In TPQOR protocol, $PR_{ovflw}(R)$ is an important element to be considered in selecting routes. A good route should not only pass through fewer hops, but also has less probability of traffic overflowing. Therefore, a variable $rt\_pri(R)$ is defined to denote the priority of each candidate route R in Eq. 3, where MAX_HOP denotes the maximum hops of routes in the network and TTL is the time to live of QRREQ packet. The value of TTL is set to MAX_HOP initially and decreased by one each time the packet is forwarded. When $Ttl = 0$, the packet will be dropped. As we know, $TTL \in [0, MAX\_HOP]$ and $rt\_pri \in [0, 2MAX\_HOP]$. TPQOR protocol tends to select a route

R with the maximum $rt\_pri(R)$ among all candidate routes for a QoS flow.

$$rt\_pri(R) = 2MAX\_HOP - (PR_{ovflw}(R) \cdot MAX\_HOP + TTL) \tag{3}$$

5 Traffic-predictive QoS on-demand routing protocol

In this section, we propose a traffic-predictive QoS on-demand routing (TPQOR) protocol for multi-channel mobile ad hoc networks. TPQOR protocol is a reactive routing protocol, which operates in two phases: route discovery and route reply. Its route discovery process depends on the flooding of QoS Routing REQuest (QRREQ) packets from the source until one of them reaches the destination. In the route reply process, a QoS Routing REPly(QRREP) packet is forwarded back from the destination to the source along a reverse path which has been built in route discovery process. In the process of route discovery, each node maintains itself a route table rt. This table is a set of rules that are used to determine where data packets will be directed. A routing table is maintained to keep the next hop information to all possible destination nodes and the previous hop information to their sources for all flows. The format of the routing table is shown in Table 2; four status {ONBUILDING, BUILT, ONREPAIRING, ERROR} are defined for a route, which are explained in Section 5.4.

5.1 Neighboring maintenance

We propose a multichannel QoS-aware routing protocol that permits a flow with the requested bandwidth and delay. The admission scheme requires the channel usage information of all interference neighbors of a node. In the proposed TPQOR protocol, a node will maintain three lists, nb_list (neighboring node list), $intf\_list$(interference node list except neighboring nodes), and CUT (channel usage table), to record the required information of channel assignment. Each node maintains its own CUT, which records the two end nodes of each assigned channel and the communication direction (receiving or transmitting) and whether the channel is reserved for a QoS flow such as shown in Fig. 2. The "Hello" detection method is used to collect the available channels of nodes in the networks. We need to find the neighboring nodes before we execute the route discovery process; the purpose of the HELLO

messages is to find the neighboring nodes and to create the neighboring node list, interference node list, and channel usage table for each node. HELLO messages are broadcasted periodically among node's one-hop neighbors. When a node v receives HELLO from its active neighbors, it updates its neighboring nodes, interference nodes, and corresponding channel usage information.

In this paper, we construct the interference node set $N_i(v)$ by discovering two-hop nodes of node v, that is, $N_i(v) = \bigcup_{u \in N(v)} N(u)$, where $N(u)$ is the neighboring node set of node u. Every Hello packet from node n contains not only information of node n itself, but also its neighbors' information, which is also known through its neighbors' Hello packets. By receiving such Hello packets, a node can indirectly obtain its two-hop neighbors' information. An example of the neighboring maintenance is shown in Fig. 2, where the neighboring node list nb_list, interference neighboring node list $intf\_list$, and channel usage table CUT are shown in its boxes. In this example, node $n3$ is the interference neighboring node of $n1$, but not its neighboring node. Although this makes it difficult for $n1$ to get $n3$'s information directly, $n1$ can get the information through $n2$'s Hello packets because $n1$ and $n3$ are both the neighboring nodes of $n2$.

5.2 Route request process

The QoS-aware routing protocol decides to accept or reject an incoming flow in the QRREQ (QoS Route REQuest) packet broadcasting process, which is based on the QoS requirement bandwidth (Bw_{req}) and delay (Del_{req}). When a source node needs to communicate with another node for which it has no route information in its table, it broadcasts a QRREQ packet to its neighboring nodes. The format of the QRREQ packet is shown in Table 3.

In TPQOR protocol, when a node receives a QRREQ packet, it updates TTL, calculates the priority of the current route by using Eqs. 2 and 3, and then decides either to rebroadcast it or drop it. A QRREQ packet will not be rebroadcasted under the following conditions: it reaches the destination node; the same packet has already been received and the new one does not have a higher route priority; the live time of the packet is more than TTL; the node, which receives the QRREQ packet, can not meet the QoS requirements.

TPQOR protocol provides the minimum bandwidth and the maximum delay guarantees for QoS routing. In a QRREQ packet, the bandwidth requirement is recorded by Bw_{req}. Intermediate node n judges whether it can meet the QoS bandwidth requirement by checking its available channel set $A_t(n) = CT - C_r(N_i(n))$ for interference-free transmitting and the available channel set $A_r(n) = CT - C_t(N_i(n))$ for interference-free receiving. Node n can meet the bandwidth requirement only if both $A_t(n)$ and

Table 2 Routing table

Sid	Did	Fid	prev_hop	next_hop	rt_pri	status

Sid source node, *Did* destination node, *Fid* flow id, *prev_hop* the previous hop of the current node, *next_hop* the next hop of the current node, *rt_pri* route priority, *status* status of the route

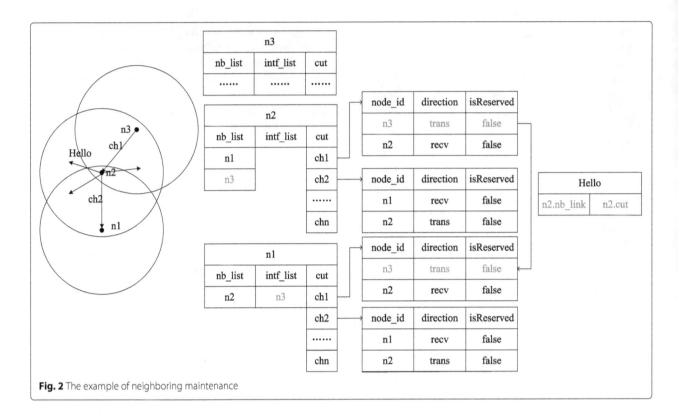

Fig. 2 The example of neighboring maintenance

$A_r(n)$ have enough channels to offer. The minimum number of channels to guarantee the QoS requirement bandwidth Bw_{req} is $\lceil Bw_{req}/Bw_{ch}\rceil$ as shown in Algorithm 1. Timeout QRREQ packets are those packets whose transmission time exceeds the delay requirement Del_{req}, that is, $Del_{req} \leq NOW\_TIME - Time$. We claim that if QRREQ packets could reach the destination within Del_{req}, so do data packets.

When a node receives a QRREQ packet, it first checks whether the entry is its own DID. If the node is the destination, it sends a QRREP packet back to the source node along the discovered path; otherwise, it broadcasts the QRREQ packet to its neighbor nodes according to Algorithm 2. Any node that receives the QRREQ message updates its neighboring node list, interference node list, and channel usage table (CUT). An example for route request is shown in Fig. 3, where node S initiates a route query to node D to establish a route with 1000 Kbps minimum request bandwidth and 0.3 s maximum delay. Here, we focus on introducing the proposed

bandwidth prediction method in the process of route request, we assume that there exist no timeout packets. In Fig. 3, we assume that $P_{ovflw}(E) = 0.7$, overflowing probability of nodes $S, A, B, C,$ and D is 0.1 respectively, $MAX\_HOP = 15$, $TTL = MAX\_HOP$ at node S. The process of QoS routing request is divided into five stages (stage.0–stage.4); we calculate the overflowing probability of the routes according to Eq. 2 and the priority of the routes according to Eq. 3 along path S-A-B-C-D and path S-E-C-D; the status items of each node are shown in the boxes:

- Stage 0: $PR_{ovflw}(S) = 0.1, rt\_pri(S) = 30 - (0.1 * 15 + 15) = 14.85$;
- Stage 1: $PR_{ovflw}(SA) = 0.19, rt\_pri(SA) = 30 - (0.19 * 15 + 14) = 13.15, PR_{ovflw}(SE) = 0.73, rt\_pri(SE) = 30 - (0.73 * 15 + 14) = 5.05$;
- Stage 2: $PR_{ovflw}(SAB) = 0.271, rt\_pri(SAB) = 30 - (0.271 * 15 + 13) = 12.935, PR_{ovflw}(SEC) = 0.757, rt\_pri(SEC) = 30 - (0.757 * 15 + 13) = 5.645$;
- Stage 3: $PR_{ovflw}(SABC) = 0.3439, rt\_pri(SABC) = 30 - (0.3439 * 15 + 12) = 12.8415, PR_{ovflw}(SECD) = 0.7813, rt\_pri(SECD) = 30 - (0.7813 * 15 + 12) = 6.2805$;
- Stage 4: $PR_{ovflw}(SABCD) = 0.40951, rt\_pri(SABCD) = 30 - (0.40951 * 15 + 11) = 12.85735$.

Although path S-E-C-D passes one hop less than its counterpart path S-A-B-C-D, its intermediate node E

Table 3 The format of QRREQ packet

Sid	Did	Fid	Type	Bw_{req}	Del_{req}	Seq	Time	PR_{ovflw}	TTL

Sid source ID; *Did* destination ID; *Fid* flow ID; *Type* the type of packet; *Bw_{req}* QoS minimum bandwidth requirement; *Del_{req}* QoS maximum delay requirement; *Seq* request packet broadcasting sequence number; *Time* QRREQ packet time stamp, it records the starting time of QRREQ packet; *Pr_{ovflw}* the probability of traffic overflowing of the current route; *TTL* the maximum time to live of the packet, whose initial value is the maximum hops of the network

---

Algorithm 2 Process of Receiving a QRREQ packet

1: **initial:** node u has received a QRREQ packet $qrreq$ from node v
2: $qrreq.TTL - -$
3: calculate overflowing probability PR_{ovflw} and priority rt_pri of the current route
4: refresh $qrreq.PR_{ovflw}$
5: $rt_item.status = $ ONBUILDING
6: **if** $rt\_item.rt\_pri < rt\_pri$ **then**
7: $\quad$ $rt\_item.rt\_pri = rt\_pri$
8: $\quad$ $rt\_item.pre\_hop = u$
9: **else**
10: $\quad$ drop $qrreq$ and return
11: **end if**
12: **if** $qrreq.TTL = 0$ **then**
13: $\quad$ $rt_item.status = $ ERROR
14: $\quad$ drop $qrreq$ and return
15: **end if**
16: **if** $|A_t(n)| < \lceil Bw_{req}/Bw_{ch} \rceil$ or $|A_r(n)| < \lceil Bw_{req}/Bw_{ch} \rceil$ or $NOW\_TIME - qrreq.Time < qrreq.del_{req}$ **then**
17: $\quad$ $rt_item.status = $ ERROR
18: $\quad$ drop $qrreq$ and return
19: **end if**
20: **if** $n! = qrreq.Did$ **then**
21: $\quad$ rebroadcast $qrreq$
22: **else**
23: $\quad$ send QRREP packet back
24: **end if**

---

suffers a high probability of traffic overflow. This makes the path S-E-C-D to have a lower route priority, though it has fewer hops. Therefore, TPQOR changes the pre-hop of node C from E to B in stage 3.

5.3 Route reply process

When a suitable path is found from a source node to its destination node, the destination node will send a QoS route reply (QRREP) packet back to the source node. The QRREP packet can be delivered to the source node through the *pre_hop* recorded in route items at nodes along the path. We define the format of the QRREP packet as shown in Table 4.

Not only reconfirming route is the purpose of route reply process, but also conflict-free channels are assigned to each link along the reverse route according to the channel assignment Algorithm 1. During the route reply phase, for a link $l = (u, v)$, QRREP packet is forwarded from downstream v to upstream u; channel information, which includes $C_t(N_i(v))$ and $C_r(N_i(v))$, is carried in the QRREP packet of node v; and the QRREP packet is forwarded along the reverse path from the destination to its source. Upon receiving a QRREP packet, each node along the route updates its routing table and channel usage table

(CUT). When node u receives unicast RREP packet from node v, it firstly extracts the interference channel sets of node v. Then, if no channel is assigned to the forwarding link (u, v), it calculates the available channel set of link $l = (u, v)$, that is, $AL(l) = CT - (C_t(N_i(v)) \cup C_r(N_i(u)))$. If $|AL(l)| * bw_{ch} \geq bw\_req$, which means there is enough bandwidth for QoS request, $m = \min\{i | i*bw_{ch} \geq bw\_req\}$ channels with the minimum number in $AL(l)$ are assigned to forwarding link (u, v). The information of the assigned channels is sent back to node v for updating the channel usage information. After the route has been established, each node along the route should have enough channels for QoS routing without channel conflicts, and then the route status is changed to *BUILT*. When a node receives a QRREP packet, the processing procedure can be summarized in Algorithm 3. An example of route reply process is shown in Fig. 4, which is according to the route request process in Fig. 3. Here, we assume that the available channel set $CT = \{ch1, ch2, ..., ch8\}$, the bandwidth of each channel is 500 Kbps, so two channels are at least needed to assign each link for 1000 Kbps QoS bandwidth requirement. The QRREP packet is forwarded from destination D to source S in five stages as shown Fig. 4. In each stage, two channels are assigned to the forwarding link. Some route items, CUT, and neighboring table are updated. The change of the related route items and assigned channels are shown in the figure.

Algorithm 3 Process of node u receiving a QRREP packet from node v

---

1: **initial:** node u has received a QRREP packet $qrrep$, rt_item is the related route item
2: for link $l = (u, v)$, $AL(l) \Leftarrow CT - (C_t(N_i(v)) \cup C_r(N_i(u)))$
3: let $m = \lceil Bw_{req}/Bw_{ch} \rceil$
4: **if** $|AL(l)| \geq m$ **then**
5: $\quad$ select m channels C_m in $AL(l)$ by using channel assignment algorithm Algorithm 1
6: $\quad$ $C_t(N_i(u)) = C_t(N_i(u)) \cup C_m, C_r(N_i(u)) = C_r(N_i(u)) \cup C_m$
7: $\quad$ $rt\_item.next\_hop = v$
8: $\quad$ $rt_item.status = $ BUILT
9: **else**
10: $\quad$ $rt_item.status = $ ERROR
11: $\quad$ return "there are no enough channels for QoS requirement"
12: **end if**
13: **if** $u \neq qrrep.Sid$ **then**
14: $\quad$ forward $qrrep$
15: **end if**

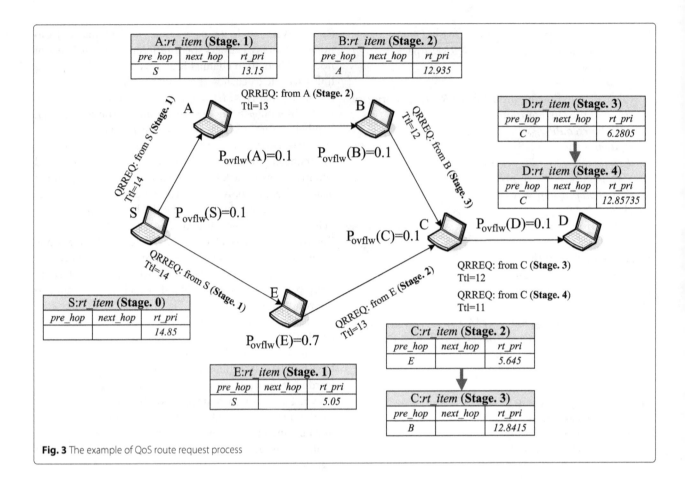

Fig. 3 The example of QoS route request process

5.4   Route maintenance

During the life cycle of each TPQOR route item, it may experience different situations. Four status, {ONBUILD-ING, BUILT, ONREPAIRING, ERROR}, are defined for a route in this paper. An ONBUILDING status denotes that a route is being built, but has not yet been confirmed through the route reply process; a BUILT status denotes a mature route for delivering data packets; an ONREPAIR-ING route is a route under rebuilding; ERROR means an invalid route and it should be abandoned. Every route item starts with ONBUILDING and ends up with ERROR. The state machine of route maintenance is indicated in Fig. 5, where the meanings of event set {A, B, C, D, E, F, G} is explained as follows: A, receiving a QRREP packet; B, destination node initiates a route reply process; C, status timed out; D, next hop is out of the transmission range; E, destination node applies for rebuilding the route; F, receiving a route error packet QERROR; G, timed out for

not receiving data packet; H, status timed out; I, source node receives a route error packet QERROR; J, receiving a fresher QRREQ packet.

A route may be broken with node mobility or unsatis-fied QoS requirements. In this case, a QoS route ERROR (QERROR) packet would be initiated for informing source node to rebuild a route. QERROR can be initiated either by the destination or intermediate node and will be forwarded to the source node. When a node receives QERROR, it should release the occupied resources. Route rebuilding process is as same as the process of building a new route. After the source receives QERROR, it will start a route rebuilding process. The format of QERROR is shown in Table 5.

6   Results and discussion

In the numerical simulation, NS-2 simulator is used to evaluate the performance of the proposed TPQOR pro-tocol and IEEE 802.11 amendment standard is used to MAC and PHY layers to support static and dynamic multi-channel access. In our design of network model, 25 static nodes are uniformly distributed in a scenario of 1000 m × 1000 m, all of which are equipped with mul-tiple interfaces and the same number of non-overlapping

Table 4 The format of QRREP packet

Sid	Did	Fid	Type	Bw_req	TTL

Sid source ID, *Did* destination ID, *Fid* flow ID, *Bw_req* QoS minimum bandwidth requirement, *Type* the type of packet, *TTL* the maximum time to live of the packet

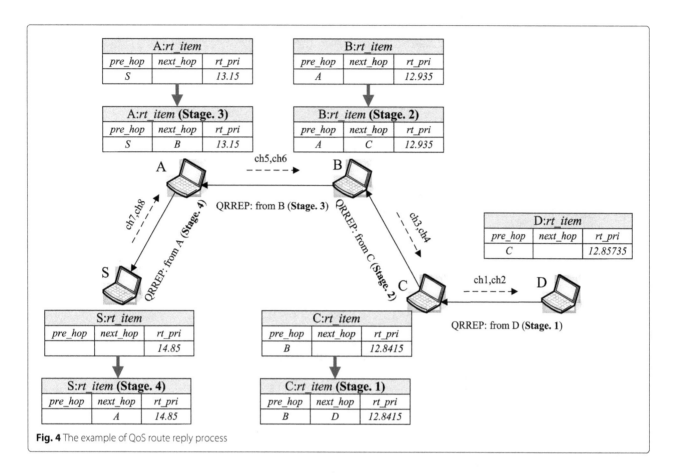

Fig. 4 The example of QoS route reply process

wireless channels. The channel maximum transmission rate is set to 1.5 Mbps. The transmission range of each node is set to 100 m, and the interference range is set to 200 m. During the simulation time of 100 s, one QoS CBR flow and other nine ordinary CBR flows with different sources and destinations are randomly chosen. We use different loads, 100 kbps, 200 kpbs, 500 kbps, 1000 kbps, and 1500 kbps, to test the performance of evaluated protocols respectively. Average network throughput and average

end-to-end delay are the average value of three times simulation. In the simulation, the following metrics are used for our performance evaluation:

Average network throughput: the average successful packet delivery over all the existing flows in the network, that is, the average number of received packets for all flows.

Packet loss ratio: the percentage that packet loss occupies over all sending packets, which is $\frac{\text{the number of lost packets}}{\text{the number of all sending packets}} \times 100\%$.

Average end-to-end delay: the average time between transmission of data packets at sources and successful reception at their receivers, which is $\frac{\sum(\text{receiving packet time} - \text{sending packet time})}{\text{the number of all received packets}}$.

6.1 Results of performance evaluation
We firstly evaluate the performance of single channel AODV protocol [20] and the proposed TPQOR protocol with different channel sets. Figure 6 compares

Table 5 The format of QERROR packet

Sid	Did	Type	Fid	Seq

Sid source ID, *Did* destination ID, *Type* the type of packet, *Fid* flow ID, *Seq* request packet broadcasting sequence number

Fig. 5 The state machine of route maintenance

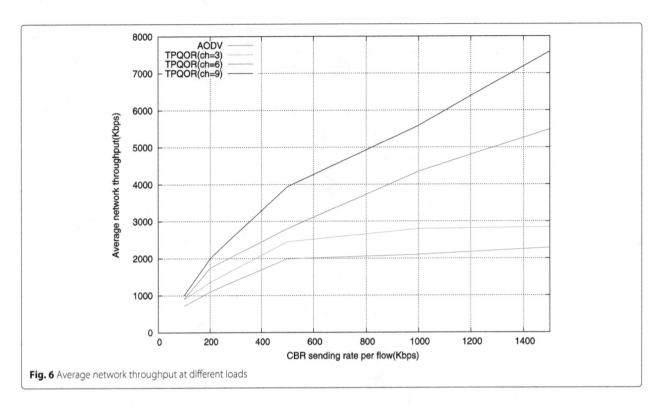

Fig. 6 Average network throughput at different loads

the average network throughput of AODV protocol to TPQOR protocol with different number of channels at different loads. It shows that the average network throughput will increase when the transmission rate of CBR flows increases. The characteristic of multi-channel makes the network throughput of TPQOR protocol increase faster than AODV when the transmission rate of CBR flows increases. Furthermore, the more channels it has, the more obvious superiority it performs. As shown in Fig. 7, the single channel AODV protocol has more packet loss rate than multi-channel TPQOR protocol, because multi-channel can decrease the packet conflict, and when CBR

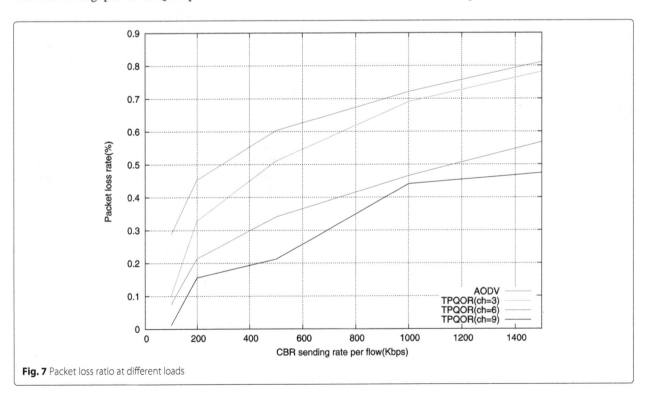

Fig. 7 Packet loss ratio at different loads

rate increases, higher load brings more conflicts and causes the packet loss ratio to increase. Figure 8 indicates that TPQOR protocol performs better than AODV protocol especially when the number of data channels increases, and as CBR rate increases, the average end-to-end delay increases as well. In Fig. 8, data does not seem as regular as data in Figs. 6 and 7. The reason is that when the packet loss ratio increases, lots of packets are dropped on the halfway, which makes the measured average end-to-end delay irregular.

In order to reveal the effectiveness of TPQOR traffic prediction scheme, a new simulation scenario is constructed: there are two paths between a source and its destination, a Pareto [21] distributed interference flow is produced along the shorter route, while no interference flow is produced along the longer one. The QoS flow throughput of our proposed TPQOR protocol with traffic prediction are shown in Fig. 9; the QoS flow with 100-Kbps bandwidth requirement starts at the simulation time 50 s. The interference flow rate is quite low at that time, TPQOR protocol chooses another path for the QoS flow at the beginning by using traffic prediction with the history traffic utilization of the interference flow, and TPQOR protocol maintains 100 Kbps end-to-end throughput for the QoS flow during the whole simulation time. The simulation results show that TPQOR protocol with bandwidth prediction can reduce the route rebuilding process and enhance the throughput of the network.

6.2 Discussion

In MANETS, nodes can communicate with each other without infrastructures and nodes are expected to forward packets for other nodes in spite of limitation of their resources. Traffic routing and channel assignment jointly play a critical role in determining the performance of MANETs. Traffic predictive has a competitive performance when traffic can be predicated accurately; a traffic-predictive QoS on-demand routing (TPQOR) protocol is proposed to support bandwidth and delay requirement for MANETs in the paper; the future traffic is predicted according to nodes' history traffic patterns. In real situation, different devices may produce different traffic and may influence each other, especially burst interference flows will affect the bandwidth requirement of the QoS flow, so the more accurate predictive scheme is needed for future research direction.

The study has the following limitations:

(1) All channels are assumed to have the same bandwidth, transmission, and interference range.

(2) The variations of wireless channels is not considered in the simulation, and the transmission data rate of all channels may change, when the wireless channel is in deep fade.

(3) The traffic prediction with the only nodes' history traffic information is not accurate, the more predictive factors are needed to be considered.

The accurate traffic prediction is a considerably different and challenging problem in MANETs, traffic

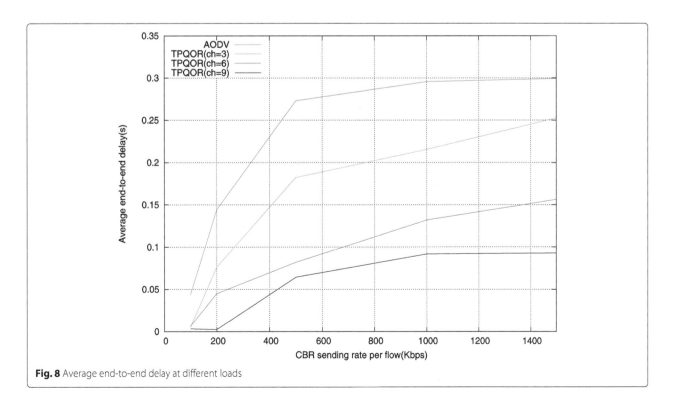

Fig. 8 Average end-to-end delay at different loads

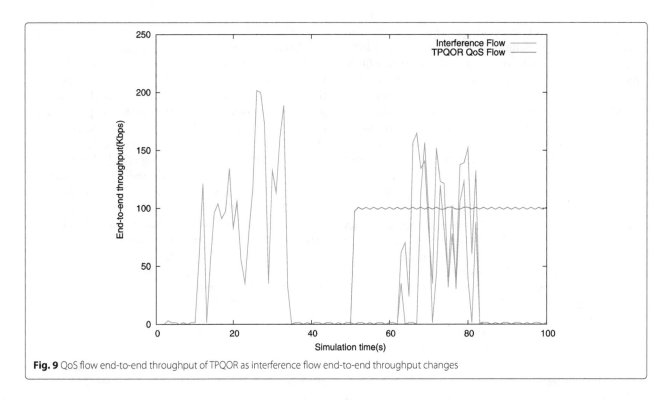

Fig. 9 QoS flow end-to-end throughput of TPQOR as interference flow end-to-end throughput changes

variations are caused by different network states at different timescales, and there are additional research opportunities to improve the proposed traffic-predictive scheme to adapt the network variations.

7 Conclusions

This paper proposes a traffic-predictive QoS on-demand routing (TPQOR) protocol to support QoS bandwidth and delay requirements. The main contributions of this work include a proposed novel channel assignment scheme and a proposed traffic prediction scheme. The proposed channel assignment scheme can efficiently express the channel usage and interference information within a certain range, which reduces interference and enhances channel reuse rate. Unlike some existing routing protocols, TPQOR protocol takes the traffic prediction as an important factor in selecting route. The simulation results show that TPQOR protocol can effectively increase throughput, reduce loss ratio as well as delay, and avoid the influences of future interference flows, as compared to AODV protocol.

Abbreviations
AODV: Ah hoc on-demand distance vector routing; CRANET: Cognitive radio ah hoc network; CUT: Channel usage table; DTN: Delay-tolerant networks; FIFO: First in first out; MANETs: Mobile ad hoc networks; NIC: Network interface card; QERROR: QoS route error; QoS: Quality-of-service; QRREP: QoS routing reply; QRREQ: QoS routing request; TPQOR: Traffic-predictive QoS on-demand routing; TTL: Time to live

Funding
This work is supported by NSFC (61373125), GDNSF (S2013020012865), and GDSTP (2013B010401016).

Authors' contributions
JZ and LL propose the TPQOR protocol cooperatively, JZ writes the manuscript, and LL contributes to the implementation of the simulation programs and collects the simulation results. HT helps to check the simulation and result analysis. All authors read and approved the final manuscript.

Competing interests
The authors declare that they have no competing interests.

References
1. O. Sahingoz, in *Proc. of 2013 International Conference on Unmanned Aircraft Systems(ICUAS), May 28–31.* Moblie networking with UAVs: opportunities and challenge (IEEE, Atlanta, 2013), pp. 933–941
2. Q. Yuan, I. Cardei, J. Wu, in *Proc. of the tenth ACM international symposium on Mobile ad hoc networking and computing (MobiHoc'09) May 18–21.* Predict and relay: an efficient routing in disruption-tolerant networks, (New Orleans, 2009)
3. P. Fazio, F.D. Rango, C. Sottile, A predictive cross-layered interference management in a multichannel MAC with reactive routing in VANET. IEEE Trans. Mob. Comput. **15**(8), 1850–1862 (2016)
4. Y. Liu, B.R. Tamma, B.S. Manoj, R. Rao, in *Proc. of IEEE INFOCOM 2010, March 15–19.* Traffic prediction for cognitive networking in multi-channel wireless networks (IEEE, San Diego, 2010)
5. J. Wellons, L. Dai, Y. Xue, Y. Cui, Augmenting predictive with oblivious routing for wireless mesh networks under traffic uncertainty. Comput. Netw. **54**, 178–195 (2010)
6. L. Dai, Y. Xue, B. Chang, Y. Cao, Y. Cui, in *Proc. of IEEE INFOCOM 2008, April 13–18.* Integrating traffic estimation and routing optimization for multi-radio multi-channel wireless mesh networks (IEEE, Phoenix, 2008)
7. K. Katsaros, M. Dianati, R. Tafazolli, R. Kernchen, in *Proc. of 2011 IEEE Vehicular Networking Conference(VNC), Nov.14-16.* CLWPR-a novel cross-layer optimized position based routing protocol for VANETs (IEEE, Amsterdam, 2011), pp. 193-146
8. D. Palma, H. Araujo, M. Curado, Link quality estimation in wireless multihop networks using Kernel based methods. Comput. Netw. **56**(16), 6629–3638 (2012)
9. Y.T. Wei, J.W. Wang, in *Proc. of 2015 27th Chinese Control and Decision Conference (2015CCDC) May 23–25.* A delay/disruption tolerant routing algorithm based on traffic prediction (IEEE, Qingdao, pp. 3253–3258

10. L. Zhang, F. Zhuo, H.T. Xu, A cross-layer optimization framework for congestion and power control in cognitive radio ad hoc networks under predictable contact. EURASIP J. Wirel. Commun. Netw. **57**, 1–23 (2018)
11. S. Shelly, A.V. Babu, Link residual lifetime-based next hop selection scheme for vehicular ad hoc networks. EURASIP J. Wirel. Commun. Netw. **23**, 1–13 (2017)
12. L. Zhang, F. Zhuo, C. Bai, H. Xu, Analytical model for predictable contact in intermittently connected cognitive radio ad hoc networks. Int. J. Distrib. Sensor Netw. **12**(7), 1–12 (2016)
13. L. Zhang, F. Zhuo, W. Huang, C. Bai, H. Xu, Joint opportunistic routing with autonomic forwarding angle adjustment and channel assignment for throughput maximization in cognitive radio ad hoc networks. Ad hoc Sens. Wirel. Netw. **38**, 21–50 (2017)
14. D. Morato, et al., in *Proc. of IEEE Tenth International Conference on Computer Communications and Networks, Oct. 15-17*. On linear prediction of Internet traffic for packet and burst switching networks IEEE, Scottsdale, 2001), pp. 138–143
15. J.K. Jayabarathan, S.R. Avaninathan, R. Savarimuthu, QoS enhancement in MANETs using priority aware mechanism in DSR protocol. EURASIP J. Wirel. Commun. Netw. **131**(1), 1–9 (2016)
16. A. Raniwala, C. Tzi-cker, in *Proc. of IEEE INFOCOM 2005, Vol. 3, March 13–17*. Architecture and algorithms for an IEEE 802.11-based multi-channel wireless mesh network (IEEE Hyatt Regency Miami, Miami, 2005), pp. 2223–2234
17. S. Kajioka, et al., A QoS-aware routing mechanism for multi-channel multi-interface ad-hoc networks. Ad Hoc Netw. **9**(5), 911–927 (2011)
18. J.P. Zhou, L.Y. Peng, Y.H. Deng, J.Z. Lu, An on-demand routing protocol for improving channel use efficiency in multichannel ad hoc networks. J. Netw. Comput. Appl. **35**, 1606–1614 (2012)
19. F. Li, S. Chen, Y. Wang, Load balancing routing with bounded stretch. EURASIP J. Wirel. Commun. Netw., 1–16 (2010)
20. C.E. Perkins, E.M. Royer, in *Proc. of the Second IEEE Workshop on Mobile Computing Systems and Applications (WMCSA'99), Feb. 25–26*. Ad-hoc on-demand distance vector routing(AODV) (IEEE, New Orleans, 1999)
21. B.C. Arnold, *Pareto distributions, Second Edition, 2015.3.3.* (Taylor&Francis Inc, Bosa Roca, 2015)

5

A cooperative V2X MAC protocol for vehicular networks

Mohamed A. Abd El-Gawad[1,2] (iD), Mahmoud Elsharief[1,3] and HyungWon Kim[1*]

Abstract

In support of traffic safety applications, vehicular networks should offer a robust Medium Access Control (MAC) layer protocol that can provide a reliable delivery service to safety-related messages. As the safety applications generally use broadcasting to propagate their messages, a reliable broadcast protocol is essential. In general, however, broadcast is considered as unreliable by nature in contrast to unicast. This paper introduces a novel MAC protocol, called a Hybrid Cooperative MAC (HCMAC), which can substantially enhance the reliability of broadcast in vehicular networks by employing a notion of channelization. HCMAC introduces a hybrid protocol that combines a time slot allocation of Time Division Multiple Access (TDMA) and a random-access technique of Carrier Sense Multiple Access (CSMA) and thus minimizes the probability of data collisions. In addition, its feedback strategy further enhances the system performance by preventing transmissions during time slots that experience collisions. Through analysis and simulations, we compare the performance of HCMAC with VeMAC, an existing TDMA protocol. The results demonstrate that HCMAC can offer substantially faster channel access and lower collision rate compared with VeMAC.

Keywords: Vehicular ad hoc networks, Reliable broadcast, TDMA, CSMA, Analytical model

1 Introduction

In recent years, road safety is receiving increasingly more attention due to the growing number of fatalities and injuries caused by road accidents. Every year, road accidents cause around 1.25 million deaths worldwide [1]. In 2015, 35,092 of 36,973 total transportation fatalities in the USA are caused by highway accidents [2]. The National Highway Traffic Safety Administration (NHTSA) estimates that safety applications provided by vehicle-to-vehicle (V2V) and vehicle-to-infrastructure (V2I) networks could prevent or mitigate 80% of crashes at intersections and during lane changes. In Dec. 2016, the U.S. Department of Transportation (DOT) announced a new proposed rule that mandates car manufacturers to incorporate V2V technologies in all new light-duty vehicles to make the road safer [2]. Consequently, vehicular networks are receiving even higher interest and support from academia, government, and automobile makers than the past years. Vehicular networks can enable us to develop a variety of road safety applications such as lane change warning (LCW), forward collision

warning (FCW), left turn assist (LTA), and intersection management assist (IMA) [3]. In addition, V2V in combination with V2I networks can offer intelligent transportation services, which includes, for example, traffic congestion control and road traffic optimization.

A vehicular network, also known as a vehicular ad hoc network (VANET), is a form of an ad hoc network, in which vehicles communicate with each other. Each vehicle uses an onboard unit (OBU) equipped with one or more radio transceivers and a global position system (GPS). OBUs based on the dedicated short-range communications (DSRC) standard communicate over a 75-MHz spectrum band starting from 5.850 to 5.925 GHz. The DSRC standard divides this spectrum to seven channels of 10 MHz: one control channel (CCH) for control information and short safety messages and six service channels (SCHs) for safety- and non-safety-related applications.

DSRC defines the framework of data transmission process between two communication ends and provides a roadmap for interoperable applications. Under this framework, a number of standards are defined by other organizations. For example, IEEE provides a set of standards such as IEEE 802.11p [4] for wireless access, IEEE 1609.3 [5] for networking services, and IEEE 1609.4 [6]

* Correspondence: hwkim@cbnu.ac.kr
[1]Department of Electronics Engineering, Chungbuk National University, Cheongju, South Korea
Full list of author information is available at the end of the article

for multichannel operations. Society of Automotive Engineers (SAE) has released the Message Set Dictionary standard (SAE J2735) [7], which defines a set of messages that represents the core functions for application development.

SAE J2735 defines the Basic Safety Message (BSM), one of the most frequently used message types for DSRC. A BSM is a broadcast message that contains critical information regarding the vehicle's state (e.g., position, speed, and heading). To realize safety applications, vehicles periodically exchange their BSM messages to track neighboring vehicles and take actions to avoid potential collisions. For instance, the IMA application relies on one-hop broadcasting of BSM messages among the vehicles within the same wireless range.

To make the safety applications acquire accurate vehicle and road information and so take timely actions, reliable transmission of BSM messages should be guaranteed with little latency overhead. Consequently, the underlying Medium Access Control (MAC).

The V2V MAC protocol specified in the current standard of IEEE-802.11p, however, does not provide a reliable broadcast service. Since it is derived from the legacy IEEE-802.11 standard, its Carrier Sense Multiple Access/Collision Avoidance (CSMA/CA) protocol offers collision avoidance mechanism only for unicast packets not for broadcast packets. In other words, the Request To Send/Clear To Send (RTS/CTS) mechanism of CSMA/CA can alleviate the hidden node problems and reduce data collisions for only unicast packets.

Another limitation resides in the Enhanced Distributed Channel Access (EDCA) of IEEE-802.11p, which provides differentiated service for urgent packets. It assigns a smaller contention window (CW) and a shorter interframe spacing (IFS) to higher priority traffic (e.g., critical safety messages). EDCA, however, cannot utilize its variable CW for broadcast packets, since the current IEEE-802.11p standard cannot detect collisions of broadcast packets. EDCA has another drawback. Although a smaller CW gives a shorter access delay, it increases the probability of collisions, especially when the number of contending vehicles increases [8].

In contrast to the contention-based protocols like IEEE-802.11p standard, other protocols have been proposed to provide more robust wireless access. Time Division Multiple Access (TDMA) or Code Division Multiple Access (CDMA) is among such protocols.

TDMA divides the access time into a group of time slots and allocates each transmitter to the time slots using distributed scheduling schemes. TDMA has been considered by several works [9–17]. RR-ALOHA [9] was originally proposed for mobile networks. It presented a distributed process that allocates time slots, to allow reliable single-hop broadcast. The authors of [10] have proposed an enhancement of RR-ALOHA, called an ADHOC MAC, which employs a slotted/framed structure. In the ADHOC MAC, each node broadcasts the status of all the slots (i.e., the slot is free or busy) in the previous frame. By advertising such information to the neighbor nodes, each node can avoid duplicate reservation of slots and eliminate the hidden node problem. It, however, still has limitations. It suffers from performance degradation for fast-moving nodes, while it does not support the multichannel operation. Thus, it is not compatible with the DSRC standard.

Another TDMA-based MAC protocol, called VeMAC [11], was also proposed for VANET. While it also utilizes the architecture of the ADHOC MAC, VeMAC supports multichannel operation and handles rapid changes in the network topology. In VeMAC, time slots are divided into disjoint sets. In two-direction roads, vehicles moving in each direction are allocated to a different set to reduce data collisions incurred by node mobility. The authors reported that VeMAC achieves higher throughput and assigns time slots much faster than ADHOC. However, it also suffers from critical limitations; access collisions can occur, as two or more vehicles might acquire the same time slot. It can completely fail to detect collisions in certain situations as to be discussed in Section 3.

STDMA [12, 13] is a time-slotted self-organizing MAC protocol which was originally proposed for the vessels industry. STDMA has been in commercial use for Automatic Identification Systems (AIS) [14]. STDMA grants a channel access for all nodes within a bounded delay, regardless of the number of neighbor nodes within the same wireless range. ETSI has evaluated the performance of STDMA by applying it to the cooperative Intelligent Transportation System (C-ITS).

As the vehicular network is highly dynamic in its nature, classical TDMA protocols suffer from frequent rescheduling demand. This limitation motivated many researchers to develop another category of TDMA schemes, called a topology transparent protocol [18, 19], which divides road segments into a set of small cells. Each cell is mapped with a group of time slots. Depending on the mapping, each vehicle selects a time slot group based on its location on the road and then acquires a unique time slot from the selected group. The topology transparent protocols, however, have critical restriction that they require a digital map [18].

CDMA is considered in some proposals due to its robustness against noise and interference [20, 21]. It, however, also suffers from its well-known shortcoming: it is very difficult to assign unique pseudo noise (PN) codes to all vehicles, especially in dense networks. If different vehicles happen to share the same PN code by distributed PN code selection processes, they can cause highly frequent collisions [21].

This paper presents a novel hybrid MAC protocol, called Hybrid Cooperative MAC (HCMAC), which is based on VeMAC [11] but further optimized for higher performance in vehicular networks. HCMAC takes advantage of the channelization scheme of TDMA and the random-access technique of CSMA. By combining the advantages of TDMA and CSMA strategies, HCMAC can substantially reduce the rate of access collisions in the CCH channel. Moreover, for the same number of contending vehicles and the available time slots, HCMAC can assign time slots to the vehicles much faster than VeMAC—a key advantage in highly dynamic networks. In addition, HCMAC introduces new techniques for more effective collision detection than conventional protocols. In summary, our contributions are as follows:

- A new MAC protocol that provides a fully distributed time slot reservation and also avoids duplicate time slot allocation
- An analytical model demonstrating that our protocol achieves faster slot reservation
- The proof of the accuracy of the analytical model through simulation results of HCMAC
- Extensive simulation results with highway and urban conditions proving that HCMAC outperforms VeMAC in all metrics: throughput, packet delivery ratio (PDR), inter-packet delay, and the rate of collisions

The remainder of this paper is organized as follows. Section 2 describes the system model, while the details of HCMAC are presented in Section 3. Section 4 analyzes the performance of the HCMAC. In Section 5, simulation results are presented, while the conclusion and future work are given in Section 6.

2 Methods
2.1 System model

In this paper, we consider the DSRC and IEEE 802.11p standard for the spectrum and basic link layer protocol, while we propose an enhancement to the link layer. The DSRC standard assigns the channel 178 (i.e., the CCH) for control messages and short safety messages (e.g., BSM messages), and other six SCHs for safety and non-safety applications as shown in Fig. 1 [22]. In our proposed protocol, transmission time over the CCH or the SCH is partitioned into frames, while each frame is divided into S equally sized slots. Each slot has two parts: part 1 consists of W equally sized backoff time units (i.e., contention window) used for carrier sensing, while part 2 represents the actual transmission time as illustrated in Fig. 1. While general vehicular networks may include road side units (RSUs), the network topology of this paper consists only of vehicles equipped with an OBU. The OBU includes two radio transceivers and one GPS receiver.

The GPS receiver provides the required synchronization among the vehicles. Hasan et al. [23] show through real experiments that the GPS synchronization accuracy is at tens of nanoseconds which is quite acceptable for the slotted channel concept. We assume that transceiver 1 is tuned to the CCH to send and receive the periodic BSM messages, while transceiver 2 is tuned to one of the service channels. We assume that all nodes have a fixed wireless range R with the same transmission power level. We also assume that the communication channel is ideal, so transmission failures can be only due to collisions, and the carrier sensing range coincides with the wireless range. Moreover, the channel is assumed to be symmetric. In other words, if node A is in the wireless range of node B, node B is also in the wireless range of node A.

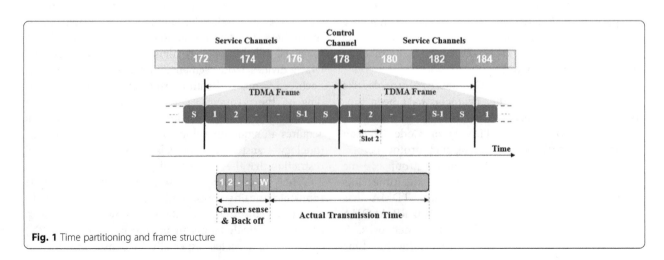

Fig. 1 Time partitioning and frame structure

To avoid the problem of hidden nodes, during each vehicle's time slot acquisition process, two sets of neighbors are defined: a one-hop neighbor list, OH_List, and a two-hop neighbor list, TH_List. Each element of the lists consists of the vehicle ID and the acquired time slot of the corresponding vehicle. Each vehicle transmits its updated one-hop neighbor list, as a part of its periodic broadcast messages. Consequently, when the other vehicles receive the one-hop neighbor list, they can determine and update their own two-hop neighbor list. In Fig. 2, for example, vehicle A has one neighbor in its wireless range, which is vehicle B, and so $OH\_List(A)$ = {(A, I), (B, J)}. As vehicle A can receive vehicle C's one-hop neighbor list via vehicle B, vehicle A can add vehicle C in its two-hop neighbor list resulting in $TH\_List(A)$ = {(A, I), (B, J), (C, K)}. By utilizing such two-hop neighbor lists, each vehicle can substantially alleviate hidden node problems during its time slot acquisition process.

3 HCMAC protocol

In this section, we present our proposed MAC protocol, HCMAC. The HCMAC is a novel hybrid MAC protocol that is aimed at improving the performance of the vehicular networks based on the conventional CSMA protocol. It combines the advantages of TDMA and CSMA. It introduces a slot acquisition process that utilizes the CSMA technique in the selected slot. In HCMAC, each node acquires a time slot in a distributed manner. Initially, the vehicle tracks all its neighboring nodes to avoid duplicate slot acquisition. Due to mobility, vehicles that have the same acquired slot might approach each other causing collisions. Even in such events, the vehicles can detect the collisions more effectively, and change their time slot that is a duplicate

selection, if the proposed collision detection techniques are employed.

In the following subsections, we describe (a) the packet format and the information exchanged among neighboring nodes, (b) the slot acquisition process of a vehicle joining the network, and (c) the collision detection mechanisms.

While we limit our discussion only to single channel case in this paper, HCMAC can operate with multi-channels, and so it conforms to the DSRC standard.

3.1 Packet format

According to IEEE 802.11p, IEEE 1609 family, and SAE J2735 standards, a BSM packet carries the sender vehicle's dynamic status including the position, heading, speed, acceleration, and brake status. As mentioned earlier, in HCMAC, to avoid the slot duplication and the hidden node problem, each vehicle checks the surrounding vehicles during the process of time slot acquisition by inspecting the two-hop neighbor list, TH_List. In HCMAC, each vehicle adds its one-hop neighbor list, OH_List, to the BSM packet's payload. The vehicles exchange the lists among all the neighbors and merge them to construct the TH_List. In order to detect collisions, we defined the slot error (SE) list, which indicates the slots that experience collisions. Figure 3 shows the proposed packet format including the new data fields for OH_List and SE list.

3.2 Slot acquisition process

By employing the packet format shown in Fig. 3, each vehicle acquires a time slot in a distributed way, and so HCMAC does not incur the overhead for expensive centralized decisions. By exchanging OH_List through periodic BSM packets, vehicles can discover the network topology, determine the time slots already occupied, and track any topology change that might occur due to vehicles' mobility.

Let V_c be the current vehicle that attempts to acquire a slot. In the start-up phase, V_c configures its radio transceiver to the listening mode and then listens to the CCH channel for S successive time slots. During this listening period, V_c receives a one-hop neighbor list, $OH\_List(V_i)$, from each neighbor V_i in its wireless range. By accumulating all OH_Lists, V_c determines its own $OH\_List(V_c)$ and $TH\_List(V_c)$, a two-hop neighbor list. V_c also determines the time slots occupied by the neighbors. Procedure 1 shows a pseudo code that describes the process of the start-up phase in details.

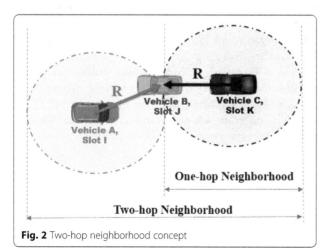

Fig. 2 Two-hop neighborhood concept

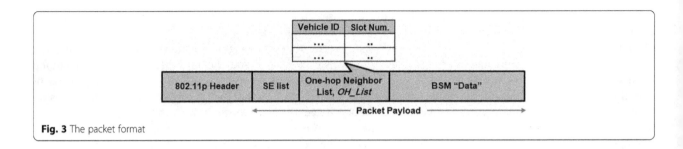

Fig. 3 The packet format

Procedure 1: Start-up Phase

Input: $V_C, S, Start\_S$ // V_C is the Vehicle ID of the current vehicle
Output: $OH\_List(V_C), TH\_List(V_C), Free_{Slots[\,]}, Ac_{Slot}$
1. $OH\_List(V_C) \leftarrow 0, TH\_List(V_C) \leftarrow 0;$
2. $Free_{Slots[\,]} \leftarrow [1{:}S];$
3.   **For** $(i \leftarrow Start\_S$ to $(S + Start\_S - 1))$ **DO**   // "Listening period"
4.   **IF** $(Pkt\_Type(V_i)$ is a BSM packet$)$ **THEN**   // i is a slot
5. $OH\_List(V_i) \leftarrow$ Neighbor list in the BSM message;
                            // V_i is the sender vehicle at slot i
6.   **FOR** every node V_k in $OH\_List(V_i)$ **DO**
7.   **IF** $(V_k == V_i)$ **THEN**
8.   Add V_k to $OH\_List(V_C)$;
9.   **END IF**
10.   Add V_k to $TH\_List(V_C)$;
11.   Remove V_k's slot from $Free_{Slots[\,]}$;
12.   **END FOR**
13.   **END IF**
14.   **END FOR**
15.   **IF** $(Free_{Slots[\,]}$ is not empty$)$ **THEN**
16. $Ac_{Slot} \leftarrow$ Select a random slot from $Free_{Slots[\,]}$;
17.   **ELSE**
18. $Ac_{Slot} \leftarrow$ Select a random slot from $[1{:}S]$;
19.   **ENDIF**
20.   Add $(V_C \,\& Ac_{Slot})$ to the $OH\_List(V_C) \,\& TH\_List(V_C);$

It takes as inputs V^C, representing the vehicle ID of the current vehicle, S indicating the total number of time slots in one TDMA frame, and $Start\_S$ denoting the slot in which the vehicle joins the network. Vehicle V starts listening to the CCH and receives its neighbors' packets. Steps 3–14 are conducted, while V receives BSM packets from its neighbors in one full sequence of slots. Upon receiving a BSM packet from V at slot i, V checks $OH\_List(V)$ which is an incremental one-hop neighbor list of the sender vehicle V. If a vehicle ID record, V in $OH\_List(V)$ matches the sender's vehicle ID, V_C this record V is stored in both the $OH\_List(V)$ and $TH\_List(V)$. Otherwise, V is stored only in the $TH\_List(V)$.

$Free_{Slots[\,]}$ is defined as the set of the available slots. Each vehicle constantly updates its own $Free_{Slots[\,]}$ by removing occupied slots. If $Free_{Slots[\,]}$ set is not empty, the vehicle randomly picks a free slot from it. Otherwise, it randomly selects any slot from the whole range, S, as described in steps 15–19 of Procedure 1. Finally, the vehicle adds V_C and the acquired slot, Ac_{Slot}, to the neighbor lists as specified by step 20 of Procedure 1.

Figure 4 illustrates an example of the start-up phase, where vehicle 3 joined two other vehicles (i.e., vehicle 1 and vehicle 2). Starting from the slot that vehicle 3 joined, it listens to the channel for S successive time slots. At slot 4, vehicle 3 receives a BSM message from vehicle 1 including the OH_List which is the incremental neighbor list of vehicle 1. Then, vehicle 3 builds its OH_List and TH_List. As shown in Fig. 4, vehicle 3 inspects its TH_List to determine that slots 1 and 4 are occupied. In this way, during the listening period, vehicle 3 discovers all the surrounding vehicles and their occupied slots, which helps vehicle 3 avoid potential collisions.

3.3 Collision detection mechanisms

We first describe how VeMAC detects collisions using its simplistic TDMA protocols, and then introduce HCMAC's enhanced collision detection later. Once the vehicle finished the listening period, it updates the neighbor lists every TDMA frame and transmits at the acquired slot until it detects a collision. To explain the collision detection process, we classify collisions into two cases.

Case 1 is called a "simultaneous-access collision," which occurs when two or more vehicles located within the same wireless range transmit at the same time. This case often occurs when the vehicles join the network at the same time and their randomly acquired slots happen to be the same. Another case is when the vehicles that have the same slot change their locations and move to the same wireless range.

Case 2 is called a "hidden-node collision," which occurs due to the hidden node problem, where a vehicle receives packets from two or more senders located in different wireless ranges. Based on the two collision cases, we define the collision detection mechanisms.

3.3.1 Detection of simultaneous-access collisions

We first describe the previous method to identify the collisions of simultaneous-access type followed by HCMAC's enhanced collision detection later.

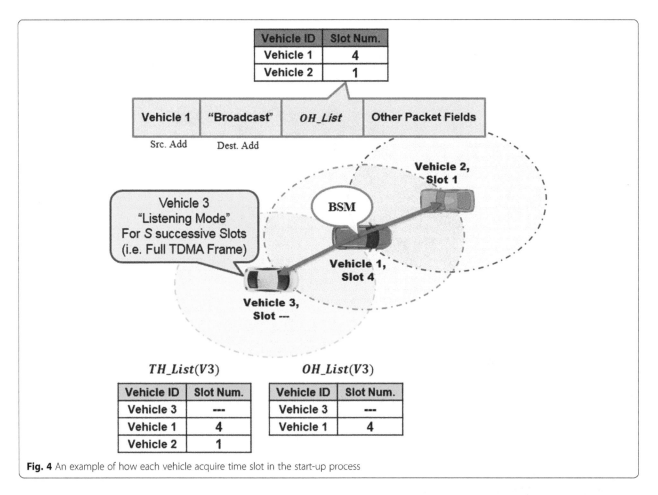

Fig. 4 An example of how each vehicle acquire time slot in the start-up process

---

Procedure 2: Collision detection in VeMAC

Input: $V_C, S, OH\_List(V_C), TH\_List(V_C), Free_{Slots[]}, Ac_{Slot}$
Output: Ac_{Slot}
1.   slot_change ← 0;
2.   **For** $(i \leftarrow (Ac_{Slot} + 1)$ to $(S + Ac_{Slot} - 1))$ **DO**
3.   **IF** $(Received\_Pkt(i) == BSM\_Pkt)$ **THEN**
4.   **IF** $(SrcAdd$ is included in the $OH\_List(V_C))$ **THEN**
5.   **IF** $((V_C$ is not in $OH\_List(V_i))$
                    AND (slot_change == 0)) **THEN**
6.   **IF** $(Free_{Slots[]}$ is not empty) **THEN**
7. Ac_{Slot} ← Select a random slot from $Free_{Slots[]}$;
8.   **ELSE**
9. Ac_{Slot} ← Select a random slot from $[1:S]$;
10.   **ENDIF**
11.   slot_change ← 1;
12.   **ENDIF**
13.   **ENDIF**
14.   Update $OH\_List(V_C)$ & $TH\_List(V_C)$;
15.   **ENDIF**
16. **ENDFOR**

To identify collisions, the previous feedback mechanism [24, 25] operates as follows: each vehicle receives a feedback from each neighboring receiver in a form of neighbor list. By inspecting the feedback, each vehicle can determine whether its packet was successfully delivered. By sending OH_List as part of a BSM message,

each message can act as an implicit acknowledgment for the previous BSMs received within the same TDMA frame. Figure 5 shows an example of the feedback mechanism for vehicle 1. According to its OH_List, vehicle 1 expects one BSM message from vehicle 2 at slot 1 and another BSM message from vehicle 3 at slot 10. If these BSMs include vehicle 1 in the OH_List field, vehicle 1 takes them as acknowledgments for the reception of vehicle 1's previous message. If these BSMs do not include vehicle 1 in OH_List, however, vehicle 1 concludes that its previous transmission has failed and selects another slot to avoid successive failures.

VeMAC protocol [11] utilizes the above mechanism for collision detection, where each vehicle needs to wait up to a full TDMA frame (i.e., up to the time it misses one acknowledgment) to check if its previous message is delivered successfully. In Procedure 2, the pseudo code explains how each vehicle detects collisions occurring in its transmitted packets. Steps 3–15 of Procedure 2 check if the received packet comes from a listed neighbor and then verify that the V_C (the vehicle ID of the receiver vehicle) is included in the OH_List. If $Free_{Slots[\ ]}$ set is not empty, the vehicle randomly picks a new slot from the free slots.

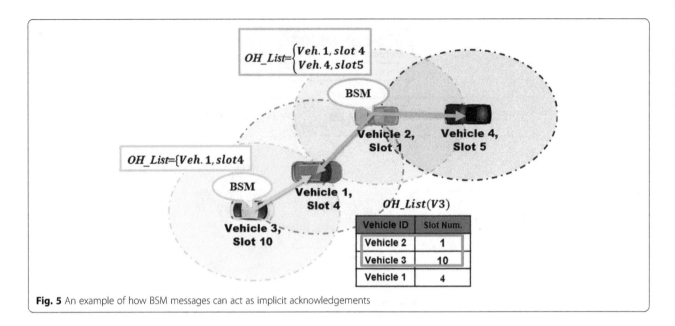

Fig. 5 An example of how BSM messages can act as implicit acknowledgements

Through analysis, we found that although the previous technique provides the benefit of detecting many collisions, it often takes more than one TDMA frame to identify a collision. Furthermore, the previous technique may completely fail to detect collisions in many scenarios. For example, Fig. 6 illustrates a common highway scenario, where vehicle 1 and vehicle 2 both are using the same slot 4. When vehicle 1 approaches and passes vehicle 2, they transmit at the same time causing a collision. Since they cannot receive any message, they cannot detect the access collision in this scenario. To address this problem, HCMAC provides far more advanced collision avoidance by inserting a contention window of fine grain backoff units in each time slot. Details of HCMAC are described later.

3.3.2 Detection of hidden-node collisions

Figure 7 shows a scenario where a hidden node collision occurs. In Fig. 7, vehicle 2 and vehicle 3 enter the wireless range of vehicle 1 in different directions. Suppose that vehicle 1 picked slot 4, while vehicle 2 and vehicle 3 picked the same slot 10. If vehicles 2 and 3 entered vehicle 1's wireless range after vehicle 1's time slot (i.e., after slot 4), vehicles 2 and 3 could not receive any information about vehicle 1. At slot 10, vehicles 2 and 3 transmit, and a collision occurs at vehicle 1. Since vehicle 2 and vehicle 3 do not have a neighbor list constructed yet at this moment, they do not expect any acknowledgment from the new neighbor vehicles leaving the collision undetected. In the next frame, at slot 4,

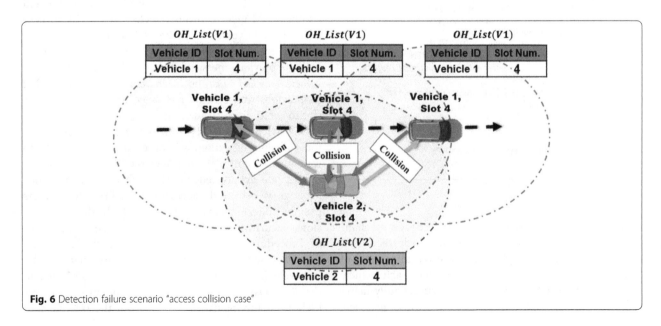

Fig. 6 Detection failure scenario "access collision case"

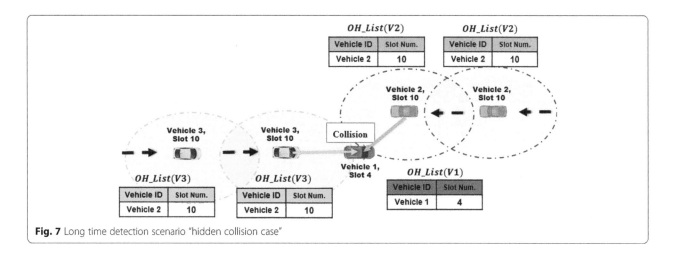

Fig. 7 Long time detection scenario "hidden collision case"

both vehicles 2 and 3 receive a BSM message from vehicle 1, but they regard vehicle 1 as a new neighbor, and thus they would not expect an acknowledgment for their previous BSM. Therefore, vehicles 2 and 3 once again transmit at slot 10, and so they waste another frame without knowing that their previous two consecutive BSMs are lost. In the next frame, finally, vehicles 2 and 3 wait for an acknowledgment from vehicle 1 at slot 4. Vehicle 1, however, transmits without including vehicles 2 and 3 in the *OH _ List*. Vehicles 2 and 3 finally can conclude that their previous packets ran into a collision, and they decide to change their slots.

Although vehicles can detect the collision eventually, it took more than one full TDMA frame. There are various other collision scenarios that can occur very often due to the mobility and the rapid change in the network topology. Therefore, the previous method relying just on the simplistic TDMA with implicit acknowledgement from the receiver nodes cannot ensure highly reliable delivery of broadcast messages like BSMs. This behavior motivates us to add two additional techniques to HCMAC in addition to the implicit acknowledgements: carrier sensing and feedback of slot error list.

3.3.3 Carrier sensing
Before transmitting the BSM at the acquired slot, each vehicle waits for a backoff interval, B, which is randomly selected out of the contention window, W. After this backoff time, the vehicle conducts carrier sensing like in CSMA. If the vehicle detected a carrier or transmitted signal exceeding the threshold of reception signal power, it defers its transmission and selects another time slot. In HCMAC, each vehicle randomly selects a pair (Ac_{Slot}, B), where Ac_{Slot} is an available time slot (or random slot if no available slot is left), while B is a random backoff interval. Utilizing these two levels of random selection of channel access time substantially reduces the

probability of multiple vehicles' transmitting in the same slot. In addition, it enables faster discovery of the potential collisions.

Figure 8 illustrates an example of HCMAC which employs the carrier sense function at the beginning of each time slot. This example configures the network with five time slots per a TDMA frame, while adding to each slot a contention window of W=5. Figure 8 shows four vehicles. Suppose that vehicle 3 and vehicle 4 joined the network almost at the same time, and they selected the same time slot, *slot*3. Due to their different backoff interval, however, vehicle 4 can transmit safely as it has a lower value, B=3. On the other hand, vehicle 3 can detect the transmission of vehicle 4 and select another free time slot.

3.3.4 Feedback of slot error list
When the physical layer cannot decode the packet, and the received data produces a Cycling Redundant Check (CRC) error while the received signal strength is higher than the receiver sensitivity, we conclude that there might be a collision. When the received signal strength indicator (RSSI) is above the receiver sensitivity, in most of the real cases, this CRC error happened because of the collision. By employing such a technique, the physical layer can deduce that a collision occurred.

By conveying the RSSI level to the MAC layer, each vehicle can construct a slot error (SE) list consisting of the slots that experienced collisions during the previous TDMA frame. By adding an SE list to the BSM packet, HCMAC allows the receiver to detect various hidden collisions; see Fig. 3 for the packet format. Figure 9 shows the same example illustrated in Fig. 7 and now shows the situation when the first collision happens. Then, vehicle 1 transmits its next packet including the SE list which notifies vehicles 2 and 3 that slot 10 had a

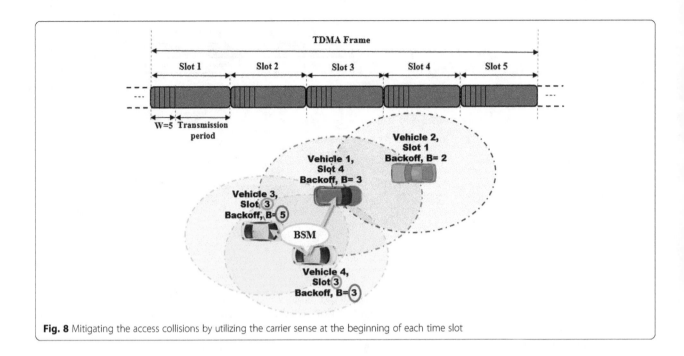

Fig. 8 Mitigating the access collisions by utilizing the carrier sense at the beginning of each time slot

collision in the previous frame. Therefore, vehicles 2 and 3 can avoid successive collisions by acquiring new slots.

The proposed sending and receiving functions of HCMAC including the new collision detection mechanism are illustrated by the pseudo codes in Procedure 3 and Procedure 4. In Procedure 3, at the acquired slot, Ac_{Slot}, the sending function observes the channel during the backoff time, B. If the channel is clear, it transmits. Otherwise, it selects another free slot. In Procedure 4, steps 4–10 check the SE list. If the SE list includes the acquired slot, Ac_{Slot}, of the receiving vehicle, the receiver selects another slot. On the other hand, steps 11–20 utilize the feedback mechanism similar to the one employed by VeMAC's Procedure 2.

Procedure 3: Sending function in HCMAC
Input: : $V_C, S, OH\_List(V_C), TH\_List(V_C), W, Free_{Slots[\,]}, Ac_{Slot}$
Output: Ac_{Slot}

1.   **For** ($i \leftarrow 1$ to S) **DO**   // Loop through the whole slot range
2.   **IF** ($i == Ac_{Slot}$) **THEN**   // when slot= the acquired slot
3. $B \leftarrow rand(\,) * W$;   // Select backoff interval
4. $CS = sense\ the\ medium\ during\ the\ B\ interval$;
5.   **IF** ($CS == 0$) **THEN**   // CS=0 if the channel is clear
6.   Send(BSM_Pkt);
7.   **ELSE**
8.   **IF** ($Free_{Slots[\,]}$ is not empty) **THEN**
9. $Ac_{Slot} \leftarrow$ Select a random slot from $Free_{Slots[\,]}$;
10.   **ELSE**
11. $Ac_{Slot} \leftarrow$ Select a random slot from $[1 : S]$;
12.   **ENDIF**
13.   **ENDIF**
14.   **ENDIF**
15.   **ENDFOR**

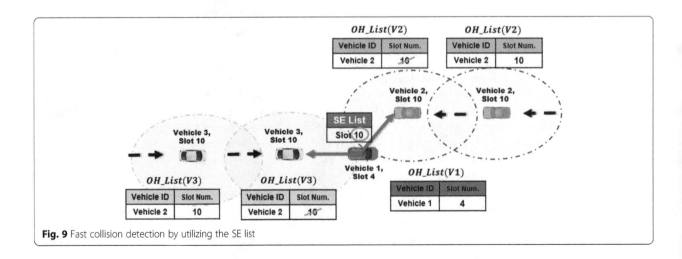

Fig. 9 Fast collision detection by utilizing the SE list

Procedure 4: Receiving function in HCMAC

Input: $V_C, S, OH\_List(V_C), TH\_List(V_C), Free_{Slots[]}, Ac_{Slot}$

Output: $OH\_List(V_C), TH\_List(V_C), Ac_{Slot}$

```
1.      slot_change ← 0;
2.      For (i ← (Ac_Slot + 1) to (S + Ac_Slot − 1)) DO
3.          IF (Received_Pkt(i) == BSM_Pkt) THEN
4.              IF (Ac_Slot is included in the SE list) THEN
5.                  IF (Free_Slots[] is not empty) THEN
6.                      Ac_Slot ← Select a random slot from Free_Slots[];
7.                  ELSE
8.                      Ac_Slot ← Select a random slot from [1: S];
9.                  ENDIF
10.             ENDIF
11.             IF (SrcAdd is included in the OH_List(V_C)) THEN
12.                 IF ((V_C is not in OH_List(V_i))
                        AND (slot_change == 0)) THEN
13.                     IF (Free_Slots[] is not empty) THEN
14.                         Ac_Slot ← Select a random slot from Free_Slots[];
15.                     ELSE
16.                         Ac_Slot ← Select a random slot from [1: S];
17.                     ENDIF
18.                     slot_change ← 1;
19.                 ENDIF
20.             ENDIF
21.             Update OH_List(V_C) & TH_List(V_C);
22.         ENDIF
23.     ENDFOR
```

---

4 Performance analysis

In this section, we introduce an analytic model to evaluate the performance of the proposed HCMAC protocol and compare it with VeMAC. We compute the average probability of successful slot acquisition when vehicles join the network. We also express the delay of slot acquisition process, which is the time it takes for all vehicles to successfully acquire unique time slots.

To develop the analytical model, we extend the probabilistic model used in [26] by introducing our parameters for the TDMA slot Ac_{Slot} and backoff interval B. To derive mathematically feasible formulation, we follow the assumptions described in [27] with our new parameters additionally considered:

a) V vehicles join the network, each of which attempts to acquire its time slot

b) All the vehicles belong to the same two-hop neighbor list, TH_List (i.e., all vehicles are within the communication range of each other). During the slot acquisition process, we assume that the TH_List does not change for a short instance of time

c) Given a set S of available time slots, each vehicle randomly chooses a time slot, Ac_{Slot}, and a backoff interval, B

d) At the end of each TDMA frame, all vehicles get informed if their trial to access the channel was successful

e) At the end of each TDMA frame, each vehicle knows the slots occupied by other vehicles during the frame, since the vehicles are within their

wireless range. Vehicles that failed to acquire a slot use the updated free slot list, $Free_{Slots[]}$, to choose another slot in the next frame

While the above assumptions are accurate with respect to the proposed HCMAC protocol, assumptions (d) and (e) intentionally impose unnecessary restrictions for the sake of simplicity of modeling. We discuss how the proposed method can further enhance the latency of collision detection by relieving (d) and (e) later.

In the slot acquisition process, $\frac{1}{s}$ is the probability that a vehicle randomly selects a slot out of s available slots with uniform distribution. Suppose that there are no transmissions during the first $(l-1)$ slots (i.e., these slots are not acquired by any vehicle), and then k out of v vehicles transmit at slot l. The probability of the above case, denoted by $P(s, l, v, k)$ ($1 \leq l \leq s$ and $1 \leq k \leq v$), can be modeled through a Bernoulli process as follows:

$$p(s, l, v, k) = \left(\left(1 - \frac{1}{s} \right)^v \right)^{l-1} * \binom{v}{k} \left(\frac{1}{s-l+1} \right)^k \left(1 - \frac{1}{s-l+1} \right)^{v-k}$$

(1)

The formula of Eq. 1 can be approximated to the formula of Eq. 2, which is originally described in [26].

$$P(s, l, v, k) = \left(1 - \frac{l-1}{s} \right)^v * \binom{v}{k} \left(\frac{1}{s-l+1} \right)^k \left(1 - \frac{1}{s-l+1} \right)^{v-k}$$

(2)

In Eq. 2, $\left(1 - \frac{l-1}{s}\right)^v$ is the probability that out of v vehicles, there is no transmission during the first $(l-1)$ slots, and $\binom{v}{k} \left(\frac{1}{s-l+1}\right)^k \left(1 - \frac{1}{s-l+1}\right)^{v-k}$ is the probability of having k out of v possible transmissions at slot l where $\frac{1}{s-l+1}$ is the probability of randomly select any slot out of the $(s - l + 1)$ upcoming slots.

While in [26] the authors model the broadcast traffic based on IEEE802.11p for the alternating mode (i.e., switching between CCH and SCH) [6], we reformulate the model to extend it for the structure of HCMAC. Thus, we define $X(s, v)$ as the mean number of successful transmissions during one TDMA frame, whereas [26] defines it as the mean number of successful transmissions during one CCH interval while taking into account the CSMA contention window size.

$$X(s, v) = \sum_{l=1}^{s} \{ P(s, l, v, 1) \, [1 + X(s-l, v-1)] \qquad (3)$$

$$+ \sum_{k=2}^{v} P(s, l, v, k) \, X(s-l, v-k) \}$$

In Eq. 3, the term $P(s, l, v, 1) \, [1 + X(s - l, v - 1)]$ expresses the success probability when we have just one successful transmission (i.e., only one vehicle acquires the lth slot). The term $X(s - l, v - 1)$ represents the mean number of successful transmissions in the remaining ($s - l$) slots. It makes Eq. 3 a recursive form with the parameter S substituted by $S - l$ indicating that the number of available free slots is reduced by l slots as Eq. 3 proceeds the recursive calculation from the first slot to the last slot to encompass one complete TDMA frame. The term $\sum_{k=2}^{v} P(s, l, v, k) \, X(s-l, v-k)$ denotes the collision cases when two or more vehicles (i.e., $k \geq 2$) acquire the same lth slot.

As described in Section 3, in HCMAC, each vehicle conducts a backoff process before transmitting its packet in the acquired slot to reduce the chances of collisions. When the first vehicle that has the lowest backoff interval transmits, all other vehicles hold their transmissions, and they reselect new slots. To derive an analytic model of the backoff process, we extend the model of Eq. 1. We derive Eq. 4 from Eq. 1 by introducing new parameters, w and b. Here, w denotes the contention window size expressed by a group of backoff units, while b indicates the backoff interval in terms of the number of backoff units. Like in Eq. 1, k is the number of the contending vehicles for a given slot. Equation 5 formulates the total probability $P(w, k, 1)$ of achieving one successful transmission when k vehicles select the same lth slot, and they contend through the contention window w. We define $Y(s, v, w)$ as the mean number of successful transmissions by incorporating our proposed backoff process as in Eq. 5.

$$P(w, b, k, 1) = \left(1 - \frac{b-1}{w}\right)^{k} * \binom{k}{1}\left(\frac{1}{w-b+1}\right)\left(1 - \frac{1}{w-b+1}\right)^{k-1} \tag{4}$$

$$P(w, k, 1) = \sum_{b=1}^{w} P(w, b, k, 1) \tag{5}$$

$$Y(s, v, w) = \sum_{l=1}^{s} \{(s, l, v, 1)[1 + Y(s-l, v-1)] \\ + \sum_{k=2}^{v} P(s, l, v, k)\,[\,P(w, k, 1) + Y(s-l, v-k)]\} \tag{6}$$

5 Evaluation of the analytical model

This section compares the performance of HCMAC with VeMAC by using the analytical model presented in Section 4. In addition, it proves the accuracy of the analytical model by extensive simulation results. By using Eqs. 2 and 3, we calculate the expected performance of the

VeMAC protocol. Similarly, by using Eqs. 5 and 6, we estimate the expected performance of the proposed HCMAC protocol. For both VeMAC and HCMAC, we compute the mean number of successfully acquired slots during the TDMA frame. To compute all the possible cases, the recursive models of Eqs. 3 and 6 start from $X(S, V)$ and $Y(S, V, W)$, and then recursively calculate all other cases, respectively. Therefore, we compute the probability of acquiring a slot in each protocol by using the following formulas:

For the VeMAC protocol,

$$P_{\text{VeMAC}} = \frac{X(S, V)}{V} \tag{7}$$

For the proposed HCMAC protocol,

$$P_{\text{HCMAC}} = \frac{Y(S, V, W)}{V} \tag{8}$$

In Fig. 10, we demonstrate the analytical results for both VeMAC and HCMAC. We compute the probability of successfully acquiring a time slot versus the number of vehicles, V, while configuring the TDMA frames with a variable number of slots, $S = [10, 20, 40]$ slots per frame, and a fixed size contention window, $W = 5$. Although the authors of VeMAC used a different analytical model [11, 27], our results obtained by Eq. 3 for VeMAC match the results originally presented in [27]. As shown in Fig. 10, HCMAC outperforms VeMAC regardless of the number of vehicles and the number of slots considered in the evaluation. HCMAC approximately achieves at $S = 10$ what can be achieved by VeMAC at $S = 20$ (i.e., the green and the blue lines in Fig. 10 are very close). Thanks to the proposed backoff and the carrier sensing processes, it is observed that HCMAC significantly improves the probability of successfully acquiring a time slot.

To verify the accuracy of the analytic model presented in Section 4, we compare the evaluation results of the analytic model with the results of a network simulator that we implemented. As both the slot acquisition and the backoff processes rely on random numbers, we have taken an average result from 50 simulation runs. Figure 11 proves that the results of the analytic model well agree with the simulation results under a wide range of S and V values. Figure 12 demonstrates the comparison results given from the simulations of VeMAC and HCMAC. Here, we configured the TDMA frames of larger sizes $S = [50, 100, 200]$, while setting the number of vehicles up to 100 vehicles and the contention window to 10 backoff units. Like in the case of

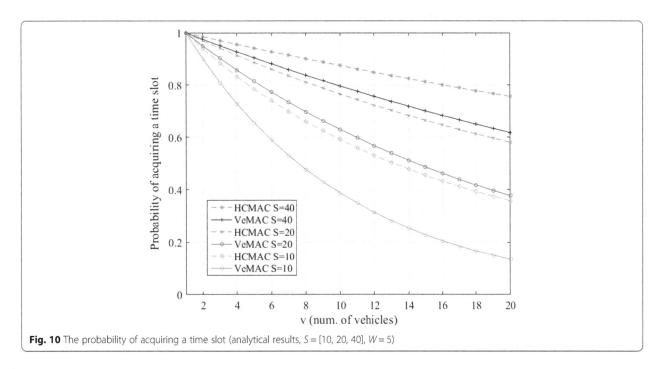

Fig. 10 The probability of acquiring a time slot (analytical results, $S = [10, 20, 40]$, $W = 5$)

Fig. 10, HCMAC achieves a substantially higher probability of acquiring a time slot compared to VeMAC.

To analyze the impact of the contention window size on the performance of the HCMAC, we used a TDMA frame of a fixed size with a contention window of various sizes. In Fig. 13, we report the analytical results for the probability of acquiring a time slot with $S = 10$ and $W = [2, 5, 10, 15, 20]$. Figure 13 shows that while the performance increases along with the contention windows of $W = [2, 5, 10, 15]$,

the performance levels off with the contention window of size beyond $W = 15$.

Similarly, Fig. 14 shows the simulation results of HCMAC obtained with $S = 50$ and $W = [2, 5, 10, 20, 40]$. It is observed that the probability of time slot acquisition well matches the results of the analytic model of Fig. 13. From Figs. 13 and 14, we can conclude that employing a contention window of a moderate size can provide a considerable improvement in performance, even when the number of vehicles is greater than the number of the

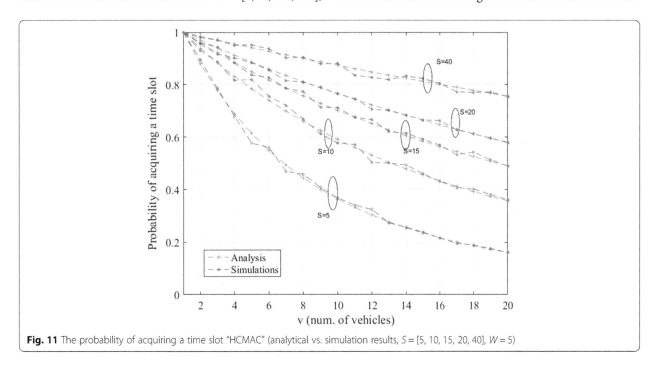

Fig. 11 The probability of acquiring a time slot "HCMAC" (analytical vs. simulation results, $S = [5, 10, 15, 20, 40]$, $W = 5$)

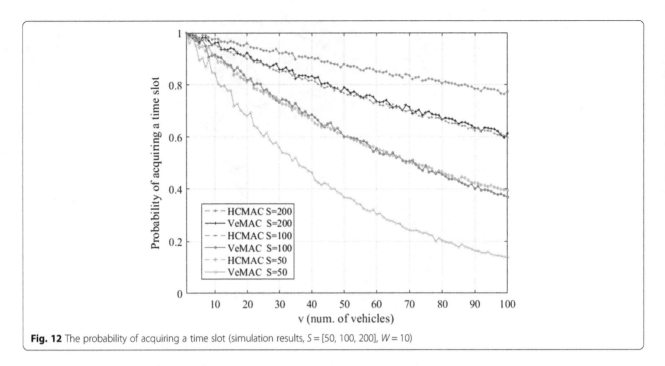

Fig. 12 The probability of acquiring a time slot (simulation results, $S = [50, 100, 200]$, $W = 10$)

available time slots. Moreover, to express how fast the vehicles can acquire a unique time slot during the slot acquisition process, we evaluate the performance in terms of the required time (i.e., number of TDMA frames) for all vehicles to successfully get a time slot. By utilizing different settings for S and V, we demonstrate the average number of vehicles acquiring a unique time slot within k TDMA frames.

In Fig. 15, the analytical results show that vehicles using HCMAC acquire a time slot faster than the

vehicles using VeMAC. For example, when $S = 15$ and $V = 15$, in HCMAC, all vehicles acquired a time slot after five TDMA frames compared to seven frames in VeMAC. To verify the analytical results, we simulated the slot acquisition phase of the HCMAC with different values of S and V. We calculated the average number of vehicles which acquire a time slot within k TDMA frames. Figure 16 shows good matching between the analytical and the simulation results. Finally, simulations have been conducted with large values of S and V to

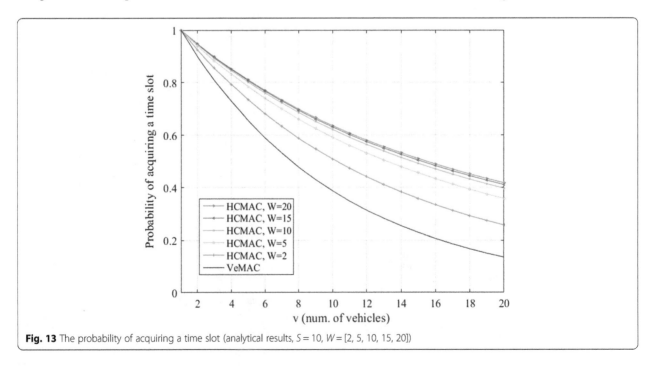

Fig. 13 The probability of acquiring a time slot (analytical results, $S = 10$, $W = [2, 5, 10, 15, 20]$)

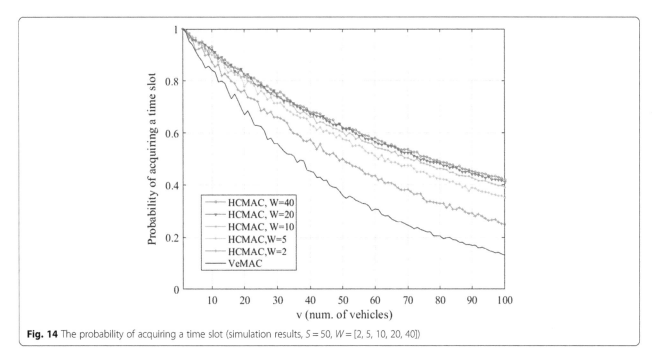

Fig. 14 The probability of acquiring a time slot (simulation results, $S = 50$, $W = [2, 5, 10, 20, 40]$)

show how fast HCMAC acquires the unique time slots compared to VeMAC. Simulation results obtained for VeMAC well match the results originally presented in [27] for all S and V values considered. Figure 17 illustrates the simulation results for HCMAC and VeMAC with different values of S and V. It is observed that, for all values of S and V, the cases of HCMAC acquire time slots faster than the cases of VeMAC, especially when the number of vehicles, V, approaches the number of the available slots, S. For example, when $V = 90$ and $S = 100$,

in HCMAC, all vehicles acquire a unique time slot within three TDMA frames compared to five TDMA frames in VeMAC.

6 Results

6.1 Implementations and simulations

In this section, we present the network simulator that implements the proposed protocol, HCMAC, and also the previous protocol, VeMAC. We evaluate the performance of HCMAC and VeMAC under various

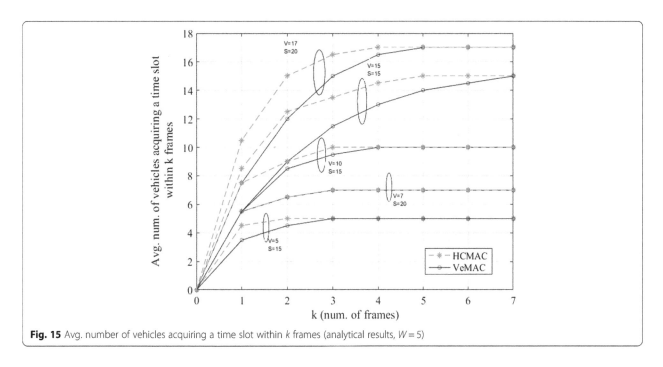

Fig. 15 Avg. number of vehicles acquiring a time slot within k frames (analytical results, $W = 5$)

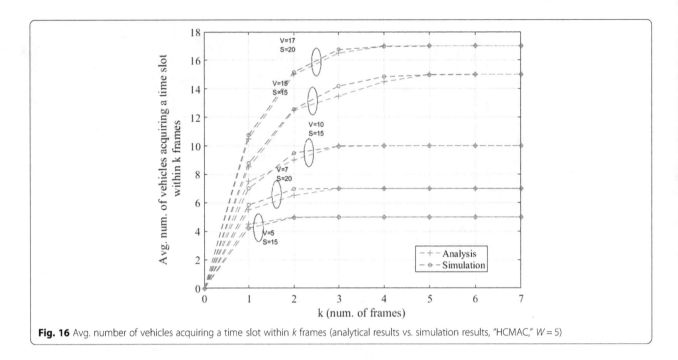

Fig. 16 Avg. number of vehicles acquiring a time slot within k frames (analytical results vs. simulation results, "HCMAC," $W = 5$)

network conditions and demonstrate remarkable performance improvement achieved by HCMAC in comparison with VeMAC.

6.1.1 Implementation of network simulator
We implemented the detailed functions of HCMAC in a network simulator using MATLAB environment. While we made restrictive assumptions of (d) and (e) in Section 4 to derive the analytic model with plausible complexity,

these restrictions are not needed in the implementation of HCMAC. We therefore implemented the network simulator without the assumptions (d) and (e), which allows early collision detection as follows. In the implementation of HCMAC, when a vehicle experiences a collision at its selected time slot, it can detect the collision within the next slot or in a few slots by analyzing BSMs received from its neighbors. It can figure out whether its previous packet was correctly acknowledged by analyzing a SE list or one-hop neighbor list in the

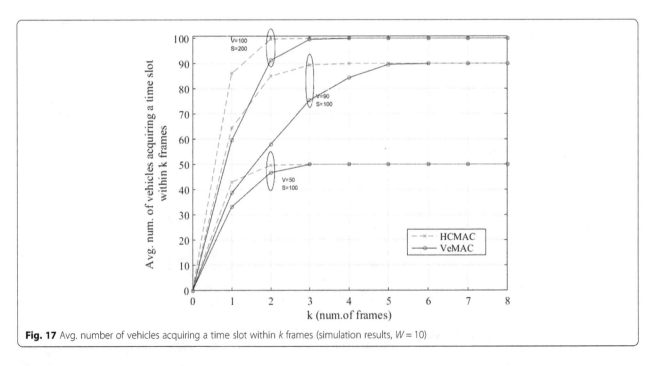

Fig. 17 Avg. number of vehicles acquiring a time slot within k frames (simulation results, $W = 10$)

BSMs. Hence, we can reduce the latency compared to the analytic model with the above restrictions.

In our simulator, we considered two propagation models for the wireless channel: ideal and two-ray ground models. For the ideal channel, vehicles can communicate with each other, if they are within a specified wireless range R. Therefore, the transmission failure can occur due to the out-of-range condition or transmission collisions. On the other hand, for the two-ray ground propagation model, a successful reception happens only if the received SINR (signal to noise and interference ratio) exceeds a certain threshold.[1] To conform to the WAVE standards, we configured the periodic safety messages with a report frequency of 10 Hz which leads to a frame length of 100 ms. We then configured each frame such that it consists of 100 slots, which makes each slot 1-ms long. According to the DSRC standard [7], each vehicle should transmit the periodic safety messages every 100 ms. Therefore, the transmission interval is an important performance metric that can judge whether the MAC protocol satisfies the delay requirement of the standard's safety applications. Each slot is further divided into a 20/80 ratio (i.e., 0.2 ms for a contention window, while 0.8 ms for data transmission). The contention window is sub-divided into 10 backoff units leading to each backoff unit of 20 µs, which is sufficient for HCMAC's carrier sensing process. Figure 14 shows that 10 backoff units represent a reasonable CW size even in the dense networks tested in this paper. The data rate of all BSM packets in our simulations is set to 12 Mbps with a packet length of 500 bytes, which is larger than a common BSM length. Although we utilize a big size packet compared to the normal size of BSM messages [7], 0.8 ms is more than enough for transmission according to the analysis made in [3].

In all the simulations presented in this paper, R was set to 150 m. We obtained reliable connections with a distance of 150 m from the measurement using commercial V2X modules in the actual road conditions. For the two-ray ground channel, vehicles transmit with a TX power of 23 dBm and the successful reception occurs when the received SINR exceeds a threshold of 15 dB. We implemented two network scenarios: a highway scenario of Fig. 18 and an urban scenario of Fig. 19. In both scenarios, we first generate the topology by randomly locating all the vehicles over the road segments. Then, the vehicles start moving and select random time slots to transmit their packets. By receiving the information that is included in the packets (i.e., neighbor list, SE list), each vehicle tries to avoid the collision by avoiding the occupied time slots and re-acquiring free time slots. With the continuous mobility, the topology continuously changes, and each vehicle keeps joining new neighbors and leaving old ones. To test the performance under different vehicle densities, we run the simulation for 2 min for each generated topology. During the simulation time, we collect all the required data to calculate the performance metrics. In the highway scenario, vehicles move on an eight-lane highway as shown in Fig. 18. Depending on the lanes, each vehicle travels with a constant speed of 60, 90, 110, and 120 km/h for lanes 1, 2, 3, and 4, respectively. A vehicle can communicate only with the vehicles located within a distance $d \le R$. To keep the same number of vehicles during the simulation time, a vehicle which reaches one end of the road re-enters from the other end. Also, this keeps changing the topology as the vehicles approaching each other from opposite directions makes each section of the road always have new arrivals. Hence, a continuous time slot rescheduling is always required.

In the urban scenario, there are three vertical and three horizontal bi-directional road segments. In between road segments, there are four building squared blocks. Each vehicle selects a random speed from the interval 40–60 km/h; then, it continues moving with the same speed during the simulation time. When a vehicle reaches to one of the intersections, it randomly turns to one of the allowable moving directions. In this way,

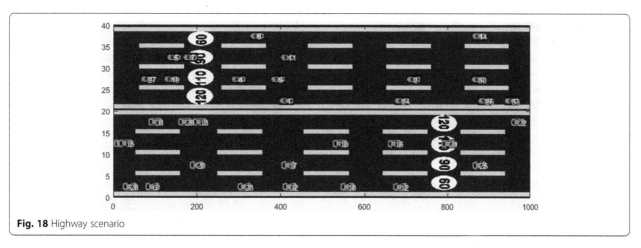

Fig. 18 Highway scenario

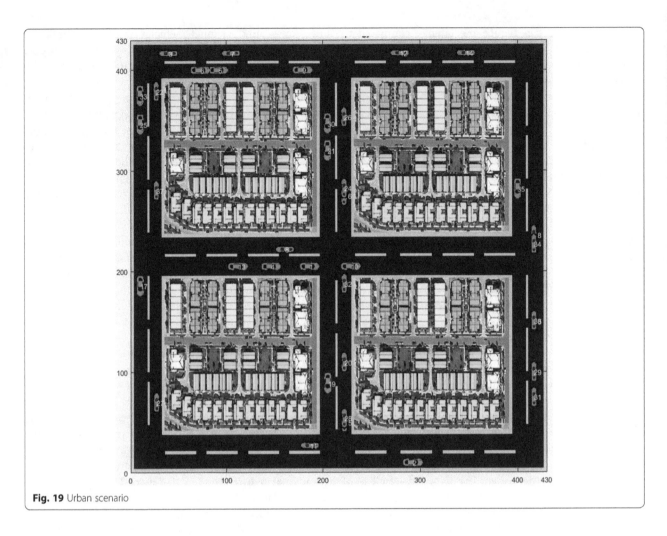

Fig. 19 Urban scenario

vehicles keep moving around, and the number of vehicles remains constant during the simulation period. We also assume that the squared building blocks can totally block the wireless signal. Therefore, away from the intersection zones, a vehicle can only communicate with the vehicles which belong to the same road segment and located at a distance $d \leq R$.

For the simulations, we chose the target network configuration such that the probability of successful slot acquisition is at least 80% with 100 slots per TDMA frame. We observed that with such network configurations the performance of the networks is acceptable for our target applications.

In Fig. 12 in Section 5, the analytic model demonstrated that the probability of successful slot acquisition becomes greater than 80% when we have no more than 45 vehicles per wireless range. For instance, in the highway scenario, 1-km road segment can have around seven adjacent wireless ranges, if each wireless range contains 45 vehicles on average. This leads to around $7 \times 45 = 315$ vehicles/km. From this perspective, we measured the performance of the considered protocols under different

vehicle densities, starting from 50 vehicles/km up to 400 vehicles/km (i.e., 400 vehicles represent around 120% of a network load). In this way, the performance can be analyzed in sparse and dense networks including the target network condition. In the same context, we tested the performance for the urban scenario starting with 50 vehicles up to 650 vehicles (i.e., with 650 vehicles, the network load is around 90%).

Table 1 summarizes the simulation parameters for the two scenarios such as the road dimension, the properties of packets and time slots, and the number of vehicles in the network. The initial positions of vehicles are uniformly distributed over the simulated road segment of 1000 m for the highway scenario and 430 m for the urban scenario, respectively.

6.1.2 Performance metrics

We evaluated the performance of various network examples using the simulator described above. We used the following performance metrics to compare the performance between HCMAC and VeMAC:

Table 1 Simulation parameters

Parameter	Highway	Urban
Road segment length	1 km	430 m
Num. of horizontal segments	–	3
Num. of vertical segments	–	3
Lane width	5 m	5 m
Num. of lanes per direction	4	1
Wireless comm. range	150 m	150 m
Vehicle speed	[60, 90, 110, 120] km/h	[40–60] km/h
Packet size	500 bytes	500 bytes
Data rate	12 Mbps	12 Mbps
TX power	23 dBm	23 dBm
SNR threshold	15 dB	15 dB
Broadcast rate	10 MHz	10 MHz
Slots per frame	100	100
Slot width	1 ms	1 ms
Contention window size	0.2 ms (10 backoff units)	0.2 ms (10 backoff units)
Number of vehicles	50–400	50–650
Number of RSUs	0	0
Simulation time	2 min	2 min

The number of collision events represents the average number of collision events per TDMA frame (two or more collided packets at the same slot are counted as one collision event). This metric considers the access collision and the hidden collision cases as defined earlier. To accommodate hidden collisions, this metric is calculated by considering a two-hop neighborhood.

Network throughput is defined as the average number of the successful transmissions per frame in the one-hop neighborhood. As the MAC protocol becomes more robust with respect to the collisions, the network throughput increases.

Packet delivery ratio (PDR) is defined as the ratio of the number of successful packet receptions to the total number of expected receivers located in the wireless range. PDR often serves as an important metric for the reliability of MAC protocols in general.

Average transmission interval is defined as the average time interval between two consecutive packet transmissions from the same vehicle, which is calculated over all vehicles at all time slots.

Maximum transmission interval is defined as the maximum value of the transmission intervals calculated for each vehicle at each selected slot.

6.2 Evaluation of simulation results

In this subsection, we present the simulation results using an extensive set of example networks under the highway and urban scenarios.[2]

6.2.1 Simulation results under highway scenarios

Figure 20 shows the number of collision events measured with the increasing number of vehicles. In the sparse networks with the number of available slots larger than the number of vehicles, both VeMAC and HCMAC exhibit very few collisions. When the network becomes denser with 250 or more vehicles, however, the number of collisions starts sharply increasing. HCMAC incurs significantly fewer collisions than VeMAC does in the dense networks. For instance, over the ideal channel, when the number of vehicles is 400, HCMAC experiences only two collisions per frame, while VeMAC has five collisions per frame. The same observation is shown for the two-ray ground channel; HCMAC experiences fewer collisions compared to VeMAC. The good performance of HCMAC is attributed to the proposed collision detection mechanisms based on CSMA and SE list.

Figure 20 also shows that with the two-ray model, each protocol has fewer collision events compared to the ideal model. That is, in the ideal channel case, receiving multiple packets from different transmitters at the same time is considered a collision event, but, on the other hand, with the employed two-ray ground model, the receiver node can still successfully capture one of the transmitted packets if the received SINR exceeds the threshold.

HCMAC's lower collision rate leads to better performance in other metrics as well.

As shown in Fig. 21, HCMAC outperforms VeMAC in the performance of network throughput. For both

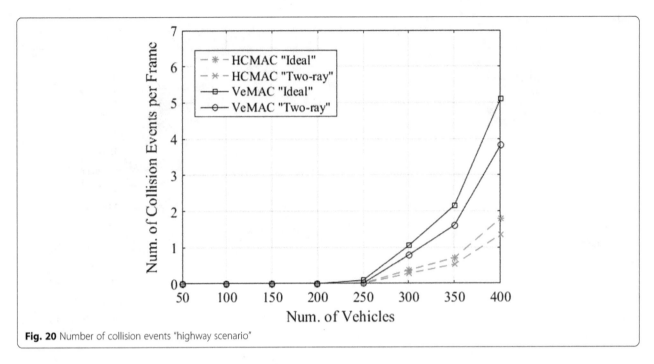

Fig. 20 Number of collision events "highway scenario"

channel models, when the number of vehicles is over 250, HCMAC experiences higher throughput compared to VeMAC. Figure 21 also illustrates that as the two-ray channel model introduces propagation losses, both protocols achieve less throughput compared to the ideal channel case. Figure 22 illustrates average PDR achieved by the two protocols. For the low densities of 250 or fewer vehicles in the network, both HCMAC and VeMAC provide very high PDR (e.g., when the number of vehicles is 150, PDR is around 99%).

As the number of vehicles increases, the PDR for VeMAC decreases. In contrast, HCMAC can achieve higher PDR even with higher vehicle densities (e.g., Over the ideal channel, when the number of vehicles is 400, HCMAC's PDR is around 96% compared to 87% for VeMAC. Also, with the two-ray ground model, HCMAC achieves a PDR of 81% compared to 74% for VeMAC.).

Finally, we measured the average transmission interval under different vehicle densities as shown in Fig. 23. HCMAC achieves lower interval in the dense networks.

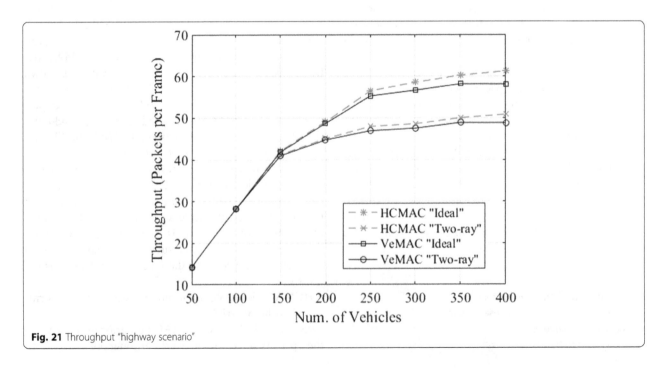

Fig. 21 Throughput "highway scenario"

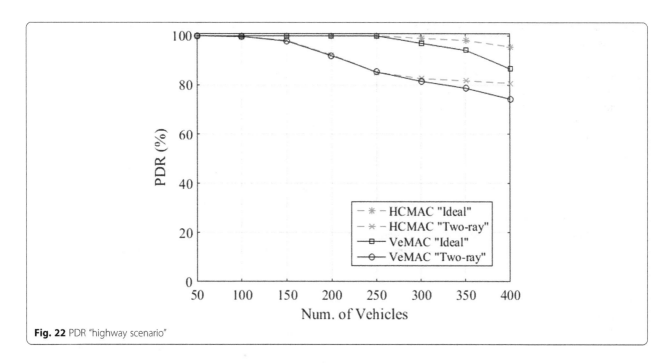

Fig. 22 PDR "highway scenario"

For example, for the ideal channel case, when the number of vehicles is 400, the average transmission interval of HCMAC is around 135 ms compared to 155 ms for VeMAC. This is attributed to the observation that VeMAC experiences more collisions while it takes more frames to detect the collisions than HCMAC. This makes the transmitter repeatedly select occupied slots leading to a longer transmission interval.

Figure 23 also demonstrates the maximum transmission interval which indicates that HCMAC's capability to reduce the transmission interval also keeps its maximum transmission interval of each vehicle bounded to much smaller value compared to VeMAC. For example, for the ideal channel case, when the number of vehicles is 400, the maximum transmission interval is around 900 ms for HCMAC compared to 1300 ms for VeMAC.

For the two-ray model, each protocol experiences a very small decrement in the average and maximum transmission intervals compared to the ideal channel case due to the small decrement in the number of

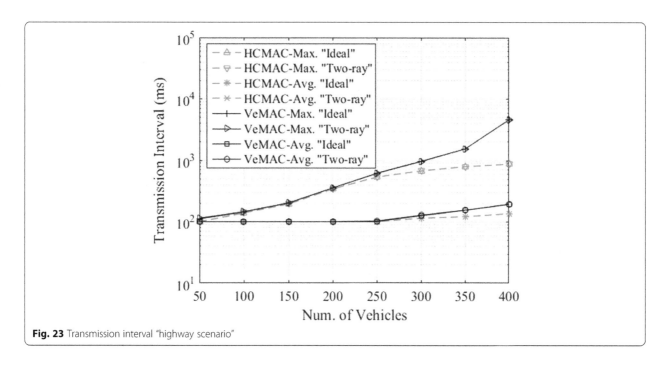

Fig. 23 Transmission interval "highway scenario"

collision events as shown in Fig. 20 (with the log scale, Fig. 23 shows completely overlapping plots).

6.2.2 Simulation results under urban scenarios

We evaluated the network performance under the urban scenarios in a way similar to the highway scenarios. Both HCMAC and VeMAC are evaluated over a wide range of vehicle densities.

Figure 24 shows the average number of collision events experienced by all the vehicles within a single TDMA frame for both channel models: ideal and two-ray ground models. As the network becomes denser with the increased number of vehicles, HCMAC's collision detection technique becomes more effective. In Fig. 24, for the ideal channel case, when the number of vehicles is 650, HCMAC has on average 3 collisions per frame, while VeMAC incurs as many as 17 collisions per frame. Similar to the highway scenario, for the two-ray channel case, each protocol has fewer collisions compared to the ideal channel case, due to the same packet capture effect that has been explained in the highway scenario.

HCMAC's capability of collision reduction also enhances the performance metrics: the throughput, the PDR, and the transmission intervals. Figure 25 compares the throughput. For the ideal channel case, when the vehicle density is 650, HCMAC achieves a throughput of around 69 successful transmissions per frame compared to a throughput of 54 for VeMAC. Figure 25 also shows that with employing the two-ray channel model, when the number of vehicles is 650, HCMAC has a throughput of around 54 packets per frame compared to a

throughput of 50 for VeMAC. Figure 26 analyzes the PDR performance. As the number of vehicles increases from 250 to 650, the PDR dramatically decreases in the VeMAC case (i.e., the PDR decreases from 97 to 65% over the ideal channel and from 97 to 60% over the two-ray channel). In the HCMAC case, in contrast, for the same range of density, the PDR decreases only from 99 to 91% over the ideal channel and from 98 to 80% over the two-ray.

Similar to the highway scenario, for both channel models, each protocol has almost the same performance of transmission interval.

Figure 27 illustrates the average transmission interval versus the vehicle density. In the dense networks, HCMAC achieves lower transmission interval than VeMAC.

For example, when the number of vehicles is 650, the average transmission interval is around 150 ms compared to 210 ms for VeMAC. The maximum transmission interval is also illustrated in Fig. 27, which demonstrates that HCMAC can substantially reduce the maximum transmission intervals for all individual vehicles. For example, when the number of vehicles is 650, the maximum transmission interval for HCMAC is around 1180 ms, a reduction of 77% compared with 3578 ms for VeMAC.

7 Conclusion and future work

In this paper, we presented a Hybrid Cooperative MAC (HCMAC) protocol for vehicular networks. HCMAC is aimed at providing a reliable broadcast service for dense vehicle-to-vehicle networks, which is regarded as highly

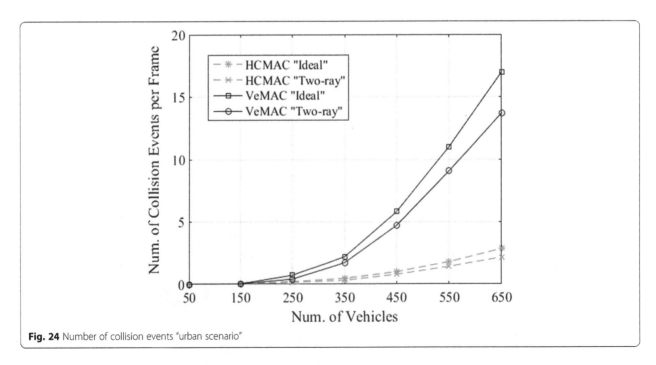

Fig. 24 Number of collision events "urban scenario"

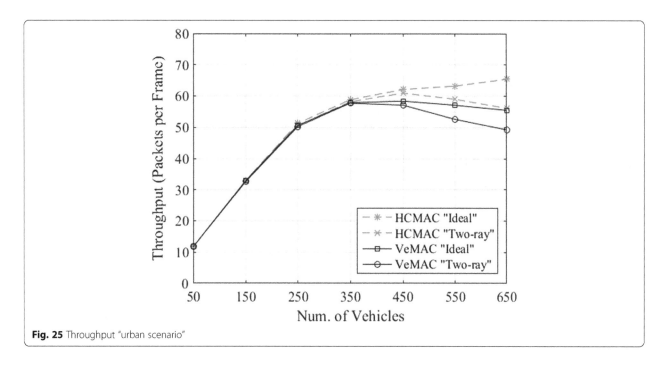

Fig. 25 Throughput "urban scenario"

challenging but is increasingly demanded for the vehicle safety applications. HCMAC integrates a TDMA protocol with a CSMA's collision avoidance strategy to provide highly efficient channel access with a substantially reduced collision. In addition, we introduced new collision detection mechanisms based on status feedback of neighbors and time slots, which further reduce the probability of collisions. We analyzed the time slot acquisition process and derived an analytic model for the collision probability of the proposed HCMAC protocol, whose accuracy is verified against network simulations. We evaluated the performance of HCMAC in comparison with VeMAC, a well-known TDMA protocol. The presented analytic model shows that HCMAC achieves substantially higher packet delivery ratio (PDR) and lower collision rate than VeMAC under various network conditions. In addition, we implemented both HCMAC and VeMAC in a realistic network simulator and conducted an extensive set of simulations with a wider range of network examples under highway and urban

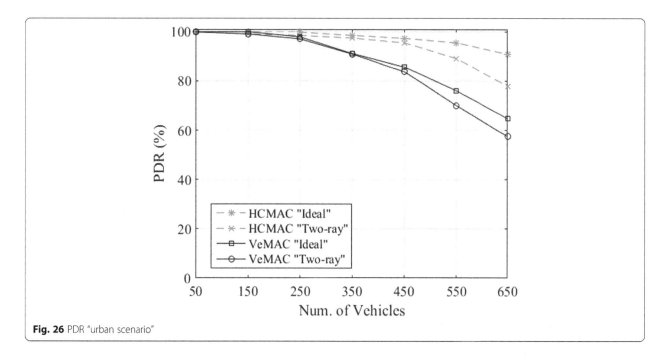

Fig. 26 PDR "urban scenario"

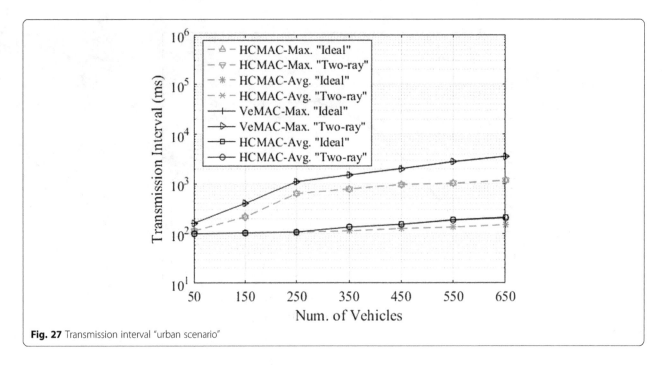

Fig. 27 Transmission interval "urban scenario"

scenarios. The simulation results demonstrated that HCMAC outperforms VeMAC in all aspect of metrics evaluated, which include collision rate, throughput, PDR, and transmission intervals.

In the future, we plan to develop a dynamic scheduling process that can determine the optimal slot number, and wireless range along with the network density changes utilizing the proposed analytic model.

8 Endnotes
[1]For a data rate of 12 Mbps, DSRC standard defines an SNR threshold of 15 dB [28].

[2]Videos of our simulator have been recorded and uploaded to [29, 30].

Abbreviations
AIS: Automatic Identification System; BSM: Basic Safety Message; CCH: Control Channel; CDMA: Code Division Multiple Access; CS: Carrier Sense; CSMA: Carrier Sense Multiple Access; CTS: Clear To Send; CW: Contention Window; DOT: U.S. Department of Transportation; DS: Detected Slots; DSRC: Dedicated Short-Range Communications; EDCA: Enhanced Distributed Channel Access; FCC: Federal Communication Commission; FCW: Forward Collision Warning; GPS: Global Position System; HCMAC: Hybrid Cooperative MAC; IMA: Intersection Management Assist; ITS: Intelligent Transportation System; LCW: Lane Change Warning; LTA: Left Turn Assist; NHTSA: National Highway Traffic Safety Administration; OBU: Onboard Unit; PDR: Packet Delivery Ratio; PN: Pseudo Noise; RSSI: Received Signal Strength Indicator; RSU: Road Side Unit; RTS: Request To Send; SAE: Society of Automotive Engineers; SCH: Service Channel; TDMA: Time Division Multiple Access; V2I: Vehicle-to-infrastructure; V2V: Vehicle-to-vehicle; VANET: Vehicular ad hoc network

Acknowledgements
This research was supported by the Center for Integrated Smart Sensors funded by the Ministry of Science, ICT& Future Planning as Global Frontier Project (CISS-2018), and it was also supported by Institute for Information & communications Technology Promotion (IITP) grant funded by the Korea government (MSIT) with grant number [R7117-16-0164], Development of wide area driving environment awareness and cooperative driving technology which are based on V2X wireless communication.

Funding
Not applicable.

Authors' contributions
MAAEG is the primary author of this work. MAAEG and ME contributed to the conception of the study and performed the system analysis. MAAEG developed the simulator environment and conducted all the simulations. MAAEG wrote the manuscript. HWK provided great comments to enhance the paper quality. All authors reviewed and approved the final manuscript.

Competing interests
The authors declare that they have no competing interests.

Author details
[1]Department of Electronics Engineering, Chungbuk National University, Cheongju, South Korea. [2]National Telecommunication Institute, Cairo, Egypt. [3]Electrical Engineering Department, Al-Azhar University, Cairo, Egypt.

References
1. The World Health Organization, "The World Health report 2015 – reducing risks, promoting healthy life." Available online: http://www.who.int/whr/ 2002/chapter4/en/index7.html. Accessed 30 June 2017.
2. U.S. Department of Transportation, Bureau of Transportation Statistics, Available online: https://www.rita.dot.gov/bts/sites/rita.dot.gov.bts/ files/ publications/national_transportation_statistics. Accessed 10 Nov 2017.
3. The CAMP Vehicle Safety Communications Consortium, "vehicle safety communications project task 3 final report," CAMP, Farmington Hills, MI, USA, Tech. Rep. DOT HS 809 859, Mar. 2005.
4. IEEE Standard for Information technology-- local and metropolitan area networks-- specific requirements-- part 11: wireless LAN Medium Access Control (MAC) and physical layer (PHY) specifications amendment 6: wireless access in vehicular environments, IEEE, pp.1–51, July 2010.
5. IEEE Standard for Wireless Access in Vehicular Environments (WAVE) -- networking services, in IEEE Std 1609.3-2016 (Revision of IEEE Std 1609.3-2010), IEEE, pp.1–160, April 2016.
6. IEEE Standard for Wireless Access in Vehicular Environments (WAVE) -- multi-channel operation, in IEEE Std 1609.4-2016 (Revision of IEEE Std 1609. 4-2010), IEEE, pp.1–94, March 2016.

7. Dedicated Short Range Communications (DSRC) Message Set Dictionary, Available online: http://standards.sae.org/j2735_201603/. Accessed 20 Oct 2017.

8. S. Eichler, Performance Evaluation of the IEEE 802.11p WAVE Communication Standard, Proc. IEEE 66th Vehicular Technology Conf. (VTC '07-Fall), pp. 2199–2203, 2007.

9. F. Borgonovo, A. Capone, M. Cesana and L. Fratta, RR-ALOHA, a reliable R-ALOHA broadcast channel for ad-hoc inter-vehicle communication networks, In *Proceedings of Med-Hoc-Net*, 2002.

10. F. Borgonovo, A. Capone, M. Cesana, L. Fratta, ADHOC MAC: new MAC architecture for ad hoc networks providing efficient and reliable point-to-point and broadcast services. Wirel. Netw **10**, 359–366 (July 2004).

11. H.A. Omar, W. Zhuang, L. Li, VeMAC: a TDMA-based MAC protocol for reliable broadcast in VANETs. IEEE Trans. Mob. Comput. **12**(9), 1724–1736 (2013).

12. L. Gallo, and H. Jérôme, Analytical study of self organizing TDMA for V2X communications. In Communication Workshop (ICCW), IEEE International Conference on, pp. 2406–2411, 2015.

13. K. Bilstrup, E. Uhlemann, E.G. Ström, U. Bilstrup, On the ability of the 802.11p MAC method and STDMA to support real-time vehicle-to-vehicle communication. EURASIP J. Wireless Commun. Netw. **2009**, 1–13 (2009).

14. Technical characteristics for universal shipborne automatic identification system using time division multiple access in the VHF maritime mobile band, Std. Recommendations ITU-R M.1371-1, 2006.

15. S. Bharati, W. Zhuang, CAH-MAC: cooperative adhoc MAC for vehicular networks. IEEE J. Sel. Areas Commun. **31**(9), 470–479 (2013).

16. J. Huang, Q. Li, S. Zhong, L. Liu, P. Zhong, J. Wang, J. Ye, Synthesizing existing CSMA and TDMA based MAC protocols for VANETs. Sensors **17**, 338–355 (2017).

17. M. Hadded, P. Muhlethaler, A. Laouiti, R. Zagrouba, L.A. Saidane, TDMA-based MAC protocols for vehicular ad hoc networks: a survey, qualitative analysis, and open research issues. IEEE Commun Surv Tutorials **17**(4), 2461–2492 (2015).

18. X. Fan, C. Wang, J. Yu, K. Xing, Y. Chen, J. Liang, A reliable broadcast protocol in vehicular ad hoc networks. Int J Distrib Sens Netw **2015**(8), 1–14 (2015).

19. J.J. Blum, A. Eskandarian, A reliable link-layer protocol for robust and scalable intervehicle communications. IEEE Trans. Intell Transp Syst **8**(1), 4–13 (2007).

20. F. Watanabe, M. Fujii, M. Itami, K. Itoh, An analysis of incident information transmission performance using MCS/CDMA scheme. Proc. IEEE Intelligent Vehicles Symp. (IV '05), 249–254 (2005).

21. H. Nakata, T. Inoue, M. Itami, and K. Itoh, A study of inter vehicle communication scheme allocating PN codes to the location on the road, Proc. IEEE Intelligent Transportation Systems Conf. (ITSC '03), vol. 2, pp. 1527–1532, 2003.

22. J. B. Kenney, Dedicated short-range communications (DSRC) standards in the United States, Proceedings of the IEEE, vol. 99, no. 7, pp. 1162–1182, 2011.

23. K.F. Hasan, Y. Feng, Y. Tian, GNSS time synchronization in vehicular ad-hoc networks: benefits and feasibility. IEEE Trans. Intell. Transp. Syst.. https://doi.org/10.1109/TITS.2017.2789291.

24. M.A. Abd El-Gawad, and H. W. Kim, Reliable broadcast protocol based on scheduled acknowledgments for wireless sensor networks, The Institute of Electronics and Information Engineers (IEIE) Conference, pp. 485–489, 2017.

25. M. Elsharief, M.A. Abd El-Gawad, H.W. Kim, Density table based synchronization for multi-hop wireless sensor networks. IEEE Access **6**, 1940–1953 (2018).

26. C. Campolo, A. Vinel, A. Molinaro, Y. Koucheryavy, Modeling broadcasting in IEEE 802.11p/WAVE vehicular networks. IEEE Commun Lett **15**(2), 199–201 (2011).

27. H.A. Omar, W. Zhuang, L. Li, VeMAC: a novel multichannel MAC protocol for vehicular ad hoc networks. Proc. IEEE INFOCOM, 413–418 (2011).

28. H. Zhou et al., Toward multi-radio vehicular data piping for dynamic DSRC/TVWS spectrum sharing. IEEE J Sel Areas Commun **34**, 2575–2588 (2016).

29. [Online]. Available: https://youtu.be/jT14WJegGSk. Accessed 5 Mar 2019.

30. [Online]. Available: https://youtu.be/PIegaReO-c8. Accessed 5 Mar 2019.

6

Performance analysis of an enhanced cooperative MAC protocol in mobile ad hoc networks

Jaeshin Jang[1]* (iD) and Balasubramaniam Natarajan[2]

Abstract

In this paper, we evaluate the performance of an enhanced cooperative MAC with busy tone (eBT-COMAC) protocol in mobile ad hoc networks via a combination of theoretical analysis and numerical simulation. Our previously proposed BT-COMAC protocol was enhanced by (1) redesigning the minislots used in the helper node selection procedure; (2) specifying complete frame formats for newly defined and modified control frames; and (3) using a new metric (the received SNR rather than the received power) in the helper node competition. In this eBT-COMAC protocol, cooperation probability is calculated based on a geometric analysis, and a Markov chain-based model is used to derive steady-state probabilities for backoff-related parameters. These results are used to analytically characterize two performance measures: system throughput and channel access delay. Numerical simulation of a mobile wireless network where all communication nodes are assumed to be uniformly distributed in space and move independently based on a random waypoint model is used to validate the analytical results and demonstrate the performance gains achieved by the proposed eBT-COMAC protocol.

Keywords: Cooperative communication, eBT-COMAC protocol, Helper node selection, Received SNR

1 Introduction

With the remarkable development of wireless technologies, 4G mobile communication systems can support peak data transmission rates up to 3 Gbps [1]. However, when mobile nodes are located around a cell boundary or when two mobile nodes in a mobile ad hoc network [2] are located far away from each other, severe fading occurs, resulting in a large number of transmission errors. This form of wireless channel impairment can be overcome with multiple input multiple output (MIMO) technology. However, it is not always possible to include multiple antennas in a small mobile node. Cooperative communication is an alternative approach for overcoming the effect of channel fading [3]. An example of cooperative communication is shown in Fig. 1. If any node is located at an intermediate position between a sender and a receiver node, for example, in the shaded area in Fig. 1,

this helper node can assist in the transmission process and help increase system throughput. In any cooperative communication scheme, finding the best helper node is critical. Helper node selection schemes are classified as two types: proactive and reactive schemes. In a proactive scheme, every mobile node maintains its relay table where wireless channel status with its neighboring nodes is stored [4–8, 23]. Each node shares a relay table with its neighboring nodes by periodically broadcasting some messages. Therefore, when a sender node wants to send a data packet to its destination node, it can find its helper node based on its relay table. In reactive schemes, the sender node begins the search for a helper node after the exchange of control frames [9, 10, 13–15, 20]. Although it takes time to select an optimal helper node, this reactive scheme guarantees that the newly selected helper node has a more conducive wireless channel than that in a proactive scheme. Initial studies in the area of cooperative medium access control (MAC) protocols focused on proactive schemes. However, reactive helper node selection schemes have gained popularity because (1) proactive schemes impose a greater load on both the network and

*Correspondence: icjoseph@inje.ac.kr
[1]Department of Electronic Telecommunications Mechanical and Automotive Engineering, Inje University, 197 Inje-ro, Gimhae, Gyeongnam 50834, South Korea
Full list of author information is available at the end of the article

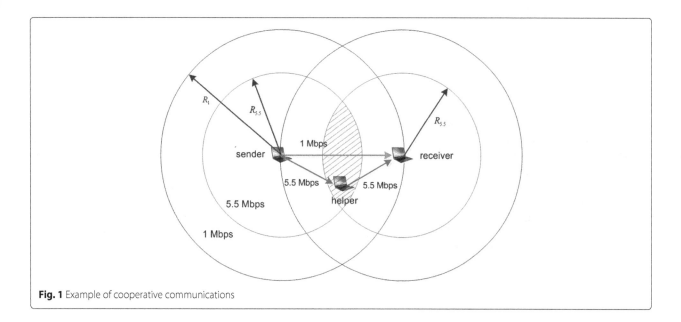

Fig. 1 Example of cooperative communications

processors within a node; and (2) there is no guarantee that the helper node chosen via a proactive scheme is optimal at data transmission time. In this work, we are interested in enhancing system performance with a new reactive helper node selection process in wireless local area networks (WLANs).

1.1 Related work

Most studies on cooperative MAC protocols follow the IEEE 802.11 WLAN design principle [11] and thus, only IEEE 802.11-based cooperative MAC protocols with link adaptation [12] are surveyed in this paper. There are three typical studies on reactive helper node selection schemes. In [13], three busy signals are used to find an optimal helper node, which is not energy efficient. A three-step helper node selection scheme was adopted in two previous studies [14, 15] consisting of GI (group indication), MI (member indication), and K minislot contention. The optimal cooperation region and system parameters were determined in [14] while an additional energy metric was used to select the best helper node in order to increase network lifetime in [15]. However, all three of these schemes use data transmission rates-related metrics for their helper node selection procedures, which has its drawbacks, as will be discussed in Section 1.2

There have also been several recent studies on cooperative MAC protocol design [16–19]. In [16], three transmisson modes are suggested where relay nodes were chosen based on proactive mechanisms: direct transmission, cooperative relay transmission, and two-hop relay transmission. Cooperative relay transmission mode is used for increasing system throughput while the two-hop relay transmission mode helps extend the service range. However, there is no suggested algorithm for choosing

an appropriate mode. In [17], a new cooperative MAC protocol based on a three-way handshake with request to send (RTS), clear to send (CTS), and relay ready to send (RRTS) is proposed. Its reactive relay node selection scheme is based on the fact that the fastest relay candidate will reply to an RRTS frame earlier. However, [17] does not consider the possibility of relay node competition and approaches to deal with collisions. In [18], a helper node initiated cooperative MAC protocol is proposed. Helper nodes are decided in advance with the help of a relay table, and they initiate cooperative communication by sending a helper clear to send (HCTS) frame when the transmission rate between sender and receiver nodes falls below a threshold. In [19], three data transmission modes similar to those suggested in [16] are discussed. In contrast to [16], an algorithm to find a suitable transmission mode is suggested in [19]. It is described that the optimal helper node is chosen via the shortest path algorithm. However, there is no detailed discussion on how to select the optimal helper node. Therefore, issues such as helper node competition and whether the shortest path can be decided without additional control frame exchanges remain unanswered. In this paper, we aim to address these issues via the design and analysis of a new cooperative MAC protocol.

1.2 Contributions

In this paper, a new cooperative MAC protocol, enhanced cooperative MAC with busy tone signal (eBT-COMAC) protocol is proposed and a mathematical analysis and simulation are carried out on it. This protocol is an enhanced version of our previously proposed MAC protocol [20]. The eBT-COMAC protocol includes a reactive helper node selection scheme with a three-step helper node selection scheme. The key difference between our proposed helper node selection scheme and prior work [14, 15] is that we use a received signal-to-noise ratio (SNR) value in the minislot contentions rather than transmission rates, which were used in two previous studies. In general, the received SNR is closely related to transmission rates. However, because the number of transmission rates is limited (i.e., in IEEE 802.11b, there are four data transmission rates: 1, 2, 5.5, and 11 Mbps), the previous schemes may have the problem that candidate helper nodes with the same transmission rates can experience continuous collisions in minislot contentions. The main contributions of our study include the following:

- The use of a new reactive helper node selection scheme with received SNR as the selection metric;
- Clear design of the packet formats for the required control frames for eBT-COMAC protocol in order to support the helper node selection scheme;
- Presentation and validation (via computer simulation) of a comprehensive mathematical analysis of the throughput and delay associated with eBT-COMAC;
- The provision of increased system throughput performance with the eBT-COMAC protocol that is 58% higher than IEEE 802.11 WLAN [11] and 6% higher than prior work [14];
- Easy extension of the entire approach to current standards, although IEEE 802.11b WLAN is the standard considered in this work.

This paper consists of five sections. A detailed explanation of the eBT-COMAC protocol is presented in Section 2; the system model and performance analysis are discussed in Section 3. The numerical results from the analysis and simulation are described in Section 4, and Section 5 presents the conclusions.

2　eBT-COMAC protocol

The frame exchange procedure for the eBT-COMAC protocol for cooperative communications is given in Fig. 2. Any sender node that has data to send begins its transmission by sending a cooperative request-to-send (CRTS) frame. When the receiver successfully receives the CRTS frame, it replies with a cooperative clear-to-send (CCTS) frame. After receiving the CRTS and CCTS frames, all mobile nodes located between the sender and receiver nodes can calculate two transmission rates, R_{SH}, R_{HR}, based on the received SNR. The direct transmission rate R_{SR} can be obtained from the physical layer convergence procedure (PLCP) header of the CCTS frame. Any candidate helper node whose two-hop effective rate (R_{e2}) is greater than the one-hop effective rate (R_{e1}) sends a short busy signal to notify all surrounding nodes that there is at least one eligible candidate helper node and thus, the

helper node selection procedure will start. The helper node selection procedure consists of three steps: harsh contention (HC), exact contention (EC), and random contention (RC). Each contention consists of several minislots or slots. The size of each HC and EC minislot is the same as slot size (σ), as shown in Table 4, and the size of RC slots is the same as the request-to-help (RTH) frame transmission time at the basic rate. If any optimal node is decided from the helper node selection scheme, this node plays the role of the helper node, at which time two-hop communication begins. The effective transmission rate represents the ratio of DATA length in bits to the required time period in seconds from the end of the busy signal to the successful reception of the acknowledgement (ACK) frame. One- and two-hop effective transmission rates are calculated as follows [20]:

$$R_{e1,2} = \frac{L_d}{T_O + T_D}, \quad 1:S\text{--}R, \ 2:S\text{--}H\text{--}R \qquad (1)$$

$$T_D = \begin{cases} \frac{L_d}{R_{SR}}, & S\text{--}R \\ \frac{L_d}{R_{SH}} + \frac{L_d}{R_{HR}}, & S\text{--}H\text{--}R \end{cases}$$

$$T_O = \begin{cases} SIFS + T_{ACK}, & S\text{--}R \\ (N_{HC} + N_{EC})\sigma + N_{RC}T_{RTH} \\ \quad + 3T_{CTH} + 3SIFS + T_{ACK}, & S\text{--}H\text{--}R \end{cases}$$

Here, L_d is the DATA length in bits; $N_{HC(EC)}$ is the number of HC (EC) minislots; N_{RC} is the number of RC slots; $T_{ACK,RTH,CTH}$ are the transmission times of control frames ACK, RTH, and clear-to-help (CTH), respectively, and $R_{SH(HR)}$ corresponds to the DATA frame transmission rates between a sender and a helper (a helper and a receiver) node; $SIFS$ is a MAC parameter representing short interframe space.

Detailed control frames used in the eBT-COMAC protocol are described in Fig. 3. The eBT-COMAC protocol is designed based on the IEEE 802.11 WLAN standard. Two control frames, the RTH and CTH frames, are newly suggested and the CRTS frame has a new field "PKT_LEN," which stands for data packet length in bytes. The CTH frame has two different formats, namely, long CTH and short CTH. The long CTH is a full-sized frame with three optional fields, helper node address (HA), and two possible transmission rates between sender and helper nodes (R_{SH}) and helper and receiver nodes (R_{HR}). This long CTH is used when helper node selection competition is successful. On the other hand, the short CTH does not have three optional fields and it is used when the helper node selection competition fails. That is, long CTH is a positive response but short CTH is a negative response for RTH transmissions in HC, EC, and RC contention.

2.1　Helper Node Selection

The proposed eBT-COMAC protocol uses a reactive helper node selection scheme. Thus, the helper node

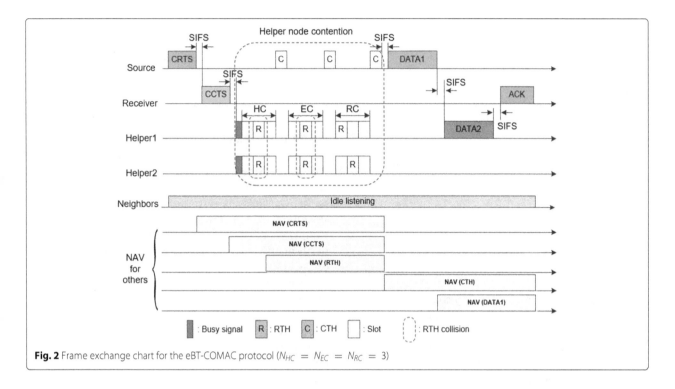

Fig. 2 Frame exchange chart for the eBT-COMAC protocol ($N_{HC} = N_{EC} = N_{RC} = 3$)

selection procedure starts after the sender and the receiver nodes exchange CRTS and CCTS frames. The helper node selection scheme in eBT-COMAC consists of three steps. The goal of HC and EC minislot contention is to find the optimal helper node, and the RC slot contention is to select one helper node on a probabilistic basis. The metric used in this contention is a utility U, corresponding to the received SNR in the dB scale, i.e., $U \equiv \log SNR_{rcvd}$. HC and EC consist of N_{HC} and N_{EC} minislots. The contention is carried out with the help of an RTH

frame transmission in the appropriate minislot. Earlier, HC and EC minislots are assigned for the candidate helper nodes with greater utility values. In HC and EC minislot contention, if any candidate helper node observes that another node has transmitted an RTH frame earlier than itself, it exits the competition. The utility window between U_{max} and U_{min} is uniformly divided, and the mapping rule between the utility values and HC and EC minislot numbers can be explained by examining Fig. 4 when $N_{HC} = N_{EC} = 3$. Here, $U_i = U_{|rmmax} - iU_{inc}, i = 1, 2,$

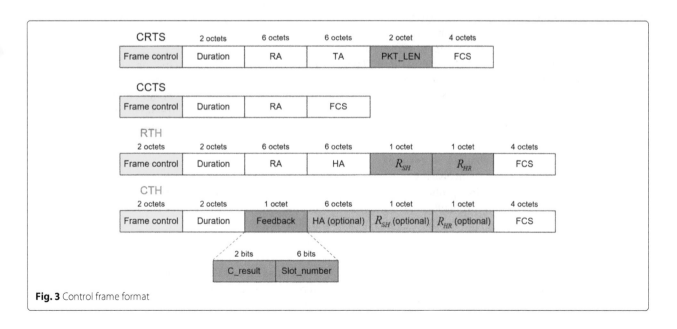

Fig. 3 Control frame format

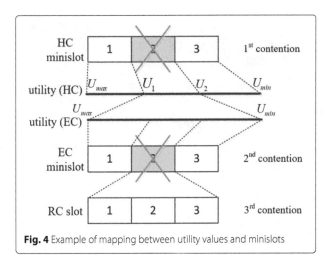

Fig. 4 Example of mapping between utility values and minislots

where $U_{inc} = \frac{U_{max}-U_{min}}{3}$. If there is a collision in minislot 2 at HC minislot contention, those candidates involved in the collision begin their contention again at the EC minislots. In this case, U_1 and U_2 in the HC contention become U_{max} and U_{min} in the EC minislot contention. If there is a continuous collision in the EC minislot contention, those candidates involved in the collision move into the RC slot contention. The RC slot contention is based on random selection. Those candidate nodes that are involved in the RTH frame collision in the EC minislot contention generate a random number between 1 and N_{RC}. Then, they transmit their RTH frame in the assigned slot. If there is more than one successful slot in this contention, the candidate that sent its RTH in the earlier slot has priority and then this candidate is chosen as the final helper node. The sender node decides the winner of the competition. If a helper node wins the competition, the sender node transmits a long CTH frame. Otherwise, the sender node transmits a short CTH frame. The "feedback" field in the CTH frame contains the competition result (C_result). "11" in "C_result" means that the competition was successful and "00" means failure in the competition. The flow chart for the operation at a candidate helper node is shown in Fig. 5.

3   Performance evaluation

Our goal is to analyze the eBT-COMAC protocol and quantify its throughput and channel access delay. The procedure to achieve this goal involves several intermediate results. First, cooperation probability and successful helper node selection probability are derived. Then, the steady-state probability for the three system state variables related to the backoff operation are evaluated. Finally, based on the calculation of average time slot size, the system throughput and channel access delay are derived. We begin by highlighting the assumptions underlying this process. First, nodes are assumed

to be uniformly distributed within the communication area. Second, to calculate the success probability in the helper node selection competition, we use an approximate approach, the classical definition of probability. Actually, it is almost impossible to derive an exact equation for the success probability because of the dynamic characteristics of helper node selection competition. Therefore, Eq. (3) has a characteristic that is sensitive to the number of minislots and the number of helper nodes, which will be described in Section 4. Next, it is assumed that all frames, including the DATA frame are susceptible to packet transmission error, which is a more realistic consideration than in previous studies [5–15, 21, 22]. For completeness, several system variables required for the performance analysis of the IEEE 802.11b CSMA/CA and eBT-COMAC protocols are defined in Table 1.

We begin the analysis of the proposed protocol with the derivation of the cooperation probability. Let us consider an example in Fig. 1 where the sender and receiver nodes are far apart and thus can communicate with each other only at a rate of 1 Mbps. In this case, a helper node, located in the shaded area, can help increase the system throughput for communication between the sender and the receiver nodes.

Lemma 1 *The cooperation probability p_h corresponds to*

$$p_h = \left\{ p_1 \frac{S_1(r_2, r_{5.5}, r_1)}{\pi r_1^2} + p_2 \frac{S_1(r_{5.5}, r_{5.5}, r_2)}{\pi r_2^2} + p_{5.5} \frac{S_1(r_{11}, r_{11}, r_{5.5})}{\pi r_{5.5}^2} \right\} p_r. \quad (2)$$

where, $r_i, p_i,$ and p_r are defined in Table 1, and $S_1(\cdot)$ represents the size of overlapping area in Fig. 1.

Proof The minimum participation criteria for cooperative communication is given in Table 2 when the relation between data transmission rates and ranges for IEEE 802.11b has those values in Table 3 [20]. The cooperation probability p_h can be approximately expressed as the weighted sum of various ratios of the overlapped area to the transmission area of the sender node when the direct transmission rate is 1, 2, and 5.5 Mbps, respectively. For example, when the direct transmission rate is 1 Mbps with the probability p_1, πr_1^2 is the transmission area of the sender node and $S_1(r_2, r_{5.5}, r_1)$ represents the overlapped area when the direct transmission rate between the sender and receiver nodes is 1 Mbps, the sender and the helper nodes transmit the DATA frame at 2 Mbps, and the helper and receiver nodes transmit the DATA frame at 5.5 Mbps. Please see Appendix 1 for the exact derivation of $S_1()$. □

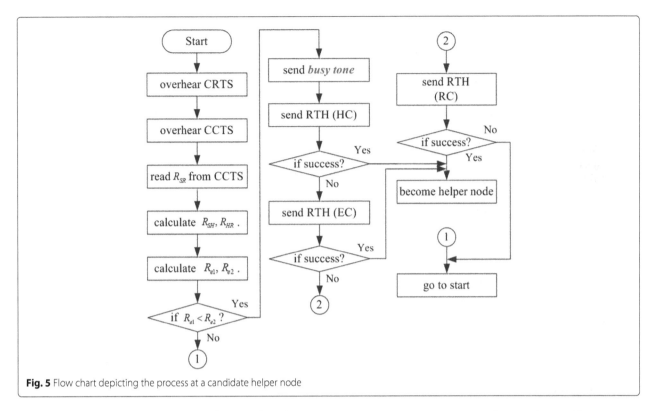

Fig. 5 Flow chart depicting the process at a candidate helper node

As described in Section 2, the helper node selection scheme consists of three steps: HC, EC, and RC competitions. The probability of successful helper node selection in each step is provided in Lemma 2.

Lemma 2 *The probability that the optimal helper node is selected successfully from three-step competitions corresponds to*

Table 1 Definition of system parameters

r	The maximum value of the backoff stage
m	The maximum value of the contention window size
$N_s(N_h)$	The number of sender (helper) nodes
τ	CRTS frame transmission probability on a wireless channel
$p_m(p_d)$	Control (DATA) frame transmission error probability due to a bad wireless channel
p_c	CRTS frame transmission failure probability due to collision
p_{sr}	Helper node selection success probability
p_{fr}	Helper node selection failure probability
R_i	Data transmission rate in Mbps for $i = 1, 2, 5.5, 11$
r_i	Maximum distance (m) for each R_i
p_i	Probability for transmitting DATA at R_i
p_h	Cooperation probability that at least one candidate helper node is in cooperation
p_r	Probability that a receiver node is located within its sender node's transmission range

$$p_{sr} = p_{s1}(1 - p_m) + \{1 - p_{s1}(1 - p_m)\}p_{s2}$$
$$\cdot (1 - p_m) + \{1 - p_{s1}(1 - p_m)\}$$
$$\cdot \{1 - p_{s2}(1 - p_m)\}p_{s3}(1 - p_m) \qquad (3)$$

Proof In the first step, let us define the possible number of candidates participating in HC the minislot contention as $M_1 = p_h N_h$. Then the probability p_{s1} that the best helper node is selected successfully in the HC minislot contention can be calculated as the ratio of the number of successful transmissions of the RTH frame to the total number of possible transmissions.

$$p_{s1} \equiv \frac{A}{(N_{HC})^{M_1}} \qquad (4)$$

$$A = \begin{cases} 0, & M_1 < 1 \\ N_{HC}, & M_1 = 1 \\ \sum_{i=1}^{N_{HC}-1} (N_{HC} - i)^{M_1-1}, & M_1 > 1 \end{cases}$$

In the second step, let us define the possible number of candidates participating in the EC minislot contention as M_2. Although an HC minislot is assigned to a candidate

Table 2 Minimum participation criteria for cooperative communication

Direct transmission	Minimum criteria for R_{SH}, R_{HR}
1 Mbps	One over 2 and the other over 5.5 Mbps
2 Mbps	All over 5.5 Mbps
5.5 Mbps	All over 11 Mbps

Table 3 Transmission rates and ranges

Data rate(R_i)	11	5.5	2	1
Distance(r_i)	≤ 48.2	≤ 67.1	≤ 74.7	≤ 100
Probability(p_i)	0.23	0.22	0.11	0.44

helper node based on its utility value, let us assume that the location of the HC minislot where the helper node competition is successful is uniformly distributed between 1 and N_{HC}. Then, it is easy to see that $M_2 = M_1/N_{HC}$. The probability p_{s2} that the best helper node is selected successfully in the EC minislot contention can be calculated as the ratio of the number of successful transmissions of the RTH frame to the total number of possible transmissions.

$$p_{s2} \equiv \frac{B}{(N_{EC})^{M_2}} \qquad (5)$$

$$B = \begin{cases} 0, & M_2 < 1 \\ N_{EC}, & M_2 = 1 \\ \sum_{i=1}^{N_{EC}-1}(N_{EC} - i)^{M_2 - 1}, & M_2 > 1 \end{cases}$$

In the third step, the RC slot contention, let us define the possible number of candidates participating in the RC slot contention as M_3, where $M_3 = M_2/N_{EC}$. Then, the probability p_{s3} that the best helper node is selected successfully in the RC slot contention can be calculated as the ratio of the number of successful transmissions of the RTH frame to the total number of possible transmissions.

$$p_{s3} \equiv \frac{C}{(N_{RC})^{M_3}} \qquad (6)$$

$$C = \begin{cases} 0, & M_3 < 1 \\ N_{RC}, & M_3 = 1 \\ N_{RC}(N_{RC} - 1)^{M_3 - 1}, & M_3 > 1 \end{cases}$$

Finally, the probability that the optimal helper node is selected successfully is the weighted sum of the successful selection of helper nodes at each step, which is provided in Eq. (3). $\qquad\square$

RTH frame transmission failure occurs when an optimal helper node is not decided from the three-step competitions. This failure probability p_{fr} corresponds to

$$p_{fr} = 1 - p_{sr}. \qquad (7)$$

The frame transmission procedure in the eBT-COMAC protocol, including the backoff operation for each station, is modeled as a Markov chain with the system state vector:

- $b(t)$: backoff stage of the sender node,
  $b(t) = 0, 1, \cdots, r$

- $c(t)$: value of the backoff counter,
  $c(t) = 0, 1, \cdots, W_{b(t)} - 1$
- $o(t)$: frame transmission phase, $o(t) = 0, 1, \cdots, 7$.

Here, the variable $o(t)$ represents the sending phase for each frame, which is shown in Fig. 7: 0 represents the sending phase of a CRTS frame; 1 refers to a CCTS frame, $2, 3, 4$, and 5 are for RTH, CTH, DATA1, and DATA2, respectively; 6 represents an ACK frame; and 7 is for DATA frame at direct transmission. We attempt to derive steady-state probabilities for this system state vector. Our mathematical analysis approach is carried out based on previous research in [21–23]. It is assumed that every sender node always has data frames to transmit in its buffer, which is known as a saturated traffic model.

The eBT-COMAC protocol uses the same retransmission scheme as IEEE 802.11b and thus, the contention window size at each retransmission is determined by the following rule:

$$W_i = \begin{cases} 2^i \cdot CW_{min}, & 0 \leq i \leq m \\ 2^m \cdot CW_{min}, & m < i \leq r. \end{cases} \qquad (8)$$

Transmission failure for the CRTS frame could occur due to collisions with other frames or a bad wireless channel. Therefore, the CRTS frame transmission failure probability is given by

$$p_f = p_c + p_m - p_c p_m. \qquad (9)$$

Let us define the steady-state probability as $\alpha_{ijk} \equiv \lim_{t \to \infty} prob.\{b(t) = i, c(t) = j, o(t) = k\}$. State transition rate diagrams for eBT-COMAC are shown in Figs. 6, 7, and 8. Figure 6 shows the total state transition rate diagram, and the detailed descriptions of $S_i, 1 \leq i \leq r$ located in the left side of Fig. 6 are shown in Figs. 7 and 8.

The balance equations for the eBT-COMAC protocol are given by

$$\alpha_{000} = \sum_{i=0}^{r-1}(1 - p_m)\alpha_{i06} + A(r) + \alpha_{r06} + p_f\alpha_{r00} \qquad (10)$$

$$\alpha_{i00} = A(i-1) + p_m\alpha_{i-106} + p_f\alpha_{i-100}, \qquad 1 \leq i \leq r \qquad (11)$$

$$\alpha_{ij0} = \frac{(W_i - j)}{W_i}\alpha_{i00}, 0 \leq i \leq r, 1 \leq j \leq W_i - 1. \qquad (12)$$

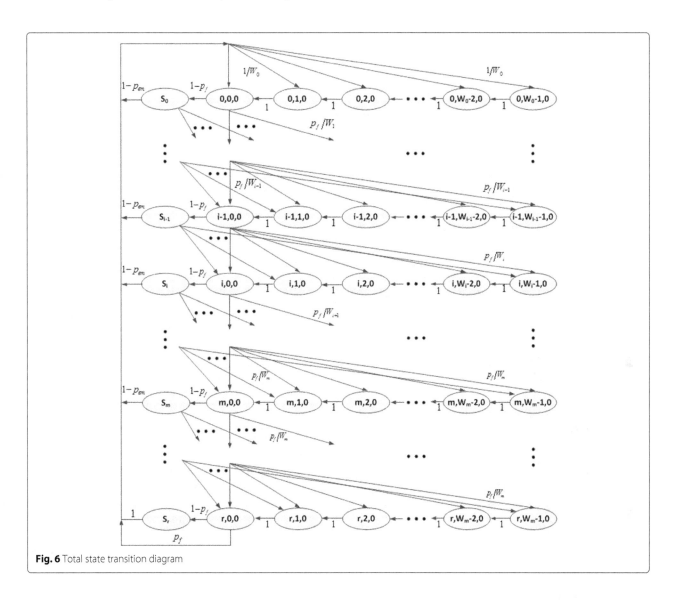

Fig. 6 Total state transition diagram

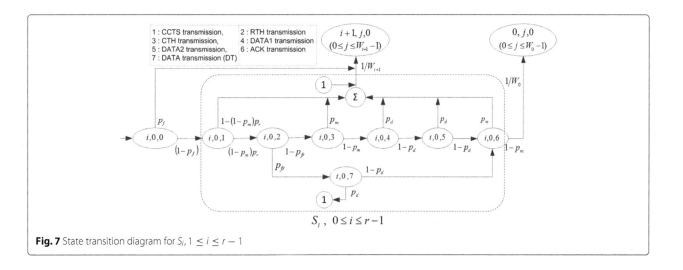

Fig. 7 State transition diagram for S_i, $1 \le i \le r-1$

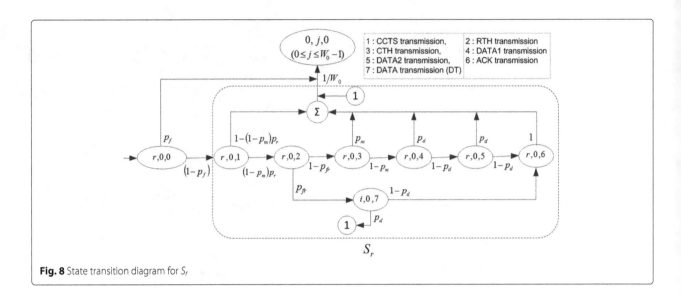

Fig. 8 State transition diagram for S_r

$$\alpha_{i0k} = \begin{cases} (1-p_f)\alpha_{i00}, & k=1 \\ (1-p_f)(1-p_m)p_r\alpha_{i00}, & k=2 \\ (1-p_f)(1-p_m)p_r(1-p_{fr})\alpha_{i00}, & k=3 \\ (1-p_f)(1-p_m)^2p_r(1-p_{fr})\alpha_{i00}, & k=4 \\ (1-p_f)(1-p_m)^2p_r(1-p_{fr}) \\ \quad \cdot (1-p_d)\alpha_{i00}, & k=5 \\ (1-p_f)(1-p_m)p_r(1-p_d)\{p_{fr} \\ \quad + (1-p_m)\,(1-p_{fr})(1-p_d)\}\alpha_{i00}, & k=6 \\ (1-p_f)(1-p_m)p_rp_{fr}\,\alpha_{i00}, & k=7. \end{cases}$$
(13)

Here, $A(i) \equiv (1 - p_r + p_m p_r)\alpha_{i01} + p_m\alpha_{i03} + p_d(\alpha_{i04} + \alpha_{i05} + \alpha_{i07})$ and $p_r = \pi r_1^2/A_c$ where A_c refers to the size of the communication area.

By the iterative method, steady-state probabilities can be derived with the additional condition that the sum of all probabilities is one. That is,

$$\sum_{i=0}^{r}\sum_{j=0}^{W_i-1}\sum_{k=0}^{7}\alpha_{ijk} \cdot \text{I}(j,k) = 1$$
(14)

where $\text{I}(j,k) = \begin{cases} 0, & j \neq 0 \;\&\; k \neq 0 \\ 1, & \text{otherwise.} \end{cases}$

In the calculation of the steady-state probabilities, the CRTS frame transmission probability τ and the transmission failure probability for a CRTS frame due to collision p_c should be expressed as

$$\tau = \sum_{i=0}^{r}\alpha_{i00} = \frac{1-D^{r+1}}{1-D}\cdot\alpha_{000}$$
(15)

$$p_c = 1-(1-\tau)^{N_s-1}.$$
(16)

Two performance measures were considered for the performance analysis. One is the system throughput in bps, and the other performance measure is the average channel access delay in seconds. In order to derive the system throughput, two types of average delays should be calculated in advance. The first one is D_S, the average time delay from the transmission of the CCTS frame to the successful reception of the ACK frame, and the second one is D_E, the average time delay from the transmission of the CCTS frame to transmission failure with the CCTS, RTH, CTS, DATA, or ACK frame.

Lemma 3 *The average time delays from the transmission of the CCTS frame to the successful reception of the ACK frame for direct and cooperative transmission, respectively, correspond to*

$$D_{S1} = D_S^1 + D_E^2 + D_S^7 + D_S^6$$
(17)
$$D_{S2} = D_S^1 + D_S^2 + D_S^3 + D_S^4 + D_S^5 + D_S^6,$$
(18)

where, D_S^k is the average time delay spent at phase k for successful frame transmission. D_E^2 is the average time delay spent at phase 2 because of frame transmission error; it will be derived in the Lemma 4.

Proof The average time delays for successful frame transmission at each phase, D_S^k, $k = 1, 2, \cdots, 7$, should be derived first. According to the sequence of frame exchanges, these delays are given by

$$D_S^1 = T_{CCTS} + SIFS$$

$$D_S^2 = T_{BT} + T_{HC} \cdot \frac{p_{s1}}{P_{ts}} + \left(T_{HC} + T_{CTH_S} + T_{EC}\right)$$

$$\cdot \frac{(1 - p_{s1})p_{s2}}{P_{ts}} + (T_{HC} + T_{CTH_S} + T_{EC}$$

$$+ T_{CTH_S} + N_{RC}T_{RTH}) \cdot \frac{(1 - p_{s1})(1 - p_{s2})p_{s3}}{P_{ts}}$$

$$D_S^3 = T_{CTH_L} + SIFS$$

$$D_S^4 + D_S^5 = T_{DATA_c} + 2 \cdot SIFS$$

$$D_S^6 = T_{ACK}$$

$$D_S^7 = T_{DATA_d} + SIFS.$$

Here, $P_{ts} = p_{s1} + (1 - p_{s1})p_{s2} + (1 - p_{s1})(1 - p_{s2})p_{s3}$. T_{HC} and T_{EC} are the average sizes of the HC and EC minislot contention periods. It is assumed that the location of the HC or EC minislot where the helper node competition is successful is uniformly distributed between 1 and N_{HC} or N_{EC}. Then, they can be expressed as $T_{HC} = \frac{N_{HC}}{2}\sigma + T_{RTH}$ and $T_{EC} = \frac{N_{EC}}{2}\sigma + T_{RTH}$, respectively. T_{BT} refers to the time period of the busy tone signal. In the above equation, $T_{DATA_{c(d)}}$ represents the data transmission period in the cooperative (direct) transmission at each data transmission rate. These values can be derived as follows:

$$T_{DATA_c} = \frac{p_1}{\pi r_1^2}\left[S_1(r_{11}, r_{11}, r_1)\left(\frac{2L_d}{R_{11}}\right)\right.$$

$$+ \{S_1(r_{5.5}, r_{11}, r_1) - S_1(r_{11}, r_{11}, r_1)\}$$

$$\cdot \left(\frac{L_d}{R_{5.5}} + \frac{L_d}{R_{11}}\right) + \{S_1(r_{5.5}, r_{5.5}, r_1)$$

$$- S_1(r_{5.5}, r_{11}, r_1)\}\left(\frac{2L_d}{R_{5.5}}\right)$$

$$+ \{S_1(r_2, r_{5.5}, r_1) - S_1(r_{5.5}, r_{5.5}, r_1)\}$$

$$\cdot \left.\left(\frac{L_d}{R_2} + \frac{L_d}{R_{5.5}}\right)\right]$$

$$+ \frac{p_2}{\pi r_2^2}\left[S_1(r_{11}, r_{11}, r_2)\left(\frac{2L_d}{R_{11}}\right)\right.$$

$$+ \{S_1(r_{5.5}, r_{11}, r_2) - S_1(r_{11}, r_{11}, r_2)\}$$

$$\cdot \left(\frac{L_d}{R_{5.5}} + \frac{L_d}{R_{11}}\right) + \{S_1(r_{5.5}, r_{5.5}, r_2)$$

$$- S_1(r_{5.5}, r_{11}, r_2)\}\left.\left(\frac{2L_d}{R_{5.5}}\right)\right]$$

$$+ \frac{p_{5.5}}{\pi r_{5.5}^2}S_1(r_{11}, r_{11}, r_{5.5})\left(\frac{2L_d}{R_{11}}\right), \quad (19)$$

$$T_{DATA_d} = \frac{L_d}{R_1}p_1\left(1 - \frac{S_1(r_2, r_{5.5}, r_1)}{\pi r_1^2}\right)$$

$$+ \frac{L_d}{R_2}p_2\left(1 - \frac{S_1(r_{5.5}, r_{5.5}, r_2)}{\pi r_2^2}\right)$$

$$+ \frac{L_d}{R_{5.5}}p_{5.5}\left(1 - \frac{S_1(r_{11}, r_{11}, r_{5.5})}{\pi r_{5.5}^2}\right)$$

$$+ \frac{L_d}{R_{11}}p_{11}. \quad (20)$$

Then, the average time delays from the CCTS transmission to successful reception of the ACK frame for direct transmission and two-hop transmission are as given in Eqs. (17) and (18). □

Lemma 4 *The average time delay from the CCTS frame to any frame transmission failure corresponds to*

$$D_E = D_E^1\frac{p_m}{P_{te}} + \left(D_S^1 + D_S^2 + D_E^3\right)\frac{(1 - p_m)}{P_{te}}$$

$$\cdot (1 - p_{fr})p_m$$

$$+ \left(D_S^1 + D_S^2 + D_S^3 + D_E^4\right)\frac{(1 - p_m)^2}{P_{te}}$$

$$\cdot (1 - p_{fr})p_d$$

$$+ \left(D_S^1 + D_S^2 + D_S^3 + D_S^4 + D_E^5\right) \cdot (1 - p_m)^2$$

$$\cdot \frac{(1 - p_{fr})(1 - p_d)p_d}{P_{te}}$$

$$+ \left(D_S^1 + D_S^2 + D_S^3 + D_S^4 + D_S^5 + D_E^6\right)$$

$$\cdot (1 - p_m)^2\frac{(1 - p_{fr})(1 - p_d)^2p_m}{P_{te}}$$

$$+ \left(D_S^1 + D_E^2 + D_E^7\right)\frac{(1 - p_m)p_{fr}p_d}{P_{te}}$$

$$+ \left(D_S^1 + D_E^2 + D_S^7 + D_E^6\right)\frac{(1 - p_m)p_{fr}}{P_{te}}$$

$$\cdot (1 - p_d)p_m, \quad (21)$$

where, the probability P_{te} *is defined for its normalization condition and it corresponds to*

$$P_{te} = p_m + (1 - p_m)\left[p_{fr}\{p_d + (1 - p_d)p_m\}\right.$$

$$+ (1 - p_{fr})p_m] + (1 - p_m)^2\left[(1 - p_{fr})\right.$$

$$\cdot \left.\{p_d + (1 - p_d)p_d + (1 - p_d)^2p_m\}\right.]$$

Here, D_E^k, $k = 1, 2, \cdots, 7$ are the average time delays spent at phase k because of frame transmission failure.

Proof The average time delays D_E^k, $k = 1, 2, \cdots, 7$, are given by

$$D_E^1 = T_{CCTS} + SIFS$$

$$D_E^2 = T_{BT} + T_{HC} + T_{EC} + N_{RC}T_{RTH} + 2T_{CTH_S}$$

$$D_E^3 = T_{CTH_L} + SIFS$$

$$D_E^4 = T_{DATA_c} + SIFS = D_E^5$$

$$D_E^6 = T_{ACK}$$

$$D_E^7 = T_{DATA_d} + SIFS.$$

For example, the phase $k = 2$ represents the transmission of RTH frames and it means helper node selection competition. Thus, D_E^2 refers to the required time delay for complete failure of the helper node selection procedure. This delay consists of a busy tone signal, HC and EC contention periods, RC slots, and two short CTH frame transmissions in the HC and EC contentions, respectively. Then, the average time delay from the CCTS frame to complete transmission failure can be expressed as a weighted sum of consumed time delays until complete transmission failure in each phase. If complete failure occurs at the phase $k = 4$, then, frame transmissions at phases 1, 2, and 3 should be successful. Thus, the time delay from the CCTS frame to DATA frame transmission failure is $D_S^1 + D_S^2 + D_S^3 + D_E^4$. Therefore, the average time delay from the CCTS frame to any frame transmission failure corresponds to Eq. (21).

$\square$

The following two theorems provide the expression for evaluating two performance measures of interest: system throughput and channel access delay.

Theorem 1 (system throughput) *The system throughput is defined as the length of successfully transmitted data in bits during a unit of time, and corresponds to*

$$TH = \frac{P_{tr}P_s(P_{a1} + P_{a2})(L_d + L_h)}{E[S]} \; [bps] \quad (22)$$

where, L_h is the sum of the MAC header and PLCP header, $E[S]$ is the average slot time, P_{tr} is the probability that there is at least one CRTS frame transmission by N_s mobile users in the considered time duration, and P_s is the probability that the transmitted CRTS frame is successfully received by the helper node without collision and transmission error. P_{a1} and P_{a2} are the probabilities that no transmission errors occur during the period from the CCTS frame to ACK frame transmission for direct and two-hop transmissions, respectively.

Proof Since CRTS frame transmission by each sender node is modeled as a Beroulli distribution with the probability τ, two probabilities P_{tr} and P_s are derived as

$$P_{tr} = 1 - (1 - \tau)^{N_s} \quad (23)$$

$$P_s = \frac{N_s \tau (1 - \tau)^{N_s - 1}}{P_{tr}}(1 - p_m). \quad (24)$$

The probabilities P_{a1} and P_{a2} correspond to

$$P_{a1} = (1 - p_m)^2 p_r p_{fr}(1 - p_d) \quad (25)$$

$$P_{a2} = (1 - p_m)^3 p_r (1 - p_{fr})(1 - p_d)^2. \quad (26)$$

Let us define S as slot time, representing the time interval between two consecutive idle slots. There are four different types of slot times. First, when there is no transmission on the channel, the slot time means the slot duration $T_I = \sigma$. Second, when the transmission of the CRTS frame results in failure due to collision or a bad wireless channel, the slot time becomes T_F. Third, when the source node does not receive the ACK frame, even after successful transmission of the CRTS frame, the slot time becomes T_E. Finally, if the total transmission scenario is successful, then this slot time is T_{S1} for direct transmission or T_{S2} for a two-hop transmission. These slot time types are indicated as

$$T_I = \sigma$$

$$T_F = T_{CRTS} + DIFS + \sigma$$

$$T_E = T_{CRTS} + SIFS + D_E + DIFS + \sigma$$

$$T_{S1} = T_{CRTS} + SIFS + D_{S1} + DIFS + \sigma$$

$$T_{S2} = T_{CRTS} + SIFS + D_{S2} + DIFS + \sigma.$$

Then, the average value of S is the weighted sum of the four different slot times and is given by

$$E[S] = (1 - P_{tr})T_I + P_s P_{tr}(P_{a1}T_{S1} + P_{a2}T_{S2})$$
$$+ P_s P_{tr}(1 - P_a)T_E + P_{tr}(1 - P_s)T_F. \quad (27)$$

The probability that the given DATA frame is transmitted successfully is the product of three probabilities derived in Eqs. (23)–(26): P_{tr}, P_s, and $P_{a1} + P_{a2}$. Because the system throughput can be expressed as the ratio of the total length of the DATA frame successfully transmitted in bits to the average time slot, it corresponds to Eq. (22). $\square$

Theorem 2 (channel access delay) *The average channel access delay, which is defined as the time period from the beginning of the backoff to the successful reception of the ACK frame, can be expressed as*

$$E[D] = \begin{cases} \sum_{i=0}^{r} \left\{ E[S] \frac{W}{2} \sum_{j=0}^{i} 2^j + T_A \right\} P_S(i), \\ \qquad\qquad\qquad\qquad\qquad\qquad r \leq m, \\ \sum_{i=0}^{m} \left\{ E[S] \frac{W}{2} \sum_{j=0}^{i} 2^j + T_A \right\} P_S(i) \\ + \sum_{i=m+1}^{r} \left\{ E[S] \frac{W}{2} \left(\sum_{j=0}^{m} 2^j \right. \right. \\ \left. \left. + \sum_{j=m+1}^{i} 2^m \right) + T_A \right\} \cdot P_S(i), \quad r > m. \end{cases}$$

$$(28)$$

Here, W represents the minimum contention window size, CW_{min} and $T_A = T_{CRTS} + SIFS + \{D_{S1}p_{fr} + D_{S2}(1 - p_{fr})\}$, and $P_S(i)$ represents the probability that one sender node receives the ACK frame successfully at the i-th backoff stage.

Proof The probability $P_S(i)$ can be approximately modeled as a geometric distribution with the parameter p_{st}. Here, the probability p_{st} represents the successful transmissions from the CRTS frame to the ACK frame and is given by

$$p_{st} = (1 - p_f)(P_{a1} + P_{a2}).$$ $$(29)$$

Then, the probability $P_S(i)$ is given by

$$P_S(i) = (1 - p_{st})^i p_{st}, \quad 0 \leq i \leq r.$$ $$(30)$$

It is assumed that when the current system state of the sender node is $\{b(t) = i, c(t) = 0, o(t) = k\}$ as in Fig. 6, the next system state is determined based on uniform distribution. That is, the next possible system state will be $(0, j, 0)$ or $(i + 1, j, 0)$ depending on whether the current frame transmission is successful or not. Here, the value j is uniformly distributed between 0 and $W_0 - 1$ or $W_{i+1} - 1$. Because the sender node stays in any state until the upcoming idle slot, the average time to stay in any state in Fig. 6 equals the average slot size $E[S]$. Thus, the channel access delay corresponds to the weighted sum of the products of $E[S]$, and the number of states where the sender node stayed from the beginning of channel sensing to any state of $(i, 0, 6), 0 \leq i \leq r$ when the ACK transmission is successful. Therefore, the expression of the channel access delay is given as Eq. (28).

$\square$

4 Network model and numerical results

4.1 Network model and environment

We consider the operation of the eBT-COMAC protocol over the IEEE 802.11b WLAN specification [11]. First, it is important to note that the proposed eBT-COMAC protocol can also be easily integrated into other current wireless networking standards. We assume that all mobile nodes are working in WLAN ad hoc mode. Therefore,

any mobile node can directly communicate with the other mobile nodes. It is assumed that all mobile nodes are uniformly distributed within a $200m \times 200m$ square communication area and mobile nodes move independently within this communication area based on a random way-point mobility model: (1) every mobile node determines its next location based on uniform distribution and then moves to its next location at the speed of v [m/sec], which is determined from the uniform distribution between 0 and V_{max}; (2) as soon as it arrives at its next location, it takes a pause during a period, which is also determined from the uniform distribution between 0 and T_{pause}; (3) after its pause, it returns to step (1) and repeats the entire process. All mobile nodes are classified into one of three types: sender, receiver, or helper nodes. It is assumed that there are N_s communication connection pairs between the sender and receiver nodes, and that these communication connections are fixed during the entire simulation period. The log-distance path loss model is used for modeling wireless channels, and the relationship between transmission distance and path loss is represented by Eq. (31) [24]:

$$L_p(d)(dB) = L_s(d_0)(dB) + 10n \log_{10}(d/d_0),$$ $$(31)$$

where d_0 is the reference distance and n is the path loss exponent (in this paper, we use $n = 3$).

The simulation code was programmed with a gnu C++ compiler using the SMPL library [25]. Computer simulation was conducted 10 times with a different seed each time, and we used the averaged data as simulation results. For simplicity, it was also assumed that transmission error probabilities for the control and DATA frames caused by a bad wireless channel were the same ($p_m = p_d$). Mathematical analysis and computer simulations were conducted and compared in order to prove the correctness of the mathematical equations. The system parameters used in the performance evaluation are described in Table 4.

Table 4 System parameters

Parameter	Value	Parameter	Value
CRTS length	176 b	SIFS	$10 \mu s$
CCTS length	112 b	DIFS	$50 \mu s$
RTH length	176 b	CWmin	32 slots
DATA length	1024 B	CWmax	1024 slots
MAC header	272 b	PLCP header	192 bits
CTH size (L/S)	136/72 b	Basic rate	1 Mbps
Slot size (σ)	$20 \mu s$	$p_m(p_d)$	Variable
N_{HC}, N_{EC}, N_{RC}	3	m	5
V_{max}	30 m/s	T_{pause}	5 s
Sim. time	1500 s	r	6

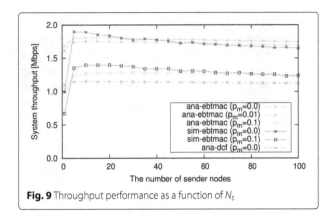

Fig. 9 Throughput performance as a function of N_s

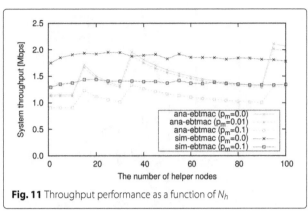

Fig. 11 Throughput performance as a function of N_h

4.2 Numerical results

First, we will explain several abbreviations used in the legends of Figs. 9, 10, 11 and 12 for notifying each numerical result. "ana-ebtmac" and "sim-ebtmac" represent the analysis and simulation results for the eBT-COMAC protocol, respectively. Figure 9 shows the comparison of throughput performances by analysis and simulation results for the eBT-COMAC protocol and analysis results for the IEEE 802.11b DCF without cooperation. These numerical results were obtained when there were 40 helper nodes in the communication area. It is shown that the eBT-COMAC provides enhanced system performance about 58% higher than IEEE 802.11 WLAN (notified as "ana-dcf" in this figure). This coincides with our expectation that cooperative communication has explicit benefits over non-cooperative communication. Simulation results were consistent with the analytical results in all ranges. It is also shown that IEEE 802.11-related MAC protocols provide the best system performance when there are about five sender nodes in the communication area.

Figure 10 shows a comparison of delay performances by analysis and simulation when there are forty helper nodes in the communication area. This figure shows that the eBT-COMAC protocol has an obvious advantage over direct communication. The analytical and simulation results are consistent; although there is a discrepancy

between the simulation and analysis of about 22% when $N_s = 100$, the difference is negligible.

Figure 11 shows a comparison of throughput performances as a function of the number of helper nodes when $N_s = 10$. According to the approximation that we used in deriving the probability p_{sr} in Eq. (3), analytical results are sensitive to the number of helper nodes. According to Eqs. (4)–(6), when $M_1 = 1$, $M_2 = 1$, and $M_3 = 1$, helper node competition becomes completely successful with $p_{sr} = 1$. In addition, when M_1, M_2, and M_3 is less than one, the probability that the best helper node is successfully decided becomes zero in Eqs. (4)–(6). Significant jumps in this figure occur when N_h is about 15, 35, and 95. These values of N_h corresponds to cases when M_1, M_2, and M_3 are slightly greater than one, respectively, and this is why there are significant jumps in this figure.

Figure 12 shows a comparison of delay performances as a function of the number of helper nodes when $N_s = 10$. The discrepancy between the simulation and analysis results may be due to our approximation when deriving Eqs. (2) and (3). However, Fig. 12 shows that the simulation results show a slight increase, although it is a little, in the section where the analysis results show a consistent increase.

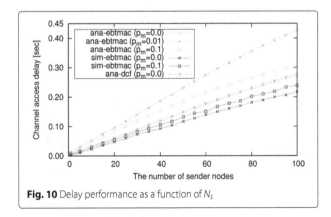

Fig. 10 Delay performance as a function of N_s

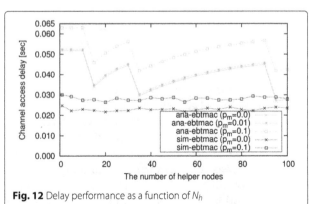

Fig. 12 Delay performance as a function of N_h

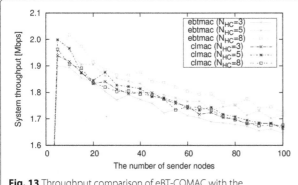

Fig. 13 Throughput comparison of eBT-COMAC with the reference [14]

Figure 13 shows a throughput comparison of the eBT-COMAC protocol with the reference [14], denoted as "clmac". When $N_{HC} = N_{EC} = 3$, it seems that the eBT-COMAC protocol provides slightly lower throughput than the reference [14]. This means that the optimal candidate helper node was not properly selected in the HC and EC minislot contentions of the eBT-COMAC protocol with this number of minislots. However, when N_{HC} and N_{EC} are greater than 5, the eBT-COMAC protocol provides approximately 6% higher system throughput performance than the reference [14] when $N_{HC} = 8$. A greater number of minislots contributes to choosing one optimal helper node because helper node candidates with different received SNR values can be classified more clearly. However, the system throughput performance results for reference [14] are almost the same when N_{HC} and N_{EC} are 3 and 5, respectively. This is because the cooperative MAC protocol in [14] uses transmission rates as a metric in the helper node selection. There are only four different transmission rates in IEEE 802.11b and thus, more minislots are useless for this MAC scheme.

5 Conclusions

In this paper, we presented for the first time, a comprehensive theoretical performance analysis of an enhanced BT-COMAC protocol and validated the analytical results via numerical simulations. The new helper node selection scheme in the eBT-COMAC protocol is based on received SNR values at each candidate node. This results in a dynamic characteristic that presents challenges in analytical modeling. In this paper, two probabilities, the cooperation probability and the probability that a helper node is successfully selected, were derived based on a geometric analysis. These probabilities, along with steady-state probabilities of backof-related parameters (derived based on a Markov analysis), are used to derive theoretical expressions for the system throughput and channel access delay of the eBT-COMAC protocol. Although the analytical results are not exact and are based on approximations that provide theoretical tractability, they are for the most part consistent with the numerical simulations. Future work will involve the design of an energy-aware eBT-COMAC protocol that can provide throughput gains while improving network lifetime.

Appendix 1: calculation of cooperative area

Let us consider two circles, the radii of which are a and b, respectively, with an overlapped area, as shown in Fig. 14. The distance between the origins of the two circles is d. Using the theory of the segment of a circle and its radius, the overlapped area S can be derived as follows:

$$S = \frac{1}{2} \left\{ a^2(\eta - sin(\eta)) + b^2(\phi - sin(\phi)) \right\}. \qquad (32)$$

According to a cosine formula, the following relation can be simply derived.

$$b^2 = a^2 + d^2 - 2ad \cdot cos\left(\frac{\eta}{2}\right). \qquad (33)$$

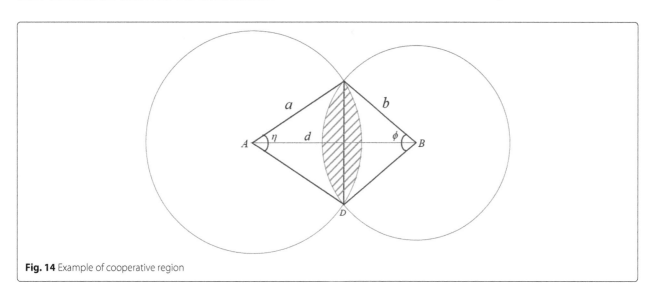

Fig. 14 Example of cooperative region

Then, combining Eq. (33) with Eq. (32), the overlapped area can be derived as follows:

$$
\begin{aligned}
S(a, b, d) = \frac{1}{2} \Bigg[& a^2 \left\{ 2cos^{-1} \left(\frac{a^2 + d^2 - b^2}{2ad} \right) \right. \\
& \left. - sin \left(2cos^{-1} \left(\frac{a^2 + d^2 - d^2}{2ad} \right) \right) \right\} \\
& + b^2 \left\{ 2cos^{-1} \left(\frac{b^2 + d^2 - a^2}{2bd} \right) \right. \\
& \left. - sin \left(2cos^{-1} \left(\frac{b^2 + d^2 - a^2}{2bd} \right) \right) \right\} \Bigg].
\end{aligned} \quad (34)
$$

As shown in Table 3, the transmission range of 1 Mbps, for example, is $74.7 \leq r_1 \leq 100\ m$. Then, the equation of $S_1(r_2, r_{5.5}, r_1)$ used in Eqs. (2), (19), (20) is calculated as the average of two transmission areas by the transmission rate R_1 and the one level higher rate R_2.

$$
S_1(r_2, r_{5.5}, r_1) = \frac{S(r_2, r_{5.5}, r_1) + S(r_2, r_{5.5}, r_2)}{2}. \quad (35)
$$

Abbreviations
ACK: Acknowledgement; BT-COMAC: Cooperative MAC with busy tone; CCTS: Cooperative clear to send; CRTS: Cooperative request to send; CTH: Clear to help; CTS: Clear to send; eBT-COMAC: Enhanced BT-COMAC; EC: Exact contention; GI: Group indication; HA: Helper node address; HC: Harsh contention; HCTS: Helper clear to send; MAC: Medium access control; MI: Member indication; MIMO: Multiple input multiple output; PLCP: Physical layer convergence procedure; RC: Random contention; RRTS: Relay ready to send; RTH: Request to help; RTS: Request to send; SIFS: Short interframe space; SNR: Signal-to-noise ratio; WLAN: Wireless local area network

Acknowledgements
The authors would like to thank the reviewers for their thorough reviews and helpful suggestions.

Funding
This research work was funded by a grant from Inje University for the research in 2016 (20160432).

Authors' contributions
JJ proposed the system model, derived the mathematical equations, and performed the simulation and manuscript writing. BN contributed in manuscript revision and correction. Both authors read and approved the final manuscript.

Author's information
Not applicable.

Competing interests
The authors declare that they have no competing interests.

Author details
[1]Department of Electronic Telecommunications Mechanical and Automotive Engineering, Inje University, 197 Inje-ro, Gimhae, Gyeongnam 50834, South Korea. [2]Department of Electrical and Computer Engineering, Kansas State University, 1701D Platt St., Manhattan, KS 66506, USA.

References
1. 3GPP LTE-Advanced specification release 10 (2009). http://www.3gpp.org/technologies/keywords-acronyms/97-lte-advanced. Accessed June 2013
2. IETF MANET Working Group. https://datatracker.ietf.org/wg/manet/about/. Accessed Nov 2016
3. A Nosratinia, TE Hunter, A Hedayat, Cooperative communication in wireless networks. IEEE Commun. Mag. 42(10), 74–89 (2004)
4. N Sai Shankar, C Chun-Ting, G Monisha, in IEEE Int. Conf. on Wireless Networks. Cooperative communication MAC (CMAC)- A new MAC protocol for next generation wireless LANs, (Hawaii, 2005), pp. 1-6
5. H Zhu, G Cao, rDCF: a relay-enabled medium access control protocol for wireless ad hoc networks. IEEE Trans. Mob. Comput. 5(9), 1201-1214 (2006)
6. P Liu, Z Tao, S Narayanan, T Korakis, SS Panwar, CoopMAC: a cooperative MAC for wireless LANs. IEEE J. Sel. Areas Commun. 25(2), 340–353 (2007)
7. K Tan, Z Wan, H Zhu, J Andrian, in 4th IEEE Conf. on Sensor, Mesh, and Ad hoc Comm. and Networks. CODE: cooperative medium access for multi-rate wireless ad hoc network, (San Diego, 2007), pp. 1-10
8. H Jin, X Wang, H Yu, Y Xu, Y Guan, X Gao, in IEEE WCNC-2009. C-MAC: a MAC protocol supporting cooperation in wireless LANs, (Budapest, 2009), pp. 1-6
9. H Shan, W Zhuang, Z Wang, Distributed cooperative MAC for multihop wireless networks. IEEE Commun. Mag. 47(2), 126–133 (2009)
10. H Shan, W Zhuang, Z Wang, in IEEE ICC-2009. Cooperation or not in mobile ad hoc networks: A MAC perspective, (Dresden, 2009), pp. 1–6
11. IEEE Standards, in Part 11: Wireless LAN medium access control (MAC) and physical layer (PHY) specifications. IEEE Std 802.11-2012, (2012)
12. G Holland, N Vaidya, P Bahl, in ACM/IEEE MOBICOM-2001. A rate-adaptive MAC protocol for multi-hop wireless networks, (Rome, 2001), pp. 236–251
13. H Shan, P Wang, W Zhuang, Z Wang, in IEEE GLOBECOM-2008. Cross-layer cooperative triple busy tone multiple access for wireless networks, (New Orleans, 2008), pp. 1–5
14. H Shan, HT Cheng, W Zhuang, Cross-layer cooperative MAC protocol in distributed wireless networks. IEEE Trans. Wirel. Commun. 10(8), 2603–2615 (2011)
15. Z Yong, L Ju, Z Lina, Z Chao, C Chen, Link-utility-based cooperative MAC protocol for wireless multi-hop networks. IEEE Trans. Wirel. Commun. 10(3), 995–1005 (2011)
16. T Zhou, H Sharif, M Hempel, P Mahasukhon, W Wang, T Ma, A novel adaptive distributed cooperative relaying MAC protocol for vehicular networks. IEEE J. Sel. Areas Commun. 29(1), 72–82 (2011)
17. J Sheu, J Chang, C Ma, C Leong, in IEEE WCNC. A cooperative MAC protocol based on 802.11 in wireless ad hoc networks, (Shanghai, 2013), pp. 416–421
18. MdR Amin, SS Moni, SA Shawkat, MdS Alam, in Int'l Conf. Computer and Information Technology. A helper initiated distributed cooperative MAC protocol for wireless networks, (Khulna, 2014), pp. 302–308
19. AFMS Shah, MdS Alam, SA Showkat, in IEEE TENCON. A new cooperative MAC protocol for the distributed wireless networks, (Macao, 2015), pp. 1–6
20. J Jang, Performance evaluation of a new cooperative MAC protocol with a helper node selection scheme in ad hoc networks. J. Inf. Commun. Converg. Eng. 12(4), 199–207 (2014)
21. G Bianchi, Performance analysis of the IEEE 802.11 distributed coordination function. IEEE J. Selec. Areas Commun. 19(3), 535–547 (2000)
22. JW Tantra, CH Foh, AB Mnaouer, in IEEE ICC-2005. Throughput and delay analysis of the IEEE 802.11e EDCA saturation, (Seoul, 2005), pp. 3450–3454
23. J Jang, SW Kim, S Wie, Throughput and delay analysis of a reliable cooperative MAC protocol in ad hoc networks. J. Commun. Netw. 5(3), 524–532 (2012)
24. TS Rappaport, Wireless communications: principles and practice. (Prentice-Hall, 2002)
25. MH MacDougall, Simulating computer systems: techniques and tools. (The MIT Press, 1992)

7

Reinforcement learning-based dynamic band and channel selection in cognitive radio ad-hoc networks

Sung-Jeen Jang[1], Chul-Hee Han[2], Kwang-Eog Lee[3] and Sang-Jo Yoo[1]* ⓘ

Abstract

In cognitive radio (CR) ad-hoc network, the characteristics of the frequency resources that vary with the time and geographical location need to be considered in order to efficiently use them. Environmental statistics, such as an available transmission opportunity and data rate for each channel, and the system requirements, specifically the desired data rate, can also change with the time and location. In multi-band operation, the primary time activity characteristics and the usable frequency bandwidth are different for each band. In this paper, we propose a Q-learning-based dynamic optimal band and channel selection by considering the surrounding wireless environments and system demands in order to maximize the available transmission time and capacity at the given time and geographic area. Through experiments, we can confirm that the system dynamically chooses a band and channel suitable for the required data rate and operates properly according to the desired system performance.

Keywords: Reinforcement learning, Cognitive radio, Ad-hoc network, Q-learning, Fairness

1 Introduction

As the demand for multimedia services increases, the problem of the frequency shortage continues to increase. The spectrum auction price is rising worldwide and passing on to users as a burden [1]. The Federal Communications Commission (FCC) had found that most of the spectrums are underutilized under its current fixed spectrum allocation [2]. The FCC had therefore proposed a new paradigm which provides an access to the spectrum resources not being used by the licensed user to resolve the increasing demand for the spectral access and inefficiency in use [3]. The cognitive radio (CR) technologies provide an opportunity for secondary users (SUs) to use spectrums that are not used by primary users (PUs), allowing the SUs to access the spectrum by adjusting their operational parameters [4, 5]. In relation to the application of CR, FCC adopted rules in April 2012 in [6] to allow license-exempt devices employing the TV white space database approach to access available channels in the UHF television bands. [7] presents the existing, emerging,

and potential applications employing CRS capabilities and the related enabling technologies, including the impacts of CR technology on the use of spectrum from a technical perspective. The U.S. Defense Advanced Research Projects Agency (DARPA) and British defense firm BAE Systems are developing a CR IC technology for next-gen communications [8]. DARPA is developing CR technologies that maintain communications under severe jamming environment by Russian electronic warfare systems from 2011 [9]. In 2016, DARPA launched the Spectrum Collaboration Challenge (SC2) to resolve the scarcity of spectrum for DoD use and a Vanderbilt team won the round 1 [10].

The CR technology enables SUs to use free spectrum holes in radio environments that vary with a time and location. When the spectrum is used by a SU, quality of service (QoS) for both the PU and SU should be maintained by ensuring the spectrum accessibility for the SU without interfering with the service for the PU through the spectrum sensing. The SU should periodically sense the channel while using the channel and switch to another channel when the PU starts accessing the current channel. In this case, when selecting a channel, it is necessary to consider the fact that the frequency resource varies depending on the time and geographical area.

* Correspondence: sjyoo@inha.ac.kr
[1]Department of Information and Communication Engineering, Inha University, 253 YongHyun-dong, Nam-gu, Incheon, South Korea
Full list of author information is available at the end of the article

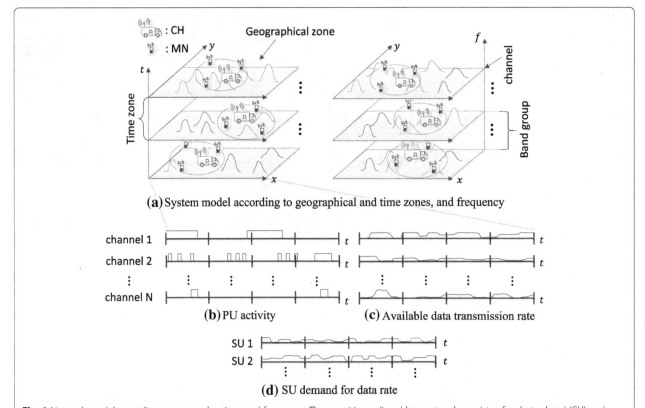

(a) System model according to geographical and time zones, and frequency

(b) PU activity   **(c)** Available data transmission rate

(d) SU demand for data rate

Fig. 1 Network model according to geography, time, and frequency. The cognitive radio ad-hoc network consists of a cluster head (CH) and mobile nodes (MNs) as shown in **a**. According to the time zone, band group, and each channel, the frequency resource is different. The PU activity, available data transmission rate, and SU demand for data rate vary according to the time and channel

Also, the CR system should consider the available data rate and possible channel acquisition time that can be achieved on each channel to guarantees the QoS of the SUs. Generally, depending on operating frequency bands such as HF (high frequency), VHF (very high frequency), and UHF (ultra-high frequency), a channel may have different channel bandwidths and the channel characteristic is different. Primary systems that are operating on different frequency bands also have diverse features and characteristics in terms of medium access mechanism, service types, and power requirements. Therefore, for choosing the best channel among the available frequency bands when the secondary CR network needs to move to another channel, several dynamic aspects such as primary system operation characteristics, radio channel conditions, frequency band characteristics, and secondary system requirements should be considered. We have to utilize a dynamic spectrum selection mechanism by considering the related environment and operational parameters to maximize the system performance. The channel access pattern of the PU, the requested data rate of the SU, and the available data rate and spectrum acquisition time can all vary dramatically according to environments. Therefore, the learning algorithm is required to dynamically solve these complex optimization problems.

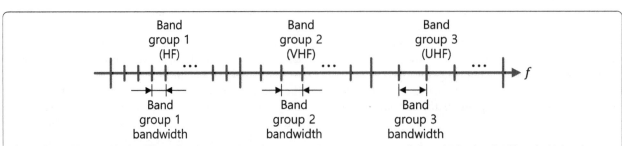

Fig. 2 Channel bandwidths for different band groups. A wireless communication system generally has a higher bandwidth and a higher data transmission rate as it goes to a higher band

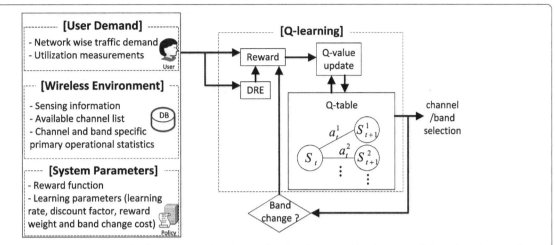

Fig. 3 Proposed system architecture. Proposed Q-learning is used to dynamically select the optimal band group and channel. As the reward function, the system considers the user demand, wireless environment and system parameters. The user demand module determines the desired data rate (DDR) of the CR ad-hoc network and measures the average utilization of the channel currently used. The wireless environment module stores the spectrum sensing results. The system parameters module is used to establish the reward function and Q-learning parameters. If the band of newly selected channel is different with the old one, the overhead for band group change is adopted to the reward function

In this paper, we propose an optimal band and channel selection mechanism in the cognitive radio ad-hoc network using the reinforcement learning. In a cluster-based CR ad-hoc network, we assumed that each member node (MN) performs a wide-band spectrum sensing periodically and reports the sensing results to the cluster head (CH) node. Based on the sensing results from the member nodes and previous channel history, the CH builds wireless channel statistic data vectors in terms of achievable data rates and average primary operational activation time (idle and busy) for each available channel of each conducted band. In addition, the CH estimates the traffic demand of the current cluster network to select a set of band and channel that provides the appropriate service to the cluster. Therefore, in CR ad-hoc networks, multiple clusters can operate in a limited area so that coexistence between ad-hoc clusters should be carefully considered in the channel selection. It is desired that if an ad-hoc cluster traffic demand is low, then the CH should select the frequency band that has relatively a narrow bandwidth (i.e., low achievable data rate). It yields the frequency band with wider bandwidth to another cluster network that needs higher traffic demand. In the proposed architecture, as a reinforcement learning, we use the Q-learning algorithm and we have designed a reward function that captures the expected consecutive operational time, affordable data rate, efficiency of spectrum utilization use, and band change overhead. In particular, the reward for channel spectrum utilization is proposed to reflect the degree of efficiency about using the supportable capacity. Using the proposed Q-learning, the CH can select an optimal band and channel that can maximize the multi-objective function of the CR network, and also, it can increase the coexistence efficiency of the overall secondary systems.

The main contributions of the proposed system architecture are as follows:

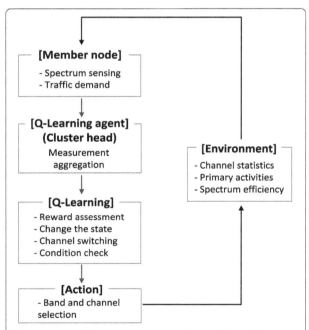

Fig. 4 Proposed Q-learning mechanism. The CH of the ad-hoc CR system is the agent of Q-learning, and the action is a selection of a tuple (band group and channel) when the PU is detected on the current band group and channel. The Q-learning agent (CH) designates the state from the information of member node and statistics of environment by the last action. From the Q-learning module, the Q-learning agent obtains the reward, change the Q-table and next action tuple

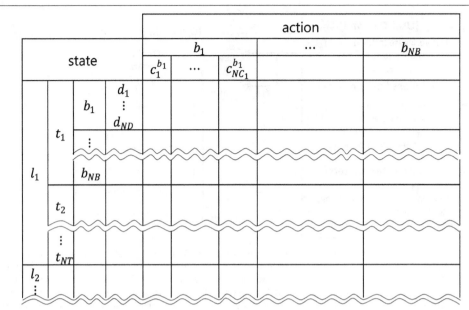

Fig. 5 Proposed Q-table structure. The column of the Q-table represents the action tuple of the band group b_q (q-th band group) and channel $c_m^{b_q}$ (m-th channel of b_q). The row of the Q-table is the state tuple of the i-th geographic location zone (l_i), j-th time zone (t_j), k-th band group (b_k), and l-th data rate efficiency level (d_l)

• We propose a new CR system architecture that maximizes the secondary user's service quality by dynamically selecting the optimal operating band and channel with consideration of the traffic demand of each CR system and the channel statistics according to the primary systems;

• We define states and actions in order to operate Q-learning considering the service state and demand of the corresponding systems, and propose a structural algorithm for it;

• We design a reward function that maximizes operating time, data rate, and channel utilization efficiency and minimizes band change overhead for secondary systems;

• The proposed system provides fairness by assigning the band and channel that are appropriate to each

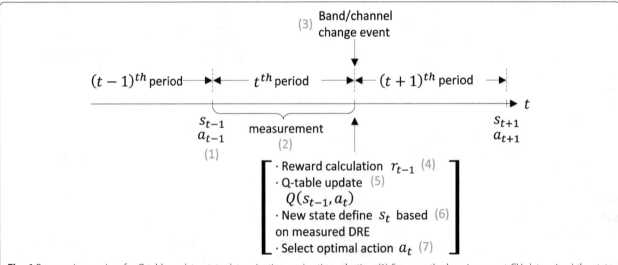

Fig. 6 Proposed procedure for Q-table update, state determination, and action selection. (1) Suppose the learning agent CH determined the state s_{t-1} and the best action a_{t-1} at the end of ($t-1$)-th time period. (2) During t-th time period, MNs and CH monitor the primary activities and channel statistics. (3) Agent CH detects the band and channel change event. (4) The CH calculates the reward r_{t-1} for the previous action a_{t-1} at state s_{t-1}. (5) The CH updates the Q-value of (s_{t-1},a_{t-1}) in Q-table. (6) The CH determines the current state s_t based on the measured DRE during t-th time period. (7) The CH selects the optimal action a_t for the next ($t+1$)-th time period. (8) Go to step 1

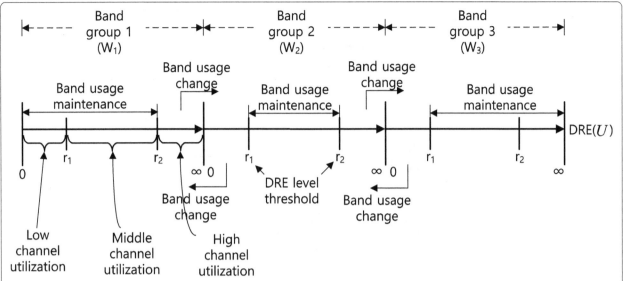

Fig. 7 Band group movement mechanism according to DRE. It is assumed the bandwidth of each channel provided by each band group follows $W_1 < W_2 < W_3$. The x-axis represents the DRE of each band group. The DRE is low, moderate, and high if it belongs to $[0, r_1)$, $[r_1, r_2)$, and $[r_2, \infty)$. If the DRE is low, the selection of the band need to be changed to lower one since the selected band supports too much bandwidth. However, the CH could not change the band in the range of $[0, r_1)$ in band group 1 since there is no band to move. If the DRE is high, the selection of the band need to be changed to a higher one since the selected band does not support the DDR. However, the CH could not change the band in the range of $[r_2, \infty)$ in band group 3 since there is no band to move. In the range of $[r_1, r_2)$ in each band group, the system does not need to change the band since the selected band supports adaptive bandwidth. Therefore, the whole DRE region is divided by the region of "Band usage maintenance" and "Band usage change"

secondary system based on its demand so that neighboring secondary systems coexist successfully.

The remainder of this paper is organized as follows. In Section 2, we describe related studies. In Section 3, we illustrate the system model and the tasks to be solved. In Section 4, we provide the proposed Q-learning algorithm to select the optimal operating band and channel.

Section 5 contains simulation results, and conclusions are given in Section 6.

2 Related works

As the CR-based ad-hoc network is often deployed in situations where resources are insufficient, it is necessary to carefully consider the frequency resource selection. In

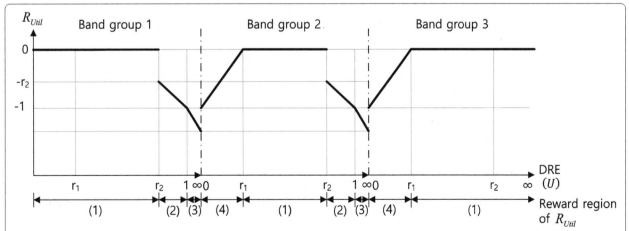

Fig. 8 Utilization efficiency reward according to DRE. To realize the mechanism in Fig. 7, the reward for channel utilization efficiency (R_{Util}) is designed as shown. The x-axis for each band group represents the DRE, and the y-axis represents R_{Util}. For the band usage maintenance range, R_{Util} is set to 0. For the band usage change range where the DRE is low, R_{Util} increases from −1 to 0 since the DRE represents better value as DRE increases. The band usage change range where the DRE is high is divided into $[r_2, 1)$ and $[1, \infty)$ to distinguish the insufficient transmission rate provided by the channel. R_{Util} represents a more rapid decrease rate in $[1, \infty)$. R_{Util} decreases from −r_2 to −1 in $[r_2, 1)$ range and from −1 to ∞ in $[1, \infty)$ range

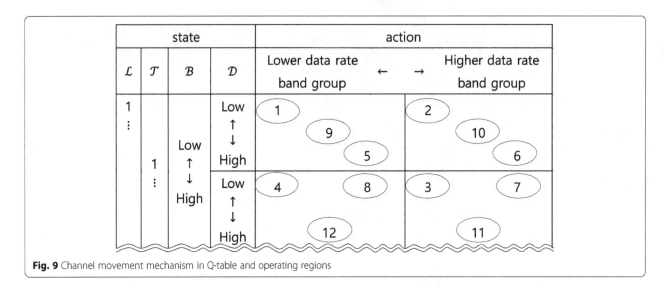

Fig. 9 Channel movement mechanism in Q-table and operating regions

this regard, studies related to the channel allocation in various fields are being conducted. Vishram et al. examined how to allocate channels using the graph coloring in the presence of homogeneous ad-hoc networks [11]. In their study, they maximized the overall performance while guaranteeing a certain grade of service to individual users with the fairness. Maghsudi and Stanczak applied the graph theory for the channel allocation in a device-to-device (D2D) environment and considered fairness by equalizing the interference for cellular users [12]. Han et al. studied channel allocation methods for maximizing the overall system performance in vehicular networks by using the submodular set function-based algorithm [13]. Li et al. investigated channel allocation methods that maximize the overall system reward using a semi-Markov decision process (SMDP) in a vehicular ad-hoc network (VANET) [14].

Other studies have considered a method of allocating channels according to either bandwidth or service characteristics. A study by Semiari et al. investigated methods of allocating a user application with dual-mode operation in the mmW and μW bands. The base station (BS) allocates a delay non-sensitive user application to the mmW while assigning a delay-sensitive user application to the μW band. Matching game theory is specifically used for channel allocation in the μW band. In non-line-of-sight (NLoS) of mmW band, the user application cannot be allocated since the wireless communication is impossible because of the frequency characteristics; therefore, channels are allocated by estimating line-of-sight (LoS) and are secured through Q-learning [15]. Liang et al. have studied a method of assigning the channel with the high transmission capacity to the vehicle-to-infrastructure (V2I) link and the channel with the high reliability to the vehicle-to-vehicle (V2V) link considering the requirements of the two types of the vehicular network links [16].

Recently, the Artificial Intelligence (AI) technology, such as machine learning, has been attracting attention in various fields [17]. Among them, the reinforcement learning is being studied in the wireless system field because it provides a solution to optimize the system parameters by learning the surrounding environment in a dynamic and complicated wireless environment [18]. The Q-learning is the representative reinforcement learning and there are also researches about using this to allocate channels in a dynamically changing environment. Asheralieva and Miyanaga studied the multi-agent reinforcement learning using rewards to maximize the signal-to-interference-plus-noise ratio (SINR) and increase the transmission capacity in D2D networks [19]. Srinivasan et al. described a way in which two BSs belonging to different operators in a cellular network can allocate channels by providing services to the nodes belonging to the all operators. They studied the reinforcement learning using the reward with the difference between quality of experience (QoE) and cost that can be obtained by providing two services [20]. Rashed et al. studied the reinforcement learning that maximizes the sum-rate of D2D users and cellular users to minimize the interference in a D2D environment [21]. Fakhfakh and Hamouda used the received SINR from the access point (AP) detected by the mobile user, QoS metrics about the channel load, and delay as the reward for choosing a WiFi over a cellular network to apply WiFi offloading and reducing the load on the cellular network [22]. Yan et al. propose a smart aggregated radio access technologies (RAT) access strategy with the aim of maximizing the long-term network throughput while meeting diverse traffic quality of service requirements by using Q-learning [23]. Maglogiannis et al. allowed the LTE system in the unlicensed band to select the appropriate muting period by using Q-learning to ensure coexistence with WiFi systems [24]. Xu et al. modeled the channel handoff

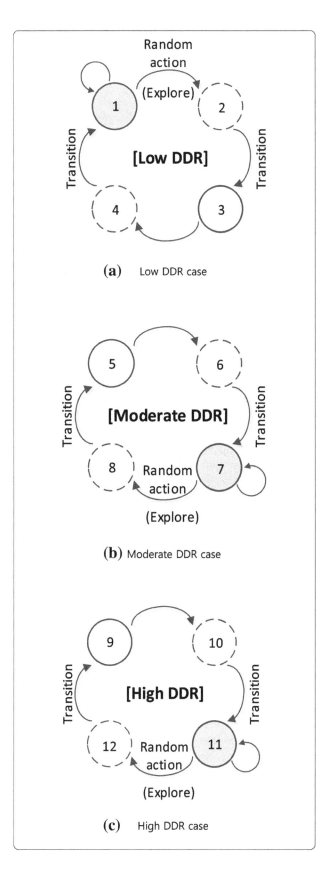

(a) Low DDR case

(b) Moderate DDR case

(c) High DDR case

Fig. 10 Channel movement example in Q-table. The update of the Q-table represents a unique pattern according to DDR by the reward for channel utilization efficiency proposed in this paper. The channel movement example of **a** low, **b** moderate, and **c** high DDR cases in Q-table is shown. The stable domain is in gray circle in each case of DDR. Each domain changes to another one by explore or natural transition

process as a partially observable Markov decision process (POMDP) and adopted a Q-learning algorithm to find an optimal handoff strategy in a long term [25]. Jang et al. proposed Q-learning based sensing parameter (sensign time and interval) control mechanism for cognitive radio networks [26]. L. Shi et al. presented optimal resource allocation for LTE-U and WiFi coexistence network using Q-learning [27].

Various studies have been carried out about selecting channels, but most studies do not consider the fairness of the channel selection between users. Even if the fairness is taken into consideration, they just allocate resources fairly regardless of the required data rate or considered it as a central manner [28]. And the central scheme is difficult to realize the realistic implementation because of the complexity, or their scheme gave the loads to the network due to the centralized control. Some distributed resource allocation mechanisms (e.g., game theory) may also cause a loss of time or resources because the channel is selected by the interaction between the systems. The fairness of the channel usage is required in order to minimize the possibility of channel resources being unnecessarily consumed by some users and unavailable for other users who require more of them. In order to reduce the load on the system, it is necessary to consider fairness within the system itself without control message exchanges. Meanwhile, the various budgets for cognitive ad-hoc networks, such as time available to the channel, transmission speed, fairness, and bandwidth conversion cost, should be considered. Moreover, these budgets must work in concert to fit an objective function with some degree of freedom about flexible operation so that the system can be operated for various purposes without altering a predetermined objective. In this paper, the reward for spectrum utilization is designed so that fairness is taken into consideration by selecting a channel suitable for the required data transmission rate. In addition, we define a reward using weighted sums for various budgets as well as a Q-learning algorithm that can operate according to the change in weights.

3 Network model and system architecture
3.1 Network model
The system considered in this paper is the cognitive radio ad-hoc network comprised of CH and MNs as shown in Fig. 1a. The channel availabilities are different with geographic locations in accordance with the primary transmitter positions, channel gain between

Table 1 Channel parameters for band groups 1 and 2

			Operation time [min]		Supportable data rate (bps)	
			Mean (T_{op})	Variance (T_{op})	Mean (D_{rate})	Variance (D_{rate})
Channel	Band group 1	1	2.1	1	10 kbps	1
		2	4.2	2	55 kbps	3
		3	8.4	1	70 kbps	2
		4	6.3	2	85 kbps	2
		5	10.5	1	100 kbps	1
	Band group 2	6	5.2	1	0.8 Mbps	2
		7	3.8	1	1.6 Mbps	1
		8	6.7	2	2.4 Mbps	3
		9	8.1	1	3.2 Mbps	1
		10	9.5	2	4 Mbps	4

primary systems and secondary users, primary activity characteristics, and so on. The characteristics of these channels for CR ad-hoc networks can be also different in time zone and frequency band groups. Therefore, in this paper, we have considered the difference of channel characteristics according to geographical zone, time zone, band group, and frequency channel. As shown in Fig. 1b and c, the primary activities (i.e., available time to access by secondary users) and possible data transmission rates are different for each channel during the same time interval. Furthermore, the desired data rates of SUs changes according to time, as shown in Fig. 1d.

In particular, when the CR system operates cross a wide frequency range, including HF, VHF, and UHF, as shown in Fig. 2, the channel bandwidth for each band group that is defined for secondary systems can be different due to the band group-specific spectrum hole nature. In general, in HF, the spectrum holes are relatively narrow in the frequency domain because the licensed spectrum of primary systems using HF band group is also usually narrow. On the other hand, the spectrum holes of UHF are comparatively wider than that of HF. Excepting for details characteristic difference of each band group, we assume that the wider channel bandwidth is used in the higher band group frequency. Therefore, if the operating frequency of band group j (BG_j) is higher than that of band group i (BG_i), then the channel bandwidth of BG_j, W_j, is wider than W_i which is the channel bandwidth of BG_i and achievable data transmission rate (i.e., capacity) of W_j is greater than that of W_i. The greater preference for the channel is given to the band group with the higher bandwidth in the

system or individual nodes. However, even though the bandwidth demanded by the secondary system can be satisfied by W_i of BG_i, if the bandwidth W_j of the higher BG_j is utilized by the secondary system, then satisfaction of the system will increase while overall spectrum resources are wasted. Because other secondary systems may exist around and their traffic demands only can be satisfied by using the bandwidth W_j of BG_j, therefore, a mechanism for adaptive allocation of band group use according to the traffic demand and bandwidth utilization efficiency of the corresponding system is required.

3.2 System architecture and problem formulation
The proposed system architecture of CH is represented in Fig. 3. The Q-learning is used to dynamically select optimal band group and channel being aware of wireless environment, network user demand, and system operation parameters. The network user demand module determines the desired data rate (DDR) of the CR ad-hoc network based on each member node's traffic demand, and it also measures the average utilization of the channel currently used. The wireless environment monitoring module stores the spectrum sensing results such as average SNR (signal to noise ratio) and primary signal detection history. Using the sensing results, this module generates band- and channel-specific statistics which includes available data rate and primary idle time. The system operator can dynamically adjust the system parameter for learning using the system parameter module. The system operator can reset the reward

Table 2 Weight parameters

Weights vector (Default)	Q-learning parameters	Reward parameters	DRE parameters
$w_1 = 0.3$, $w_2 = 0.3$,	Learning rate (α) = 0.3,	overhead (η) = 0.01,	$r_1 = 1/6$,
$w_3 = 0.3$, $w_4 = 0.1$	Discount factor (γ) = 0.7	$\delta = 2$	$r_2 = 5/6$

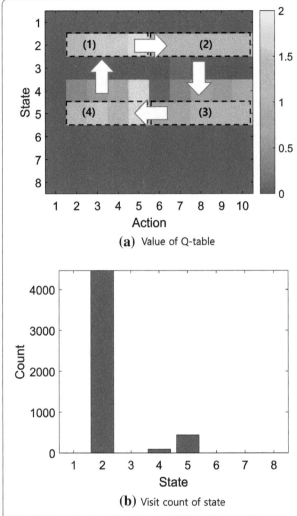

(a) Value of Q-table

(b) Visit count of state

Fig. 11 The **a** Q-value of Q-table when DDR = 40 kbps, and **b** number of state visits. Similar to Figs. 9 and 10a, the Q-value of Q-table shows the channel movement of low DDR case. The visit count of state is high in the state of low DRE and low CBG because the stable domain is in that region

function by learning parameters for Q-learning. Based on all information, the Q-learning module determines which band group and channel can meet the data rate demand and maintain effective utilization level.

The Q-learning module changes from one channel to another when a PU appears on the current channel being used by the secondary system. The reward function is used to update the Q-table, and channel and band group selection is performed based on the current Q-table. The reward function proposed in this paper captures user demand, wireless environment, data rate efficiency (DRE), and band change overhead cost. The DRE is an evaluation metric to determine how much the ad-hoc network efficiently utilizes the data rate supported by the current channel.

In the proposed algorithm, we design the Q-learning reward function to satisfy the following criteria:

- Maximize the secondary system operational time;
- Satisfy the desired data rate of the CR ad-hoc secondary network;
- Provide the coexistence and fairness between secondary systems;
- Consider the overhead of band change for system reconfiguration;
- Guarantee operational flexibility and adaptability to meet the desired purpose.

4 Reinforcement learning for dynamic band and channel selection

4.1 Action, state, and Q-table design for Q-learning

The Q-learning is one of the model-free reinforcement learning techniques. It is able to compare the expected utility of the available actions for a given state without requiring a specific model of the environment. An agent tries to learn the optimal policy from its history of interaction with the environment, in which an agent applies a specific action at the current state and receives a response as a form of a reward from the given environment. The Q-learning eventually finds an optimal policy, in the sense that the expected value of the total reward return over all successive iterations is the achievable maximum one. The problem space consists of an agent, a set of states $\mathcal{S}$, and a set of actions per state $\mathcal{A}$. By performing an action $a \in \mathcal{A}$, the agent can move from state to state.

Figure 4 shows the Q-learning mechanism of the proposed method. The CH of the CR ad-hoc system is the agent of Q-learning, and the action of the CH is a selection of a new tuple (band group, channel) for the CR system operation when the primary signal is detected on the current band group and channel. The structure of the Q-table is expressed by rows of states and columns of actions. In this paper, the set of action $\mathcal{A}$ is given by:

$$\mathcal{A} = \mathcal{B} \times \mathcal{C}_k \qquad (1)$$

where $\times$ is the Cartesian product; $\mathcal{B} = \{b_1, b_2, ..., b_{NB}\}$ expresses the set of channel band groups; NB is the number of band groups; $\mathcal{C}_k = \{c_1^{b_k}, c_2^{b_k}, ..., c_{NC_k}^{b_k}\}$ represents set of available channels in k-th band group (b_k); NC_k is the number of channels of band group b_k; and $c_j^{b_k}$ is $j - th$ channel of band group b_k.

In this paper, a multi-layered state is defined, in which it is composed of geographic location $(\mathcal{L})$, time zone $(\mathcal{T})$, channel band group $(\mathcal{B})$, and data rate efficiency level (D). The state of space in this system is defined as follows.

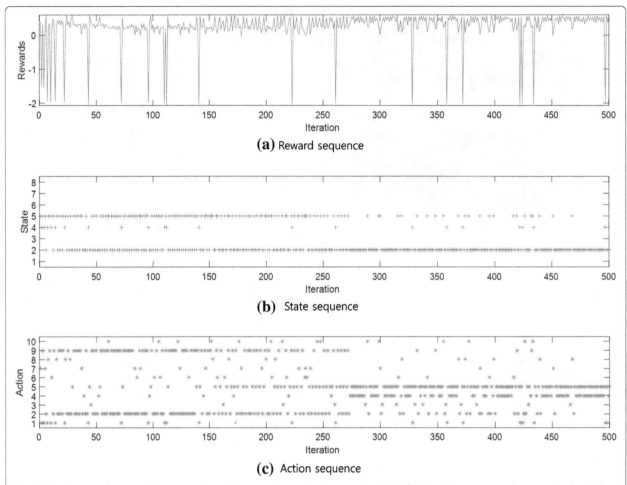

Fig. 12 The change of **a** rewards, **b** states, and **c** actions according to iteration at a low DDR of 40 kbps. The temporary low reward value is due to the random action of Q-learning exploration. The agents visits the state 2 more often than the states 4 and 5 over time as seen in Fig. 11b. As shown in Fig. 11a, the action in **c** mainly visits channel 2 or 3 and is adaptive to the DDR at the latest possible moment, even if a channel from band group 2 is selected or a channel from band group 1 offering a high data rate is selected. **c** represents how the agent selects the channels in band group 1 suitable for the DDR over time. Therefore, we can see that the agent operates according to the designed mechanism

$$\mathcal{S} = \mathcal{L} \times \mathcal{T} \times \mathcal{B} \times \mathcal{D} \qquad (2)$$

where $\mathcal{L} = \{l_1, l_2, ..., l_{NL}\}$, $\mathcal{T} = \{t_1, t_2, ..., t_{NT}\}$, and $\mathcal{B} = \{b_1, b_2, ..., b_{NB}\}$ represent the sets of geographic location zones, time zones, and band groups, respectively. NL, NT, and NB are the number of location zones, time zones, and band groups, respectively. In this paper, we have defined a new additional state dimension $\mathcal{D}$ to represent the operational state of the secondary system in terms of how much the CR system effectively utilize the given channel of the selected band group. $\mathcal{D} = \{d_1, d_2, ..., d_{ND}\}$ indicates the set of DRE levels, and ND is the predefined number of DRE levels. The DRE is the ratio of the DDR of the secondary network to the average supportable data rate of the current channel of the selected band group. Therefore, the current state is defined as a form of (l_i, t_j, b_k, d_l) tuple and it represents the current location zone, time zone, operation band

group, and DRE level. For the given geolocation area and time period, the secondary CR system needs to select the next operational band group and channel whenever a channel switching is required. The current band group (CBG) and DRE capture the dynamic goodness of the selected band group and channel in terms of spectrum efficiency and support of the desired rate. The CH selects the best action for the current state using the current Q-table.

Figure 5 shows the proposed Q-table structure. At the current state (l_i, t_j, b_k, d_l), the CH selects the best action $(b_q, c_m^{b_q})$, i.e., the next band group b_q and m-th channel of b_q, which shows the maximum Q-value in the current Q-table. It needs be noted that the candidate channels of each band group as possible actions should be available channels at the current time as a result of spectrum sensing.

Figure 6 shows the procedure of the proposed Q-table update, state determination, and action selection. It is

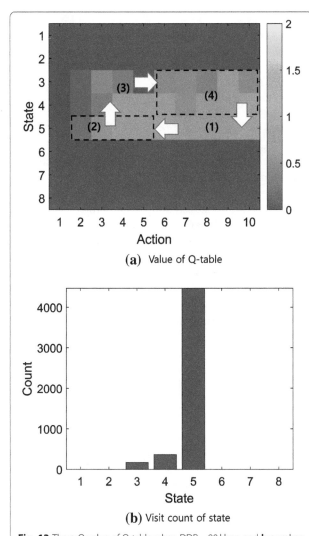

Fig. 13 The **a** Q-value of Q-table when DDR = 90 kbps, and **b** number of state visits. Similar to Figs. 9 and 10b, the Q-value of Q-table shows the channel movement of moderate DDR case. The visit count of a state is high in the state of low DRE and high CBG because the stable domain is in that region

assumed that the MNs transmit the average channel operation time and average supportable data rate to the CH through sensing and channel use report.

1. Suppose the learning agent CH determined the state s_{t-1} and the best action a_{t-1} at the end of $(t-1)$-th time period.
2. During t-th time period, MNs and CH monitor the primary activities and channel statistics.
3. Agent CH detects the band and channel change event.
4. The CH calculates the reward r_{t-1} for the previous action a_{t-1} at state s_{t-1}.
5. The CH updates the Q-value of (s_{t-1}, a_{t-1}) in Q-table.

6. The CH determines the current state s_t based on the measured DRE during t-th time period.
7. The CH selects the optimal action a_t for the next $(t + 1)$-th time period.
8. Go to step 1.

The Q-learning updates the Q-value for each pair of state and action (s, a) visited through these series of processes. The Q-value reflects the value that the system can accept when selecting action a in state s.

The Q-value update of the Q-table can be represented by:

$$\mathcal{Q}(s_t, a_t) \leftarrow (1-\alpha)\mathcal{Q}(s_t, a_t) + \alpha\left\{ r_t + \gamma \max_{a_{t+1}} \mathcal{Q}(s_{t+1}, a_{t+1}) \right\} \quad (3)$$

where α and γ denote the learning rate and discount factor, respectively. The learning rate $\alpha \in [0, 1]$ is used as a weight to reflect the $\mathcal{Q}(s_t, a_t)$ accumulated from the past, the newly obtained reward, and the expected reward value for the next action. If α is low, it increases the weight of the past experience so the system takes an extended time to learn, but the fluctuation of the reward sequence is low. If α is high, the learning speed is increased by assigning a high weight to both the present and future values. However, an extremely high α causes instability in the system, while a fairly low α prevents the system from reaching a satisfactory reward at the desired time. The discount factor $\gamma \in [0, 1]$ is the weight for how much the Q-value of a_{t+1}, the future reward, should be reflected in the Q-value of action a_t in the Q-table of the current action and state. A high γ has a high contribution on the Q-value of the future expected reward, and a low γ weights the reward according to the current action a. That is, when Q-learning reflects the immediate reward and the future tendency in the Q-value of the action and corresponding state, γ is a weight that takes into account whether to further consider the volatility of the current action or to reflect the future value predicted from past trends of the Q-table.

If the CH only uses the updated Q-values to select actions, it may fall into local optimum. Therefore, we use ε-greedy policy to add randomness to selecting of actions that are explorative in the learning algorithm, as follows:

$$a = \begin{cases} \underset{\tilde{a}\in\mathcal{A}}{\arg\max}\ \mathcal{Q}(s, \tilde{a}), & \text{with probability } 1-\varepsilon \\ \text{random } a\in\mathcal{A}, & \text{with probability } \varepsilon \end{cases} \quad (4)$$

where, $\varepsilon \in [0, 1]$ is the probability of choosing a random action. If ε is high, it is more likely that new information will be added to the already accumulated information while searching for the next action; if it is low, the next action is selected using only the accumulated information. ε

starts with a specific value and lowers this value for each iteration, so that the Q-table can operate stably after a certain time. However, when the value of ε decreases continuously, a considerable amount of time is required for adapting to the changing environment by updating the Q-table. Therefore, a lower limit of ε is required.

The overhead of Q-learning can arise from the memory size for the use of Q-tables. It depends on the level of the actions (the number of channels and bands) and resolution level of states, and it increases linearly with each level. If you set the number of level too low, the system cannot use the Q-table for learning dynamic environments. Otherwise, the system takes a long time to learn the surrounding environment using the Q-table. Therefore, the selection of appropriate level is required.

4.2 Reward function design
The reward that the CR system obtained by using the selected set of (band group, channel) is composed of the system operation time, average data transmission rate,

channel utilization efficiency, and overhead required for the system to change the frequency band. The reward for the action a is expressed as follows:

$$R(a) = w_1 \frac{T_{op}}{\max\left(T_{op}^{cbg}\right)} + w_2 \frac{E[D_s]}{\max\left(E\left[D_s^{cbg}\right]\right)} + w_3 R_{Util} - w_4 BC_{(a,a')} \tag{5}$$

where T_{op} is the consecutive channel operation time for the secondary system after the channel is selected, in which if a channel shows high T_{op} value, then it indicates that once the secondary system takes this channel it can use the channel relatively long time before the primary appears. $E[D_s]$ is the average supportable data rate of the selected channel. R_{Util} represents how the secondary system utilizes the channel effectively. $BC_{(a,a')}$ is the overhead for band group change. The operation time and average supportable data rate are normalized to their maximum values for the current band group. a and a' are the current action and previous action, respectively.

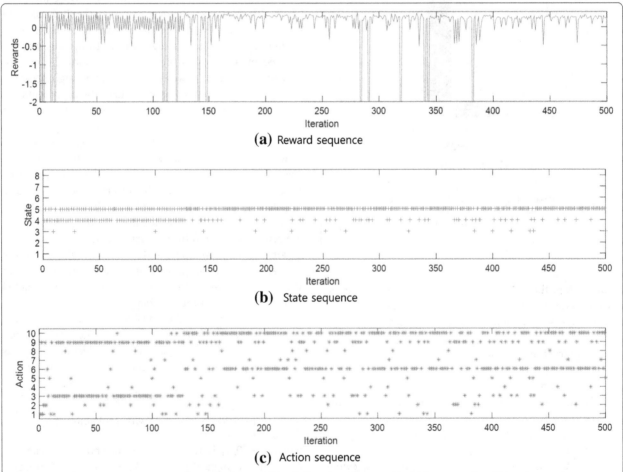

(a) Reward sequence

(b) State sequence

(c) Action sequence

Fig. 14 Rewards, states, and actions according to iteration at DDR = 90 kbps. In **a,** the reward is stable at more than 10 iterations, and we can see that the reward is temporally low in the overall interval by random action, similarly to Fig. 12. As shown in **b** the agent mainly visits the state 5. **c** Reveals that actions in band group 2 are selected mostly

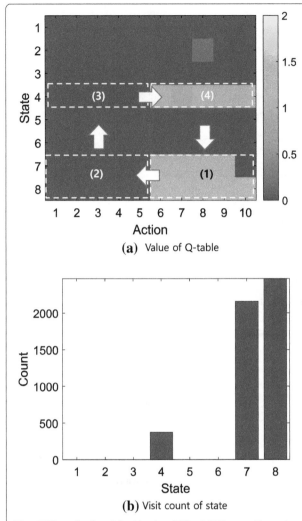

(a) Value of Q-table

(b) Visit count of state

Fig. 15 The **a** Q-value of Q-table when DDR = 3.5 Mbps, and **b** number of state visits. Similar to Figs. 9 and 10c, the Q-value of Q-table shows the channel movement of high DDR case. The visit count of state is high in the state of high DRE and high CBG because the stable domain is in that region

$max(T_{op}^{cbg})$ and $max(E[D_s^{cbg}])$ are the maximum channel operation time and maximum expected supportable data rate value from all channels in the current band group, respectively. w_i is the weight for i-th reward component and $\sum_{i=1}^{4} w_i = 1$. The first and the second term are normalized by each maximum value of the operation time and average supportable data rate in each channel group so that the relative value to the other channels can be reflected in the reward. The third term is described in (7) and serves to adjust the reward to select a channel suitable for the desired data rate. The fourth term represents the cost due to an overhead when a band group change occurs, which is described in (6). All the terms are linearly coupled to allow the system designer or user to operate the system for a specific purpose through weighting changes.

The overhead for band group change, $BC_{(a,a')}$, in (5) is to capture the required additional time and energy for reconfiguring some system operational parameters when the band group is changed. In most cases, different band groups have different channel bandwidths and wireless characteristics so that communication system may need to reconfigure radio frequency (RF) front-end, modulation method, and medium access control (MAC) layer components whenever it changes its operation band group. The overhead is represented as in (6).

$$BC_{(a,a')} = \begin{cases} \eta, & \text{channel a and a' belong to different band groups.} \\ 0, & \text{else} \end{cases}$$
(6)

where η is the cost when the current channel and the previous channel belong to different band groups. In this paper, we do not consider the channel switching overhead inside the same band group.

R_{Util} of (5) is defined as a function of DRE. The DRE is defined in this paper as in (7)

$$DRE = \frac{DDR}{E[D_s]}$$
(7)

where DDR is the desired data rate (DDR) of the secondary CR network. To design R_{Util} function, first we considered the desired system operation in terms of band group selection depending on the current DRE value.

Figure 7 shows the example of the desired mechanism by which the channel selection is performed according to the current DRE. It is assumed that the bandwidth of each channel provided by each band group follows $W_1 < W_2 < W_3$, in which we have three band groups. The x-axis is divided by band group, and the parts represent the DRE for each band group. DRE (U) indicates that the utilization ratio of the channel is low when it belongs to $[0, r_1)$ of band group 1, and that ratio is moderate when it belongs to $[r_1, r_2)$. If U belongs to $[r_2, 1)$, it denotes a high channel utilization ratio so that some time instances of the network may not be able to meet the user traffic demand. The range of $[1, \infty)$ means the channel cannot support enough bandwidth for the system. A low channel utilization ratio means that the possible transmission rate provided by the selected channel of the current bad group is much higher than the desired CR network data rate so that most of spectrum resource is wasted after it satisfies the desired data transmission rate. It is therefore necessary to move to a channel that provides a lower bandwidth and data rate (i.e., change to the lower band group channel). Furthermore, if U shows a higher channel utilization ratio than the defined r_2, which means that the possible data transmission rate provided by the selected channel is not

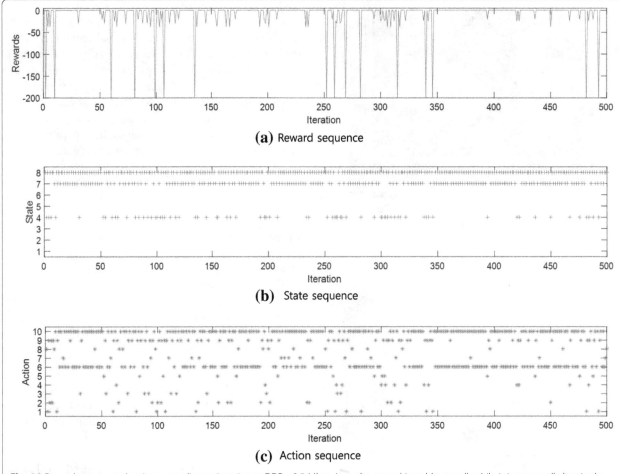

Fig. 16 Rewards, states, and actions according to iteration at DDR = 3.5 Mbps. In **a,** the reward is stable overall, while it is temporally low in the overall interval by random action, similar to Figs. 12 and 14. In **b,** the system visits state 4 to a degree, but it mainly remains in states 7 and 8. **c** shows that actions in band group 2 are selected mostly

likely to support the desired data rate of the system, then it is necessary to move to a channel band group that can provide a wider bandwidth and a higher data rate. However, in Fig. 7, even though DRE is in $[0, r_1)$ for band group 1, the secondary system does not have any band group that has narrower channel bandwidth so that it needs to keep the current band group. On the other hand, in case of band group 3, when DRE is in $[r_2, \infty)$, the system does not have any band group that has wider channel bandwidth so that it has to find other best channel in the same band group. In each band group, $[r_1, r_2)$ is the band usage maintenance interval, because the selected band group channel provides an appropriate transmission rate.

Based on the band group selection movement mechanism in Fig. 7, the proposed utilization efficiency reward function R_{Util} of Eq. (5) is shown in Fig. 8, in which we assume that there are three band groups. The x-axis for each band group represents DRE (U), and the y-axis represents R_{Util}. For the band usage maintenance

range in Fig. 7, R_{Util} is set to 0 in $[r_1, r_2)$ DRE range for all band groups, which represents the medium channel utilization efficiency ratio. In $[0, r_1)$ DRE range, the selected band group channel can support much larger data rate than the desired data rate so that channel utilization efficiency is low. Therefore, as the value of DRE goes from 0 to r_1, R_{Util} increases from −1 to 0 except the in first band group 1. Any ad-hoc CR secondary systems that has its current DRE value in $[0, r_1)$ need to move to the band group channel that has narrower channel bandwidth and lower supportable data rate. It makes the secondary system to yield the current band group channel to other secondary systems that requires more data rate. For the first band group 1, there is no other narrower band group so that R_{Util} is maintained at 0 in $[0, r_1)$ DRE range. The range $[r_2, \infty)$ is divided into $[r_2, 1)$ and $[1, \infty)$ to distinguish the insufficient transmission rate provided by the channel, with R_{Util} representing a more rapid decrease rate in $[1, \infty)$ range except in the last band group 2. In $[r_2, 1)$ DRE range, the band group

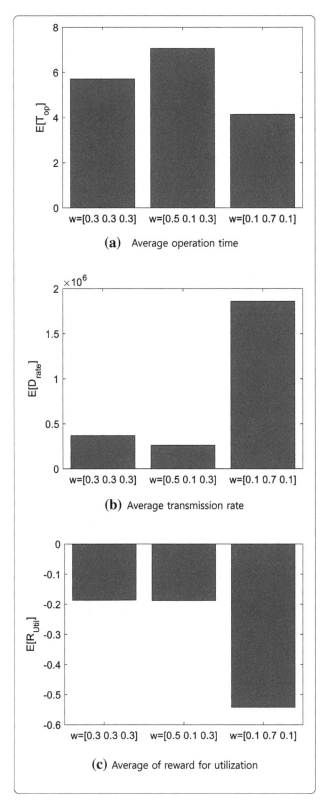

(a) Average operation time

(b) Average transmission rate

(c) Average of reward for utilization

Fig. 17 a Average operation time, **b** average transmission rate, and **c** reward of utilization according to weight change. The average operation time, average data rate, and reward for channel utilization by changing the weight assignment for DDR to 40 kbps. Since the reward function is composed of the weighted sum of the objective functions, the Q-learning can be operated according to the desired objective function by adjusting the weight. Therefore, if the weight of the operation time is increased, the average operation time is increased, and if the weight of the data transmission rate is increased, the average transmission rate is increased. Finally, increasing the weight of reward for utilization increases the average of reward for utilization

channel supportable data rate may not be enough to guarantee the desired rate in some time instances so that R_{Util} decreases from $-r_2$ with a slope of -1 as DRE increases. In $[1, \infty)$ DRE range, the current band group channel cannot support desired data rate so that R_{Util} decreases with a slope of $-\delta$ ($\delta > 1$). For the last band group 3, there is no other wider band group so that R_{Util} is maintained with 0 in $[r_2, \infty)$ DRE range.

Figures 9 and 10 show how this intentional mechanism is supported in the Q-table. Figure 9 shows the Q-table where the state is divided into geographic zones and time zones, and again into band groups and discrete DRE levels. The columns of the Q-table are divided into bands, which are then divided into selectable channels as possible actions. In the action shown in Fig. 9, a channel that represents a narrower band is shown as a lower data rate toward the left, and a channel that can use a wider band appears as a higher data rate toward the right. Figure 10 represents how the Q-value updating area changes in the Q-table of Fig. 9 through the example of three DDR cases. First, Fig. 10a depicts a case where the DDR is low. In Fig. 9, suppose that the secondary system is operating in action domain 1 (low CBG, low DRE) and by the Q-learning ε-greedy policy it may randomly select action domain 2 band group channel. As a result, the DRE is significantly lowered, and the system gets low R_{Util} because a high channel band provides a high data rate and it results in low reward for channel utilization efficiency. Therefore, after updating the Q-value of domain 2, the system operating area changes to domain 3 by the change of the state which represents (high CBG, low DRE). Because selecting a channel which supports a high data rate makes the DRE low, the Q-table then updates the Q-value of domain 3 and the Q-learning agent will select the best action of domain 4 because selecting a channel with a low band gives a high R_{Util}. After selecting the low-band channel, the transition to domain 1 (low CBG, low DRE) is performed. In this case, R_{Util} does not have a negative value because there is no longer a lower channel to select, as in the $[0, r_2)$ of band group 1 seen in Fig. 8, and no value is subtracted from the total reward. Therefore, in the

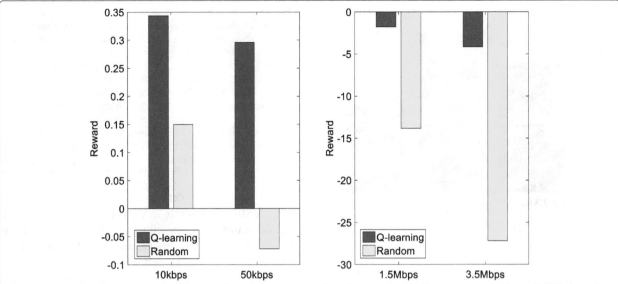

Fig. 18 Average reward comparison for Q-learning channel selection vs. random channel selection. The average reward for each DDR depending on the method of the channel selection. For all of the DDR cases, Q-learning band and channel selection has more reward value than random selection. The reward for a DDR of 1.5 or 3.5 Mbps (e.g., more than medium or high DDR case) is lower than that of 10 and 50 kbps. The ε-greedy policy in case of a high DDR causes very low reward for channel utilization R_{Util} due to the selection of a low-band channel which support insufficient data rate and these effects are accumulated in the Q-table

case of a low DDR, the preferred domain in the Q-table is not domains of 2, 3, and 4 where high band is selected by occasional ε-greedy policy for exploration but domain 1 (stable operating domain). Cases of both moderate and high DDR, as shown in b and c of Fig. 10, can be similarly explained.

5 Simulation results and discussion

The simulation environment in this paper assumes five channels for each of the two band groups, as listed in Table 1. The channels of band group 2 provides higher supportable data rates than those of the band group 1 while the operation time available for the transmission is not significantly different. We use a Gaussian distribution to determine the operation time and supportable data rate of each channel in each band group based on the mean and variance values provided in Table 1. The other simulation parameters are shown in Table 2.

In this paper, the action is defined as the selection of the channel in each band group as shown in Table 1. The index number of the action corresponds to the number of the channel in Table 1, and the total number of action is 10. We define the state as the combination of (band group, DRE) where the domain of DRE is divided as $[0, r_1), [r_1, r_2), [r_2, 1), [1, \infty)$. The domain of DRE corresponds to each band group so the total number of state is 8.

The ε-greedy policy for action exploration is as follows:

$$p(n) = \begin{cases} p_0(0.999)^n, & p(n) > p_{low} \\ p_{low}, & p(n) \le p_{low} \end{cases} \tag{8}$$

where $p_0 = 0.3, p_{low} = 0.1$

where n represents the iteration sequence number with time. In applying the ε-greedy policy, when the wireless environment of the system is changed, p_{low} is required to set a lower limit for a random value in order to maintain a certain degree of exploration. Otherwise, the Q-table cannot adaptively operate in the changed environment.

The overall simulation configuration starts by looking at the operation of Q-learning for each DDR, 40 kbps – 90 kbps – 3.5 Mbps, and confirming the change of average operation time and average transmission data rate according to the weight of the reward. Next, we compare the results of Q-leaning and random channel selection according to the reward, operation time, data rate, and utilization. Finally, we compare the change of DDR according to iteration for Q-learning and random channel selection.

5.1 Adaptive channel selection according to DDR

In this section, we identify our proposal adaptively selects the channel according to each DDR (e.g., low, moderate and high) as described by Figs. 9 and 10 in Section 4.2. Figure 11a shows the value of the Q-table in scaled colors when the DDR is 40 kbps, and b shows the number of visits to each state in the Q-table. The DDR of

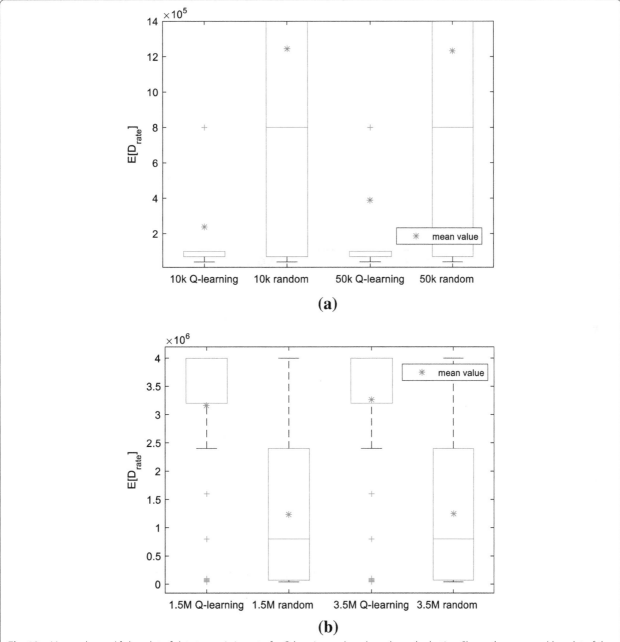

Fig. 19 a Mean value and **b** boxplot of data transmission rate for Q-learning and random channel selection. Shows the mean and boxplot of the data transmission rate for Q-learning and random channel selection for each DDR. In cases of the random channel selection, the average data transmission rate of all DDR cases is the same as the average value of the data rates for all channels in Table 1 belonging to band groups 1 and 2. The boxplots of all DDR cases for the random channel selection have a similar distribution. For the DDR of 10 kbps and 50 kbps, the mean of the Q-learning selection is lower than the random selection and the Q-learning has more narrow distribution. For the DDR of 1.5 and 3.5 Mbps, the mean of the Q-learning selection is higher than the random selection and the Q-learning has more narrow distribution

40 kbps is the low data rate comparing the data rate of channels in Table 1. Therefore, if the CH selects the channel of band group 1, the DRE belongs to almost $[0, r_2)$ of band group 1 in Fig. 8 comparing the channels in band group 1, and R_{Util} does not give any effect on total reward. However, if the CH selects the channel of band group 2, the DRE belongs to $[0, r_1)$ of band group 2 in

Fig. 8 and R_{Util} has an impact on the total reward linearly according to DRE. Figure 11a represents the process of changing the channel (action) selected by the ad-hoc CH (agent). When the CH selects the channel of band group 1, the process of updating the Q-value in the Q-table takes place in domain (1), which represents the channel selection of band group 1 and the low DRE

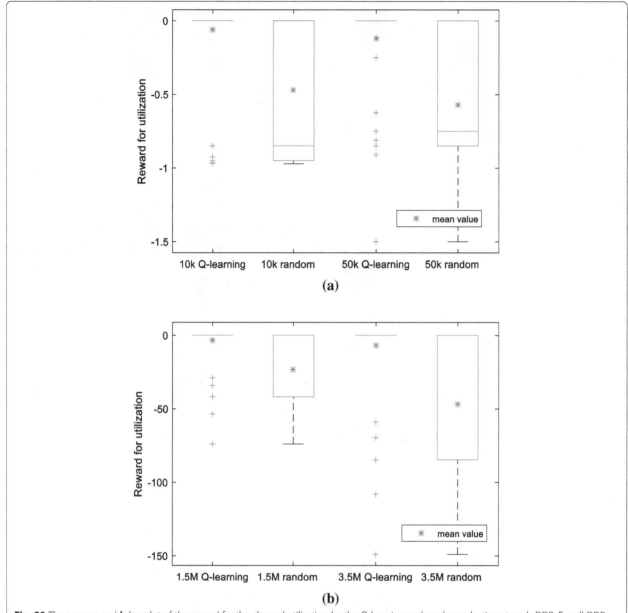

Fig. 20 The **a** mean and **b** boxplot of the reward for the channel utilization by the Q-learning and random selection at each DDR. For all DDRs, the boxplot of Q-learning has denser distribution and higher values than that of the random selection, and it has a higher average value

of band group 1. If the CH selects the channel of band group 2 by the explore policy of the Q-learning, the Q-value of the domain (1) is changed and the update process moves to the domain (2) which represents the channel selection of band group 2 maintaining the current state. Since the CH selected the channel of band group 2 that provides a high data rate for low DDR, the state is changed to the low DRE of band group 2. Thus, the update process of the Q-table moves to domain (3) and the Q-value of domain (2) changes. If CH selects channel of band group 1 in state 5, the state is maintained and update process moves to domain (4) after the Q-value change in domain (3). Finally, the state is

changed to the state 2, which represents the band group 1 and low DRE, and the update process moves to domain (1) after the Q-value of domain (4) changes. The Q-value of Fig. 11a is the highest in domain (1), which represents the low DRE of band group 1 same with the low DDR case of Fig. 10a by the reward for utilization in Fig. 8. As a result, Fig. 11b represents that the number of visits in state 2 is the highest which corresponds to domain (1).

Figure 12 shows the change of rewards, states, and actions according to iteration at a low DDR of 40 kbps. The temporary low reward value is due to the random action of Q-learning exploration. The agents visit the

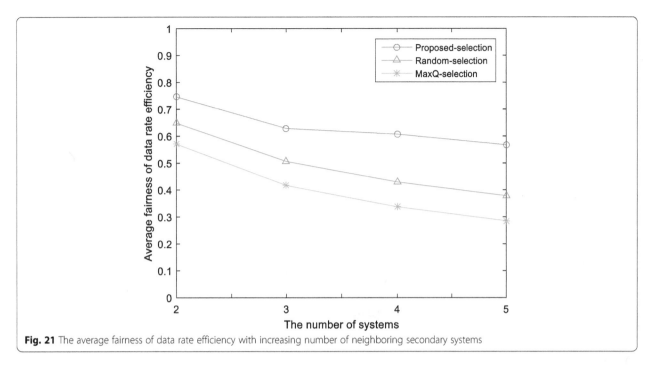

Fig. 21 The average fairness of data rate efficiency with increasing number of neighboring secondary systems

state 2 more often than the states 4 and 5 over time as seen in Fig. 11b. As shown in Fig. 11a, the action in Fig. 12c mainly visits channel 2 or 3 and is adaptive to the DDR at the latest possible moment, even if a channel from band group 2 is selected or a channel from band group 1 offering a high data rate is selected. Figure 12c represents the agent selects the channels in band group 1 suitable for the DDR over time. Therefore, we can see that the agent operates according to the designed mechanism.

Figure 13a shows the value of the Q-table in scaled colors when the DDR is 90 kbps and b shows the number of visits for each state in the Q-table. If the DDR is 90 kbps and the CH chooses a channel in the band group 1, the DRE belongs to $[r_2, \infty)$ in Fig. 8. Meanwhile, the DRE belongs to $[0, r_1)$ in Fig. 8 if a channel is chosen from the band group 2. Therefore, the supportable data rate by the channels in band group 1 is not enough in comparison with the channels in band group 2, as seen in Table 1, since the channel selection from band group

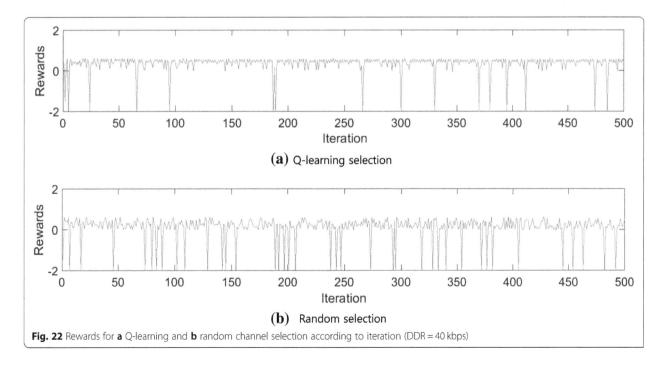

Fig. 22 Rewards for **a** Q-learning and **b** random channel selection according to iteration (DDR = 40 kbps)

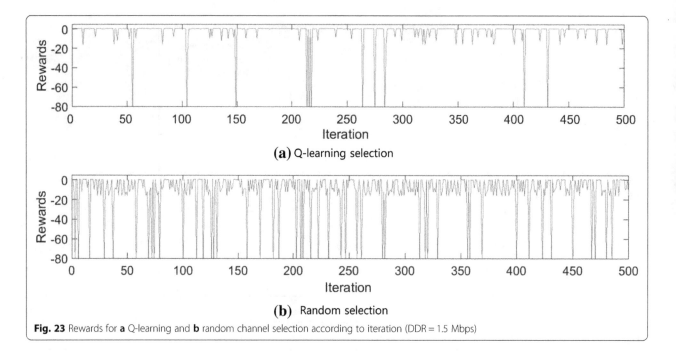

Fig. 23 Rewards for **a** Q-learning and **b** random channel selection according to iteration (DDR = 1.5 Mbps)

2 offers better R_{Util} than that of band group 1. We assume that the process of updating the Q-value starts from domain (1) which represents the channel selection of band group 2 and the DRE is low in band group 2. After selecting the channel in band group 1 by random channel selection, the Q-value of domain (1) is renewed and the update process moves to domain (2). If the CH selects channels from band group 1, the renewal process

of the Q-table changes to domain (3) due to the high DRE which means the selected channel does not support a high enough transmission data rate after the renewal of domain (2). The process of updating moves to domain (3) by the change of state then transfers to domain (4) by the random or best selection. Because the channel selection in band group 2 provides more reward for utilization by Fig. 8, the Q-table in Fig. 13b has the

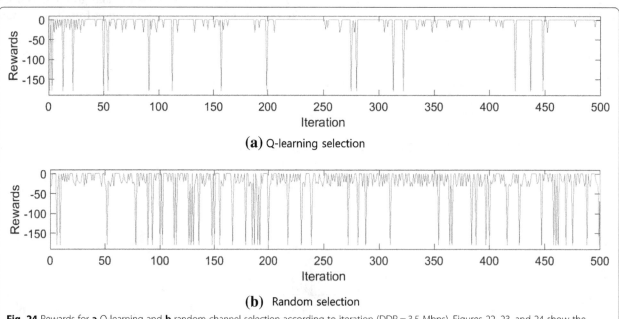

Fig. 24 Rewards for **a** Q-learning and **b** random channel selection according to iteration (DDR = 3.5 Mbps). Figures 22, 23, and 24 show the rewards for Q-learning and random band and channel selection according to the iteration for each DDR. In the case of Q-learning for all DDRs, the fluctuation decreases over time and the system operates with the intended reward design. In the random selection, there are more notches and fluctuation than Q-learning selection

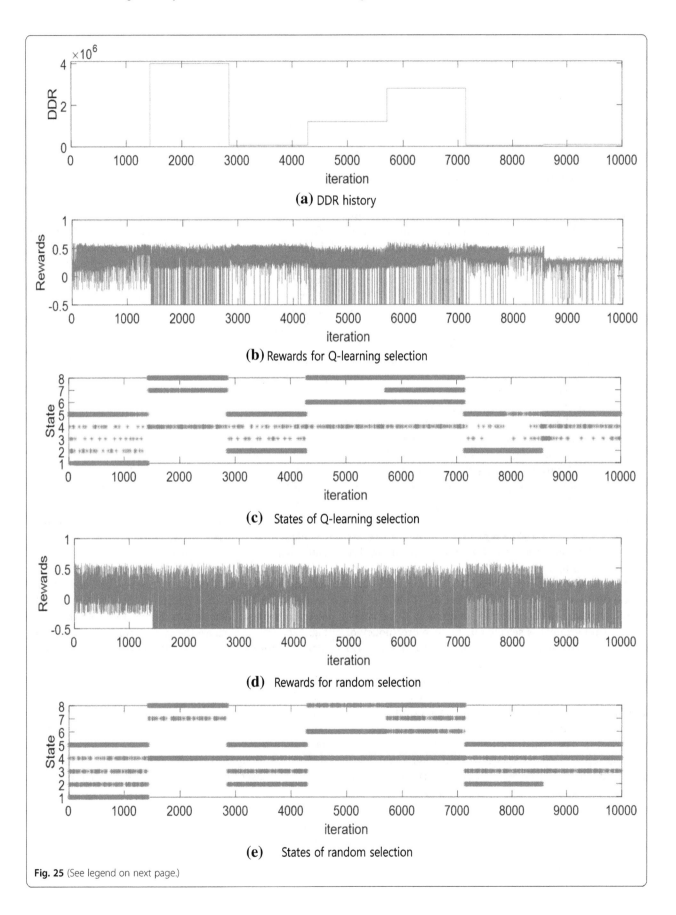

(a) DDR history

(b) Rewards for Q-learning selection

(c) States of Q-learning selection

(d) Rewards for random selection

(e) States of random selection

Fig. 25 (See legend on next page.)

(See figure on previous page.)
Fig. 25 Rewards for Q-learning and random channel selection according to DDR change. Represents the rewards and visits of states according to the changes in DDR as shown in **a**. Comparing **b** and **d,** the rewards of Q-learning selection are more stable than that of the random selection. From **c** and **e,** we can see that the ad-hoc CH selects a low data rate channel and Q-learning visits the state of low DRE in band group 1 when the DDR is low. Furthermore, we can see that the Q-learning visits a state of low DRE of band group 2 when the system selects a high data rate channel by the explorer policy. When the DDR is high, the Q-learning tries to select the channel of band group 2 mainly which provides higher data rates so that the sates of band group 1 are visited less frequently. However, in random channel selection, the visits of states are distributed evenly in various DREs when the DDR is low or high. **c** and **e** shows that the visiting states of Q-learning and random channel selection are the same for a particular DDR. However, since the Q-learning channel selection tries to select a channel adaptive for the specific DDR, Q-learning mainly visits the state of the band group 1 when the DDR is low and visits the state of the band group 2 when the DDR is high. From these results, the proposed Q-learning selects the appropriate channel even if the DDR changes

highest Q-value in domain (1) which represents a low DRE for band group 2. As a result, the number of visits to state 5 belonging to the low DRE of band group 2 is the highest, as shown in Fig. 13b.

Figure 14 shows the change of rewards, states, and actions according to iteration with a medium-level DDR at 90 kbps. In Fig. 14a, the reward is stable at more than 10 iterations, and we can see that the reward is temporally low in the overall interval by random action, similarly to Fig. 12. As shown in Fig. 14b, the agent mainly visits the state 5. Figure 14c reveals that actions in band group 2 are selected mostly.

Figure 15a shows the value of the Q-table in scaled colors when the DDR is 3.5 Mbps, and b shows the number of visits to each state in the Q-table. Comparing with the supportable data rate of channels in Table 1, the DDR of 3.5 Mbps makes the DRE belong to $[1, \infty)$ of the band group 1 in Fig. 8 when the CH selects the channel from band group 1, and belongs to $[r_1, \infty)$ when a channel is selected from band group 2. However, the reward for utilization R_{Util} remains as 0 since there are no other channels to move out. As illustrated in Fig. 10c about the example of high DDR case, the same explanation can be given about Fig. 15a. At first, update process is assumed starting from domain (1) in Fig. 15a by the channel selection from band group 2. If the channel of band group 1 is selected by the explorer policy of Q-learning, the update process moves to domain (2) after the change the Q-value of domain (1). The state changes to the state 4 which represents high DRE in the band group 1 by the given DDR and the channel selection of band group 1. Therefore, after the Q-value of domain (2) is updated, the update process moves to domain (3). The update process selects the channel of band group 2 through the best or random channel selection and could be moved to domain (4), thereby the Q-value of domain (3) is updated. Finally, since the channel selection of band group 2 changes state to high state of band group 2, the update process moves to domain (1) and the Q-value of domain (4) is updated. As described for high DDR case in Fig. 10c, the CH of Fig. 15 also tends to select the channel of band group 2 and stay on the state which has high DRE of band group

2 since the channels of this band group gives no harmful effect on R_{Util}. In the Q-table of Fig. 15a, the Q-value of domain (1) showing high DRE in band group 2 is the topmost, and this is also shown in Fig. 15b as the high visit count of states 7 and 8.

Figure 16 represents the change of rewards, states, and actions according to iteration at a high level DDR of 3.5 Mbps. In Fig. 16a, the reward is stable overall, while it is temporally low in the overall interval by random action, similar to Figs. 12 and 14. In Fig. 16b, the system visits state 4 to a degree, but it mainly remains in states 7 and 8. Figure 16c shows that actions in band group 2 are selected mostly.

These results demonstrate that the proposed system can select an appropriate channel according to the DDR required by ad-hoc CR users.

5.2 Reward reconfiguration with weights

Figure 17 shows the average operation time, average data rate, and reward for channel utilization by changing the weight assignment for DDR = 40 kbps. In the reward calculation, if weights for [operation time, supportable data rate, reward for channel utilization] are assigned to [0.5, 0.1, 0.3], then it increases the importance for the operation time. As a result, it has the best average increase in operation time, as shown in Fig. 17a, and the least average of data transmission rate, as shown in Fig. 17b. This is because the system wants to reserve the highest priority for operation time and the least for data transmission rate. If weights are assigned to [0.1, 0.7, 0.1], then it increases the data transmission rate. However, it results in the lowest average operation time and reward for channel utilization because they are less important for consideration. This weight shows the highest average transmission data rate in Fig. 17b. Therefore, it is possible to operate the CR system according to the purposes of user by changing the weight assignment.

5.3 Performance comparison for the proposed Q-learning

In this section, we compare the channel selection performance between the proposed Q-learning mechanism and random selection from the available channel lists in terms of obtained rewards, average data rate of the secondary systems, and channel utilization efficiency. We also consider the fairness of selfish channel selection without considering channel utilization efficiency.

Figure 18 shows the average reward for each DDR case. When the DDR is 10 kbps or 50 kbps, the random channel selection has a lower reward than Q-learning because the random channel selection causes a waste of channel resources and obtains the low reward for channel utilization R_{Util}. In case the DDR is 1.5 Mbps or 3.5 Mbps (e.g., more than medium or high DDR case), a channel providing a sufficient data rate is not selected adaptively by random channel selection, leading to a lower reward than Q-learning channel selection. The rewards for a DDR of 1.5 Mbps and 3.5 Mbps is lower than those of 10kbps and 50 kbps. As shown in DRE range of $[r_2, 1)$ and $[1, \infty)$ in Fig. 8, the ε-greedy policy in cases of a high DDR causes very low reward for channel utilization R_{Util} due to the selection of a low-band channel which support insufficient data rate and these effects are accumulated in the Q-table. These results show that the Q-learning channel selection adaptively selects the channel for the overall DDR.

Figures 19 and 20 show the boxplot and mean value for the data rate and reward for channel utilization resulting in Q-learning and random channel selection. The red line represents the median value, and a star denotes the mean value of the data. Figure 19 shows the mean and boxplot of the data transmission rate for Q-learning and random channel selection for each DDR. In cases of the random channel selection, the average data transmission rate of all DDR cases is the same as the average value of the data rates for all channels in Table 1 belonging to band groups 1 and 2. The boxplots of all DDR cases for the random channel selection have a similar distribution, as well. The DDR of 10 kbps and 50 kbps by Q-learning channel selection have similar distributions, and the mean for 50 kbps Q-learning is higher than that of 10 kbps, since a higher DDR attempts to choose the channel supporting higher data transmission rate. In case of a DDR for 1.5 Mbps and 3.5M bps, as in Fig. 19b, the distribution and average value of the data transmission rate by the Q-learning channel selection are higher than that of 10 kbps and 50 kbps since the channels are mainly selected from band group 2.

We can see that the Q-learning channel selection can select the channel which provides higher data transmission rate when the DDR is 3.5 Mbps than 1.5 Mbps from the mean values of each DDR case.

Figure 20 shows the mean and boxplot of the reward for the channel utilization by the Q-learning and random channel selection at each DDR. The reward for channel utilization mainly operates as a harmful value in the total reward function when the ad-hoc CH chooses an appropriate channel for its DDR. For all DDRs, the boxplot of Q-learning is denser and distributed at higher values than that of the random channel selection, and it has a higher average value since Q-learning tries to choose the channel that does not create harm in terms of R_{Util}. Outlier values of Q-learning cases are generated by random selection.

Figure 21 shows the average fairness of data rate efficiency with increasing number of neighboring secondary systems. To compare the fairness between secondary systems, two compared methods are considered in this experiment: (i) the random selection, in which the operating band and channel are selected randomly by each secondary system from its available channels and (ii) MaxQ-selection [29–32], in which each secondary system selects the channel that has the maximum supportable data rate. As we can see in Fig. 21, the proposed method provides the highest fairness because it selects the band and channel based on the desired traffic demand and current channel utilization efficiency. Therefore, if a secondary system needs relatively low data rate, then it will select the band that has a low channel bandwidth in the proposed system and it yields the bands with wider channel bandwidth to the neighbor secondary systems that require higher data rates.

Figures 22, 23, and 24 show the rewards of Q-learning and random channel selection according to an iteration for each DDR. In the case of Q-learning for all DDRs, the fluctuation decreases over time and the system operates with the intended reward design. In the random selection, there are more notches and fluctuation than Q-learning channel selection.

Figure 25 represents the rewards and visits of states according to the changes in DDR as shown in Fig. 25a. Comparing Fig. 25b and d, the rewards of Q-learning selection are more stable than those of the random selection. From Fig. 25c and e, we can see that the ad-hoc CH selects a low data rate channel and Q-learning visits the state of low DRE in band group 1 when the DDR is low. Furthermore, we can see that the Q-learning visits a state of low DRE of band group 2 when the system selects a high data rate channel by the explorer policy. When the DDR is high, the Q-learning tries to select the channel of band group 2 mainly which provides higher data rates so that the states of band group 1 are visited less frequently. However, in random channel selection, the visits of states are distributed evenly in various DREs when the DDR is low or high. Figure 25c and e show that the visiting states of Q-learning and random channel selection are the same for a particular DDR. However, since the Q-learning channel selection tries to select a channel adaptive for the specific DDR, Q-learning mainly visits the state of the band group 1 when the DDR is low and visits the state of the band group 2 when the DDR is high. From these results, the proposed Q-learning selects the appropriate channel even if the DDR changes.

6 Conclusions

In this paper, we propose a band group and channel selection method considering the consecutive channel operation time, data transmission rate, channel utilization efficiency, and cost of the band group change for a cognitive radio ad-hoc network composed of CH and MNs. The proposed method uses the Q-learning in order to operate in a channel environment that varies dynamically according to the geographical region, time zone, band group, channel, and primary user's activity. As the core of the Q-learning operation, a Q-table and reward function consisting of an action and state are designed to consider various parameters related to the channel selected by the CR ad-hoc system. In particular, the reward for channel utilization is designed to select the appropriate band and channel so that the frequency resources are not wasted and a CR ad-hoc system can coexist with other CR systems with fair resource utilization efficiency. The simulation results represent how the proposed system selects an adaptive band and channel for the required data rate and also explain the principle of operation through the change of action and state in Q-table. It also can be confirmed that the system operates according to the intended purpose through the weight change, and the channel is selected adaptively when the required transmission rate is changed. These simulations clearly demonstrate these advantages of the proposed method.

7 Methods/experimental

The purpose of this paper is to select the band and channel for a cognitive ad-hoc system to move when the primary user appears in the channel used by the CR system and is to consider the fairness with other systems in selecting the channel. The characteristics of frequency resources such as an available transmission opportunity and data rate vary depending on the time zone, geographical location, and band group, and the activity of the primary user and the desired data rate of the secondary user are also different according to them. Therefore, considering such a complicated environment, it is necessary to select a band and a channel that can maximize the performance of the system. In this paper, the Q-learning is used to dynamically select the band and channel according to the complex surrounding environment which is time-varying. The reward function of the Q-learning is designed considering the available channel use time, data rate, utilization efficiency of the selected channel, and cost for band change. Each of the considered terms is combined with a weight sum so that the

performance related to the preferred parameters can be properly realized according to the adjustment of the weights. In particular, we designed a reward for utilization in the reward function so that the CR ad-hoc system does not choose a channel that provides unnecessarily high data rate and other system has the opportunity of selecting adaptive channel which supports adaptive high data rate. The Q-table is designed so that the reward function of Q-learning works properly. The state of the Q-table is composed of time zone, geographical zone, band group, and data rate efficiency (DRE) so that the proposed Q-learning can operate well.

Experimental results in this paper had been performed using MATLAB R2015b on Intel® Core i7 3.4 GHz system. The Gaussian random function to generate the operation time and supportable data rate of each channel over time and Q-table matrix for Q-learning can be made by constructing appropriate MATLAB code.

Abbreviations
BS: Base station; CBG: Current band group; CH: Cluster head; CR: Cognitive radio; D2D: Device-to-device; DDR: Desired data rate; DRE: Data rate efficiency; FCC: Federal Communications Commission; HF: High frequency; LoS: Line-of-sight; MN: Mobile node; NLoS: Non-line-of-sight; PU: Primary user; QoE: Quality of experience; QoS: Quality of service; RF: Radio frequency; SINR: Signal-to-interference-plus-noise ratio; SMDP: Semi-Markov decision process; SNR: Signal to noise ratio; UHF: Ultra-high frequency; V2I: Vehicle-to-infrastructure; V2V: Vehicle-to-vehicle; VANET: Vehicular ad-hoc network; VHF: Very high frequency

Dataset of simulations
The simulation was performed using MATLAB in Intel Core i7 (32bit). The operation time and supportable data rate is made of the mean and variance in Table 1 by Gaussian function using MATLAB. The Q-table is made up of tables as defined in the paper, and it works according to the Q-table update equation.

Funding
This work was supported by a grant-in-aid of Hanwha Systems and the Agency for Defense Development (ADD) in the Republic of Korea as part of the Contract UC160007ED.

Authors' contributions
All authors contribute to the concept, the design and developments of the theory analysis and algorithm, and the simulation results in this manuscript. All authors read and approved the final manuscript.

Authors' information
- Prof. Sang-Jo Yoo, PhD (Corresponding author): Sang-Jo Yoo received the B.S. degree in electronic communication engineering from Hanyang University, Seoul, South Korea, in 1988, and the M.S. and Ph.D. degrees in electrical engineering from the Korea Advanced Institute of Science and Technology, in 1990 and 2000, respectively. From 1990 to 2001, he was a Member of the Technical Staff in the Korea Telecom Research and Development Group, where he was involved in communication protocol conformance testing and network design fields. From 1994 to 1995 and from 2007 to 2008, he was a Guest Researcher with the National Institute Standards and Technology, USA. Since 2001, he has been with Inha University, where he is currently a Professor with the Information and Communication Engineering Department. His current research interests include cognitive radio network protocols, ad-hoc wireless network, MAC and routing protocol design, wireless network QoS, and wireless sensor networks.

- Mr. Sung-Jeen Jang: Sung-Jeen Jang received a B.S degree in electrical engineering from Inha University Incheon, Korea, 2007. He received his M.S. degree in Graduate School of Information Technology and Telecommunication, Inha University, Incheon Korea, 2009. Since March 2009, he has been pursuing a Ph.D. degree at the Graduate School of Information Technology and Telecommunication, Inha University, Incheon Korea. His current research interests include cognitive radio network protocols and machine learning applied wireless communications.
- Dr. Chul-Hee Han: Chulhee Han received the B.S. degree in Electronic Engineering from Chung-ang University, Korea, in 1997, and M.S. and Ph.D. degrees in Electrical and Electronic Engineering from Yonsei University, Korea, in 1999 and 2007, respectively. Currently, he is working at Hanwha Systems, Korea, as a chief engineer. He was involved in various projects including tactical mobile WiMAX system and tactical LOS PMP radio. His research interests include Tactical Broadband Communications, Combat Network Radio, and Cognitive Radio for Military Applications.
- Dr. Kwang-Eog Lee: Kwang-Eog Lee received the B.S. and M.S. degrees in electronic engineering from Kyungpook National University, Daegu, South Korea, in 1988 and 1990, respectively. He has been working in Agency for Defense Development since 1990. From 2007 to 2008, he was an exchange scientist in CERDEC (Communications-Electronics Research, Development and Engineering Center) U.S. Army. Currently, he is a principal researcher and his research interests include cognitive radio and terrestrial and satellite tactical communication.

Competing interests

The authors declare that they have no competing interests.

Author details

[1]Department of Information and Communication Engineering, Inha University, 253 YongHyun-dong, Nam-gu, Incheon, South Korea. [2]Hanwha Systems 188, Pangyoyeok-Ro, Bundang-Gu, Seongname-Si, Gyeonggi-Do 13524, South Korea. [3]Agency for Defense Development, P.O.Box 35, Yuseong-Gu, Daejeon, South Korea.

References

1. R. Marsden, B. Soria, H.M. Ihle, *Effective Spectrum Pricing: Supporting Better Quality and more Affordable Mobile Services* (GSMA, London, UK, 2017) Tech. Rep
2. FCC, ET Docket No 03-222, Notice of Proposed Rule Making and Order. (2003)
3. FCC, ET Docket No 03-237, Notice of Proposed Rule Making and Order. (2003)
4. J. Miltola, *Cognitive radio: An Integrated Agent Architecture for Software Defined radio* (Doctor of Technology, Royal Inst. Technol. (KTH), Stockholm, 2000)
5. I.F. Akyildiz et al., Next generation/dynamic spectrum access/cognitive radio wireless networks: a survey. Comp. Networks J. **50**(13), 2127–2159 (2006)
6. Unlicensed operation in the TV broadcast bands and additional spectrum for unlicensed devices below 900 MHz and in the 3 GHz band, Third Memorandum Opinion and Order, (FCC 2012), http://transition.fcc.gov/Daily_Releases/Daily_Business/2012/db0405/FCC-12-36A1.pdf. Accessed 25 Jun 2018
7. Report ITU-R M.2330–0 (11/2014) ITU-R M.2330–0 Cognitive radio Systems in the Land Mobile Service
8. DARPA Developing Cognitive radio IC Technology for Next-Gen Communications, Radar, and Electronic Warfare, (IDST 2018) www.darpa.mil/program/computational-leverage-against-surveillance-systems. Accessed 25 Jun 2018
9. DARPA, DARPA-BAA-11-61, Computational Leverage against Surveillance Systems (CLASS), (2011)
10. The Spectrum Collaboration Challenge. (DARPA, 2018), https://spectrumcollaborationchallenge.com. Accessed 25 Jun 2018
11. M. Vishram, L.C. Tong, C. Syin, A channel allocation based self-coexistence scheme for homogeneous ad-hoc networks. IEEE Wireless Commun. Lett. **4**(5), 545–548 (2015)
12. S. Maghsudi, S. Stańczak, Hybrid centralized–distributed resource allocation for device-to-device communication underlaying cellular networks. IEEE Trans. Veh. Technol. **65**(4), 2481–2495 (2016)
13. Y. Han, E. Ekici, H. Kremo, O. Altintas, Throughput-efficient channel allocation algorithms in multi-channel cognitive vehicular networks. IEEE Trans. Wirel. Commun. **16**(2), 757–770 (2017)
14. M. Li, L. Zhao, H. Liang, An SMDP-based prioritized channel allocation scheme in cognitive enabled vehicular ad hoc networks. IEEE Trans. Veh. Technol. **66**(9), 7925–7933 (2017)
15. O. Semiari, W. Saad, M. Bennis, Joint millimeter wave and microwave resources allocation in cellular networks with dual-mode base stations. IEEE Trans. Wirel. Commun. **16**(7), 4802–4816 (2017)
16. L. Liang, J. Kim, S.C. Jha, K. Sivanesan, G.Y. Li, Spectrum and power allocation for vehicular communications with delayed CSI feedback. IEEE Wireless Commun. Lett. **6**(4), 458–461 (2017)
17. P. Simon, *Too Big to Ignore: The Business Case for Big Data* (Wiley, (2013)
18. M.L. Littman, Reinforcement learning improves behavior from evaluative feedback. Nature **521**(7553), 445–451 (2015)
19. A. Asheralieva, Y. Miyanaga, An autonomous learning-based algorithm for Joint Channel and power level selection by D2D pairs in heterogeneous cellular networks. IEEE Trans. Commun. **64**(9), 3996–4012 (2016)
20. M. Srinivasan, V.J. Kotagi, C.S.R. Murthy, A Q-learning framework for user QoE enhanced self-organizing spectrally efficient network using a novel inter-operator proximal spectrum sharing. IEEE J. Sel. Areas Commun. **34**(11), 2887–2901 (2016)
21. K. Salma, S. Reza, G.S. Ali, Learning-based resource allocation in D2D communications with QoS and fairness considerations. Eur. Trans. Telecommun. **29**(1), 2161–3915 (2017)
22. E. Fakhfakh, S. Hamouda, Optimised Q-learning for WiFi offloading in dense cellular networks. IET Commun. **11**(15), 2380–2385 (2017)
23. M. Yan, G. Feng, J. Zhou, S. Qin, Smart multi-RAT access based on multiagent reinforcement learning. IEEE Trans. Veh. Technol. **67**(5), 4539–4551 (2018)
24. V. Maglogiannis, D. Naudts, A. Shahid, I. Moerman, A Q-learning scheme for fair coexistence between LTE and Wi-Fi in unlicensed spectrum. IEEE Access **6**, 27278–27293 (2018)
25. N. Xu, H. Zhang, F. Xu, Z. Wang, Q-learning based interference-aware channel handoff for partially observable cognitive radio ad hoc networks. Chin. J. Electron. **26**(4), 856–863 (2017)
26. S.J. Jang, S.J. Yoo, *Reinforcement learning for dynamic sensing parameter control in cognitive radio systems* (IEEE ICTC, 2017), pp. 471–474
27. Y.Y. Liu, S.J. Yoo, *Dynamic resource allocation using reinforcement learning for LTE-U and WiFi in the unlicensed spectrum* (IEEE ICUFN, 2017), pp. 471–475
28. L. Shi, S.J. Yoo, Distributed fair resource allocation for cognitive femtocell networks. Wireless Personal Commun. **93**(4), 883–902 (2017)
29. Fairness measure, (Wikipedia 2018) https://en.wikipedia.org/wiki/Fairness_measure. Accessed 25 Jun 2018
30. Q. Zhang, Q. Du, K. Zhu, *User Pairing and Channel Allocation for Full-Duplex Self-Organizing Small Cell Networks* (WCSP, Nanjing, 2017), pp. 1–6
31. M. Yan, G. Feng and S. Qin, Multi-RAT access based on multi-agent reinforcement learning. IEEE GLOBECOM, 1–6 (2017)
32. F. Zeng, H. Liu and J. Xu, Sequential channel selection for decentralized cognitive radio sensor network based on modified Q-learning algorithm. ICNC-FSKD, 657–662 (2016)

8

Methods of increasing two-way transmission capacity of wireless ad hoc networks

Kan Yu[1], Guangshun Li[2*], Jiguo Yu[2] and Lina Ni[1]

Abstract

Two-way communication is required to support control functions like packet acknowledgement and channel feedback. Most previous works on the transmission capacity of wireless ad hoc networks, however, focused on one-way communication; reverse communication from the destination to the source was ignored. In this paper, we first establish mathematical expression for two-way transmission capacity under the fixing transmission distance (i.e., the distance between the source and the destination is a constant), by introducing the concept of two-way outage and setting different rate requirements in both directions. Next, based on the concept of guard zone and cooperative communication, methods of increasing two-way transmission capacity are proposed. Simulation results show that the proposed methods can improve two-way transmission capacity significantly.

Keywords: Two-way transmission capacity, Guard zone, Cooperative communication

1 Introduction

Recently, more and more attentions have focused on Internet of Things (IoT) applications. The coexistence of a massive number of IoT devices poses a challenge in maximizing the successful transmission capacity of the overall network alongside reducing the multi-hop transmission delay in order to support mission critical applications [1, 2]. As one of the most important and critical part of the IoTs, considerable progress has been made in the field of wireless ad hoc networks, particularly with respect to improving their transmission performance [3]. Two-way communication is one of the fundamental communication methods, such as state feedback among IoT devices, data acknowledgement or route initiation, and update requests. Two-way transmission capacity of a wireless ad hoc network refers to the maximum number of successful transmissions existing per unit network area, constrained by two-way outage probability. The transmission capacity provides a framework to derive closed-form bounds for the interference distribution by using stochastic geometry when the locations of nodes form a Poisson point process

(PPP). Most of the previous works on deriving transmission capacity only focused on one-way transmission capacity (reverse communication from the destination to the source is ignored), which considered the effect of various physical and medium access layer techniques such as successive interference cancellation [4], multiple antennas [5–8], guard zone-based scheduling [9], and cooperative relaying [10, 11].

In the landmark work [12], Truong et al. developed the concept of transmission capacity of two-way communication in wireless ad hoc networks with the concept of a two-way outage. Specifically, they derived an upper bound and an approximation for two-way transmission capacity, which are shown to be relatively tight for small outage probability constraints. Finally, they concluded that the two-way capacity loss is considerable by numerical and simulation results.

First, we consider two-way transmission in a wireless ad hoc network, where each source destination pair has data to exchange from each other and their locations are modeled as a PPP. The success probability with two-way transmission is the probability that the communication in both directions (from source to destination and from destination to source) is successful simultaneously. Then, the two-way outage probability is 1—success probability.

*Correspondence: guangshunli@sohu.com
[2] School of Information Science and Engineering, Qufu Normal University, Rizhao, 276826 Shandong, China
Full list of author information is available at the end of the article

However, the interference received in both directions is correlated, since the distance between the source and the receiver of interfering transmission and another distance between the destination and the sender of interfering transmission are not independent from each other. Moreover, explicitly computing the correlation between the above two kinds of distances is also a hard problem. In [12], Truong et al. assumed that the interference received in both directions is independent to derive the exact outage probability and transmission capacity of two-way simply.

Instead, to get a meaningful insight into two-way transmission capacity without the simple assumption of independence, we derive lower and upper bounds for outage probability and transmission capacity of two-way. The main contributions of this paper can be summarized as follows:

Considering correlation of interference in both directions and using tools of stochastic geometry, we derive lower and upper bounds for outage probability and transmission capacity of two-way. Specifically, the difference between the lower and upper bounds of two-way transmission capacity is only constrained by a constant $\frac{1}{2} + \frac{1}{\alpha}$, which is consistent with the result in [13], where α is the path-loss exponent.

To increase two-way transmission capacity, we introduce the concept of guard zone, which can be modeled as a disc of radius φ centered at the receiving node; interfering transmissions cannot be allowed to exist within this disc. Specifically, the difference between the lower and upper bounds of two-way transmission capacity in this case is still $\frac{1}{2} + \frac{1}{\alpha}$. By setting properly a guard zone size, we can ignore safely the interference outside φ.

Next, combining cooperative communication with guard zone, two-way transmission capacity can further be increased. Under decode-and-forward relaying scheme, we give a method to find the optimal relay node to achieve maximum successful transmitting nodes per unit area to satisfy outage probability and data rates.

Finally, theoretical analyses are evaluated by simulation results.

2 Methods

For the purpose of deriving expression of two-way transmission capacity in a wireless network, considering guard zone and cooperative communication strategies, we apply FKG inequality, Cauchy-Schwarz inequality to derive outage probability, and two-way transmission capacity. Finally, we utilize the simulator MATLAB to evaluate the performance of guard zone and cooperative communication.

The results of this paper in part have been presented in [12] and [13]. The differences between [12] and [13] and the present paper are as follows. For simplification

of analysis, [12] assumed that the successful reception events in two directions are independent. The independence assumption was removed in [13] and the upper and lower bounds on two-way transmission capacity were shown to be tight. Compared to [13], the present paper introduces the concept of guard zone and cooperative communication to quantify the increment in bidirectional transmission capacity. In addition to this, this paper offers more additional simulation results for more insights into the effects of two-way communication.

The rest of this paper is organized as follows. We summarize the related work in Section 3. In Section 4, we introduce network model and definitions. In Sections 5, 6, and 7, we give the expression of two-way transmission capacity of a wireless ad hoc network. Simulation results are shown in Section 8. Finally, the conclusion and future work are given in Section 9.

3 Related work

In the past few years, most of previous works focused on the transmission capacity of one way (e.g., [4, 5, 9, 14–20]). The concept of one-way transmission capacity was first introduced in [14] and [15] as a way to evaluate the performance of specific communication strategies and different MAC protocols. This performance metric can be used to characterize a decentralized wireless ad hoc network under an outage constraint [14].

By using tools of stochastic geometry to quantify the interference among multiple nodes in the network, in [15], Weber et al. determined the relationship between the optimal spatial density and success probability of transmissions in the network and presented tight upper and lower bounds on transmission capacity via lower and upper bounds on outage probability of one-way. Finally, they applied these results to show how transmission capacity can be used to better understand scheduling, power control, and the deployment of multiple antennas in a decentralized network.

However, these prior works all concentrated on investigating transmission capacity in one-way ad hoc networks (communication from the destination to the source is ignored). Specifically, two-way communication is needed, such as state feedback, packet acknowledgement, and update request. In [12], Truong et al. first developed the concept of transmission capacity of two-way communication in wireless ad hoc networks. An improved version was extended in [13]. In [13], Vaze et al. derived the lower and upper bounds of two-way transmission capacity. The obtained bounds are used to derive the optimal solution for bidirectional bandwidth allocation that maximizes two-way transmission capacity, which is shown to perform better than allocating bandwidth proportional to the desired rate in both directions. Specifically, they showed that an intuitive strategy that allocates the bandwidth in proportion to the desired rate in each direction is optimal only for symmetric traffic (same rate requirement in both directions) and performs poorly for asymmetric traffic in comparison to the optimal strategy. However, they did not consider the problem of how to increase two-way transmission capacity.

Cooperative communication is one of the popular ways to increase transmission capacity. In [10], Lee et al. analyzed the transmission capacity for dual-hop relaying in a wireless ad hoc network in the presence of both co-channel interference and thermal noise. Specifically, they first presented the exact outage probability for amplify-and-forward and decode-and-forward protocols in a Poisson field of interferers, and then, they derived transmission capacity of such networks. For cognitive radio networks, in [11], Jing et al. proposed a cooperative framework in which a primary sender, being aware of the existence of the secondary network, may select a secondary user that is not in transmitting or receiving mode to relay its traffic. The feasible relay location region and optimal power ratio between the primary network and the secondary network are derived in the underlay spectrum sharing model. Based on the optimal power ratio, they derived the maximum achievable transmission capacity of the secondary network under the outage constraints from both the primary and the secondary network with or without cooperative relaying. However, reverse communication from the destination to the source is ignored.

In fact, two-way communication is closely related to full-duplex scheme. In other words, a successful probability of a half-duplex transmission considers one-way success probability from a sending node to a receiving node, which is not suitable for full-duplex transmission cases; both of these successful probabilities of sending and receiving messages are considered. In [21] , Tong and Haenggi considered a wireless network of nodes with both half-duplex and full-duplex capabilities and derived an optimal throughput by using tools of stochastic geometry. In [22], Marašević et al. introduced a new realistic model of a small form-factor (e.g., smartphone) full-duplex receiver and quantified the rate gain as a function of the remaining self-interference and SNR values by considering the multi-channel case. However, their successful probability of a transmission was based on half-duplex rather than full-duplex within the same frequency band and time.

4 Models and definitions

First, we give the definition of Poisson point process (PPP) as follows.

Definition 1 [23] *The PPP* Φ *of intensity measure* Λ *is defined by means of its finite-dimensional distributions:*

$$Pr\left[\Phi(A_1)=n_1, ..., \Phi(A_k)=n_k\right] = \prod_{i=1}^{k}\left(e^{-\Lambda(A_i)}\frac{\Lambda(A_i)^{n_i}}{n_i!}\right),$$

for every $k = 1, 2, ...$ *and all bounded, mutually disjoint sets* A_i *for* $i = 1, ..., k$. *If* $\Lambda(dx) = \lambda dx$ *is a multiple of Lebesgue measure (volume) in* R^d, Λ *is a locally finite non-null measure on* R^d, *we call* Φ *a homogeneous PPP, and* λ *is its intensity parameter.*

It is notable that the Poisson network model can nicely capture the random geometric properties of networks and enable the analytical modeling of network interference statistics in general [24].

4.1 Network model

Consider a wireless ad hoc network consisting of N transmissions, where the sender and its receiver want to exchange data between each other. All senders and receivers construct sender set Φ_T and receiver set Φ_R, respectively. We assume that each node has a single antenna and applies full-duplex scheme. We consider a slotted Aloha random access protocol, where at any given time, the pair of sender-receiver transmits data to each other with an access probability p_a, and the distance between them is fixing value, i.e., d.

The set Φ_T is modeled as a homogenous PPP on a two-dimensional plane with intensity λ_0, similar to [13] and [25]. Due to one-to-one correspondence relationship of each sender-receiver pair, the set Φ_R is also a homogenous PPP on a two-dimensional plane with intensity λ_0. According to the assumed Aloha random access protocol, locations of the active senders and receivers Φ_T^a and Φ_R^a are homogenous PPPs on a two-dimensional plane with intensity $\lambda = p_a\lambda_0$.

4.2 The residual self-interference (RSI)

The existence of self-interference is a main challenge in applying full-duplex scheme. Since the interference cancellation is a challenging problem and the number of self-interference cancellation is often related to the wireless network capacity, the residual self-interference has negative effects on the network capacity. Although previous results generally assumed that self-interference can be completely eliminated, in the realistic applications, the self-interference only can be eliminated to the level of noise in the best case. In this paper, we will describe the residual self-interference as a constant fraction of the transmission power P, that is $RSI = gP$, where g is a constant fraction [21] and P is transmission power of all transmitting nodes.

4.3 Interference model

Specifically, we consider the interference-limited wireless networks where the noise power is ignored. Signal prorogation between senders and receivers is considered under the Rayleigh fading channel. Namely, the strength of received power at receiver r_i transmitted by sender s_i can be represented as Ph_{ii}/d^α, α denotes the path-loss exponent, and h_{ii} denotes the channel fading gain between s_i and r_i, which is an exponentially distributed random variable with unit mean, i.e., $h_{ii} \sim \exp(-1)$ [25, 26]. Furthermore, the received signal-to-interference ratios (SIRs) for the transmissions from s_i to r_i and from r_i to s_i are respectively

$$\gamma_i^f = \frac{Ph_{ii}/d^\alpha}{\sum_{t_j \in \Phi_T^a} Ph_{ji}/d_{ji}^\alpha + gP} \tag{1}$$

and

$$\gamma_i^r = \frac{Ph_{ii}'/d^\alpha}{\sum_{r_j \in \Phi_R^a} Ph_{ji}'/d_{ji}'^\alpha + gP}, \tag{2}$$

where d_{ji} and h_{ji} are the distance and channel fading gain between sender s_j and receiver r_i, respectively. Similarly, d_{ji}' and h_{ji}' are the distance and channel fading gain between receiver r_j and sender s_i, respectively. gP is the remaining self-interference.

The network employs frequency duplexing to support two-way communication, i.e., two separate frequency carriers, of which the bandwidths W_f and W_r are used to transmit data in two directions between each pair of nodes. We have used the subscripts "f" and "r" to indicate the directions, namely *forward direction* and *reverse direction*, respectively. We also refer to the nodes sending information in forward direction as *senders* and their partners as *receivers* [12].

We denote the SIR thresholds in two directions β_f and β_r. The relationship between a decoding threshold and the corresponding transmission rate, i.e., the *forward transmission rate* R_f and the *reverse transmission rate* R_r, can be given by the following equations [12]:

$$R_f = W_f \log_2(1 + \beta_f) bits/sec, \tag{3}$$

$$R_r = W_r \log_2(1 + \beta_r) bits/sec. \tag{4}$$

In this study, we employ the following performance metrics: the probability of failure for two-way transmissions, denoted as the probability that the signal reception is failed in at least one direction of a two-way communication, is given by [12, 13]

$$p_{out} = 1 - \Pr\left[\gamma_i^f \geq \beta_f, \gamma_i^r \geq \beta_r\right];$$

the concept of two-way transmission capacity, as shown in the following definition.

Definition 2 [13] *Two-way transmission capacity, denoted by τ, is defined as*

$$\tau = (1 - \epsilon)p_{out}^{-1}(\epsilon)\left(\frac{R_f + R_r}{W_{total}}\right) bits/sec/Hz/m^2, \tag{5}$$

where ϵ is outage probability constraint, $p_{out}^{-1}(\epsilon)$ denotes the inverse, the maximum spatial density of simultaneously successful two-way links subject to an outage probability constraint of ϵ, and

$$\mu_2(\epsilon) = \max\{\lambda | p_{out} \leq \epsilon\} = p_{out}^{-1}(\epsilon),$$

$W_{total} = W_f + W_r$ is the total available bandwidth.

5 Analysis of two-way transmission capacity

In this section, we derive the upper and lower bounds of two-way transmission capacity.

Lemma 1 *The two-way outage probability of a two-way transmission can be upper-bounded and lower-bounded by*

$$1 - \exp\left[-gd^\alpha(\beta_f + \beta_r) - \lambda\pi C(\alpha)\left(s_f^\delta + s_r^\delta\right)\right],$$

and

$$1 - \exp\left[-gd^\alpha(\beta_f + \beta_r) - \lambda\pi\left(\frac{1}{2} + \frac{1}{\alpha}\right)C(\alpha)\left(s_f^\delta + s_r^\delta\right)\right],$$

respectively, where $C(\alpha) = \Gamma(1 + \delta)\Gamma(1 - \delta)$, $\Gamma(a) = \int_0^{+\infty} t^{a-1}e^{-t}dt$ *is the gamma function,* $s_f = d^\alpha\beta_f$, $s_r = d^\alpha\beta_r$, *and* $\delta = \frac{2}{\alpha}$.

Proof From Slivnyak's theorem [23], the distribution of a point process is unaffected by adding a reference receiver at the origin, from which the sender is d distance away. Therefore, the interference measured at the reference receiver under this conditional point process is the same as the one measured at any place under a homogeneous PPP. Thus, for a forward direction, the received SIR at the reference receiver is given by $\qquad\square$

$$\gamma_o^f = \frac{Ph/d^\alpha}{\sum_{t_j \in \Phi_T^a} Ph_{jo}/d_{jo}^\alpha + gP}, \tag{6}$$

where h is the channel fading gain between the sender and the reference receiver and h_{ko} and d_{ko} are respectively the channel fading gain and the distance between an interferer k the reference receiver.

By shifting the entire point process so that the corresponding sender of the reference receiver lies at the origin, the received SIR at this sender for reverse direction is given by

$$\gamma_o^r = \frac{Ph'/d^\alpha}{\sum\limits_{r_j \in \Phi_R^a} Ph'_{jo}/d_{jo}^{'\alpha} + gP}. \tag{7}$$

The derived process for upper bound of two-way outage probability is given in Formula (8).

$$
\begin{aligned}
p_{out}(\beta_f, \beta_r) &= 1 - \Pr\left[\gamma_i^f \geq \beta_f, \gamma_i^r \geq \beta_r\right] \\
&= 1 - \Pr\left[\gamma_i^f \geq \beta_f, \gamma_i^r \geq \beta_r\right] \\
&= 1 - \Pr\left[\frac{Phd^{-\alpha}}{\sum\limits_{t_j \in \Phi_R^a} Ph_{jo}/d_{jo}^\alpha + gP} \geq \beta_f, \frac{Ph'd^{-\alpha}}{\sum\limits_{r_j \in \Phi_R^a} Ph'_{jo}/d_{jo}^{'\alpha} + gP} \geq \beta_r\right] \\
&= 1 - \Pr\left[h \geq d^\alpha \beta_f \left(\sum_{t_j \in \Phi_T^a} h_{jo}d_{jo}^{-\alpha} + g\right), h' \geq d^\alpha \beta_r \left(\sum_{r_j \in \Phi_R^a} h'_{jo}d_{jo}^{'-\alpha} + g\right)\right] \\
&\overset{(*)}{=} 1 - \exp\left(-gd^\alpha(\beta_f + \beta_r)\right) \mathrm{E}\left[\exp\left(-s_f \sum_{t_j \in \Phi_T^a} h_{jo}d_{jo}^{-\alpha}\right) \cdot \exp\left(-s_r \sum_{r_j \in \Phi_R^a} h'_{jo}d_{jo}^{'-\alpha}\right)\right] \\
&\overset{(**)}{\leq} 1 - \exp\left(-gd^\alpha(\beta_f + \beta_r)\right) \mathcal{L}_{\Phi_T^a, I}(s_f) \cdot \mathcal{L}_{\Phi_R^a, I}(s_r) \\
&= 1 - \exp\left(-gd^\alpha(\beta_f + \beta_r)\right) \exp\left[-\lambda\pi\Gamma(1+\delta)\Gamma(1-\delta)\left(s_f^\delta + s_r^\delta\right)\right]
\end{aligned}
\tag{8}
$$

where $\mathcal{L}_{\Phi_T^a, I}(s_f)$ and $\mathcal{L}_{\Phi_R^a, I}(s_r)$ denote the Laplace transform of the interference evaluated at s_f and s_r, respectively, $s_f = d_0^\alpha \beta_f$ and $s_r = d_0^\alpha \beta_r$.

Based on the fact that d_{jo} and d'_{jo} are not independent [13], although we cannot establish an equation to evaluate the expectation with respect to d_{jo} and d'_{jo} in Formula (8), we derive an upper bound by applying FKG inequality [27], i.e., inequality (**) holds in Formula (8). Moreover, term (*) holds since $\Pr[h_{xy} \geq z] = \exp(-z)$ for an exponentially distributed random h_{xy} with unit mean [26].

Using the Cauchy-Schwarz inequality, we derive the lower bound of two-way outage probability, as shown in Formula (9).

$$
\begin{aligned}
p_{out}(\beta_f, \beta_r) &\overset{(*)}{=} 1 - \exp\left(-gd^\alpha(\beta_f + \beta_r)\right) \mathrm{E}\left[\exp\left(-s_f \sum_{t_j \in \Phi_T^a} h_{jo}d_{jo}^{-\alpha}\right) \cdot \exp\left(-s_r \sum_{r_j \in \Phi_R^a} h'_{jo}d_{jo}^{'-\alpha}\right)\right] \\
&= 1 - \exp\left(-gd^\alpha(\beta_f + \beta_r)\right) \mathrm{E}\left[\prod_{t_j \in \Phi_T^a} \exp\left(-s_f h_{jo}d_{jo}^{-\alpha}\right) \prod_{r_j \in \Phi_R^a} \exp\left(-s_r h'_{jo}d_{jo}^{'-\alpha}\right)\right] \\
&\overset{(**)}{\geq} 1 - \exp\left(-gd^\alpha(\beta_f + \beta_r)\right) \left\{\mathrm{E}\left[\prod_{t_j \in \Phi_T^a} \left[\exp\left(-s_f h_{jo}d_{jo}^{-\alpha}\right)\right]^2\right]\right. \\
&\quad \left. \mathrm{E}\left[\prod_{r_j \in \Phi_R^a} \left[\exp\left(-s_r h'_{jo}d_{jo}^{'-\alpha}\right)\right]^2\right]\right\}^{\frac{1}{2}} \\
&= 1 - \exp(-gd^\alpha(\beta_f + \beta_r)) \exp\left[-\lambda\pi\left(\frac{1}{2} + \frac{1}{\alpha}\right)s_f^\delta \Gamma(1-\delta)\Gamma(1+\delta)\left(s_f^{\frac{2}{\alpha}} + s_r^{\frac{2}{\alpha}}\right)\right]
\end{aligned}
\tag{9}
$$

where inequality (**) holds according to the Cauchy-Schwarz inequality, and calculating process is given in Eq. (11).

For the Laplace transform of the interference in Formula (8), we flip the order of integration and expectation, and $\lambda'(r) = 2\pi\lambda r$ is the intensity function of PPP Φ_T^a. Then, we calculate the integral, corresponding calculating process given in Eq. (10).

$$
\begin{aligned}
\mathrm{E}\left[\exp\left(-s_f \sum_{t_j \in \Phi_T^a} h_{jo}d_{jo}^{-\alpha}\right)\right] &= \exp\left\{-\mathrm{E}\left[\int_0^\infty \left(1 - \exp(-s_f h r^\alpha)\right)\lambda'(r)dr\right]\right\} \\
&= \exp\left\{-\mathrm{E}\left[\lambda\pi\int_0^\infty \left(1 - \exp(-s_f h x^{-1})\right)\delta x^{\delta-1}dx\right]\right\}, \ \delta = \frac{2}{\alpha}, x = r^{\frac{1}{\delta}} \\
&= \exp\left\{-\lambda\pi\mathrm{E}\left[(s_f h)^\delta \int_0^\infty t^{(1-\delta)-1}\exp(-t)dt\right]\right\}, \ t = s_f h x^{-1} \\
&= \exp\left\{-\lambda\pi\mathrm{E}\left[(s_f h)^\delta \Gamma(1-\delta)\right]\right\} = \exp\left\{-\lambda_T \pi s_f^\delta \mathrm{E}[h^\delta]\Gamma(1-\delta)\right\} \\
&\overset{(*)}{=} \exp\left[-\lambda\pi s_f^\delta \Gamma(1+\delta)\Gamma(1-\delta)\right],
\end{aligned}
\tag{10}
$$

where (*) holds due to $\mathrm{E}[h^\delta] = \Gamma(1+\delta)$ under Rayleigh fading [28].

$$
\begin{aligned}
\mathrm{E}\left[\prod_{t_j \in \Phi_T^a} \exp\left(-(s_f h_{jo}d_{jo}^{-\alpha})^2\right)\right] &\overset{(*)}{=} \exp\left[-\lambda\int_{R^2} 1 - \left(\frac{1}{1 + s_f x^{-\alpha}}\right)^2 dx\right] \\
&= \exp\left[-2\pi\lambda\int_R \left(\frac{s_f^2 x^{-2\alpha+1} + 2s_f x^{-\alpha+1}}{(1 + s_f x^{-\alpha})^2}\right)dx\right]^{\frac{1}{2}}, \ \delta = \frac{2}{\alpha}, x = r^{\frac{1}{\delta}} \\
&\overset{(**)}{=} \exp\left(-\lambda\pi\left(\frac{1}{2} + \frac{1}{\alpha}\right)s_f^\delta \Gamma(1-\delta)\Gamma(1+\delta)\right),
\end{aligned}
\tag{11}
$$

where (*) functional of PPP [29], (**) holds according to results in [13].

Theorem 1 *Two-way transmission capacity is lower and upper-bounded by*

$$\tau \geq (1 - \epsilon) \frac{\ln\left(\frac{1}{1-\epsilon}\right) - gd^\alpha(\beta_f + \beta_r)}{\pi C(\alpha)\left(s_f^\delta + s_r^\delta\right)} \left(\frac{R_f + R_r}{W_{total}}\right)$$

$$\tau \leq (1 - \epsilon) \frac{\ln\left(\frac{1}{1-\epsilon}\right) - gd^\alpha(\beta_f + \beta_r)}{\pi C(\alpha)\left(\frac{1}{2} + \frac{1}{\alpha}\right)\left(s_f^\delta + s_r^\delta\right)} \left(\frac{R_f + R_r}{W_{total}}\right),$$

respectively.

Most importantly, we can see that the upper and lower bounds of two-way transmission capacity only differ by a constant.

6 Two-way transmission capacity with guard zone

In ad hoc networks, it may be helpful to suppress interfering transmissions around the receiving nodes in order to increase the probability of successful communication. In this section, we introduce the concept of a guard zone, defined as the region around each receiving node where interfering transmissions are inhibited. Using stochastic geometry, the relationship between guard zone size and two-way outage probability (or two-way transmission capacity) is established.

Define the guard zone of a receiving node as a disc of radius φ, denoted by D_φ; potential transmissions inside this disc are inhibited. Before transmitting data, nodes with some fraction of transmission power broadcast message *stop* to interrupt transmissions within their guard zone. Thus, for a forward direction, the received SIR at the reference receiver is given by

$$\gamma_{0,\varphi}^f = \frac{Ph/d^\alpha}{\sum\limits_{t_j \in \Phi_T \cap \bar{D}_\varphi} Ph_{jo}/d_{jo}^\alpha + gP},$$

where $t_j \in \Phi_T \cap \bar{D}_\varphi$ denotes the set of nodes transmitting simultaneously while potential senders inside the disc D_φ are inhibited, where $\bar{D}_\varphi$ denotes the area out of the guard zone.

The only difference on calculating outage probability is the lower bound of integration, and we get Eq. (12).

$$\mathbb{E}\left[\exp\left(-s_f \sum_{t_j \in \Phi_T^a} h_{jo}d_{jo}^{-\alpha}\right)\right] = \exp\left\{-\mathbb{E}\left[\int_\varphi^\infty \left(1 - \exp\left(-s_f h r^\alpha\right)\right) \lambda'(r) dr\right]\right\}$$

$$= \exp\left\{-\lambda\pi \mathbb{E}\left[\int_{\varphi^{1/\delta}}^\infty \left(1 - \exp\left(-s_f h x^{-1}\right)\right) \delta x^{\delta-1} dx\right]\right\} \quad \delta = \frac{2}{\alpha} \text{ and } x = r^{1/\delta}$$

$$= \exp\left\{-\lambda\pi \mathbb{E}\left[(sh)^\delta \int_{sh\frac{1}{\varphi^{1/\delta}}}^\infty t^{(1-\delta)-1} \exp(-t) dt\right]\right\} \quad t = shx^{-1}$$

$$= \exp\left[-\lambda\pi s_f^\delta \Gamma(1+\delta) \Gamma\left(1-\delta, s\frac{1}{\varphi^{1/\delta}}\right)\right],$$

(12)

where $\Gamma(s,x) = \int_x^\infty t^{s-1}e^{-t}dt$ being the upper incomplete gamma function.

Lemma 2 *Using the guard zone, the two-way outage probability of a two-way transmission can be upper-bounded and lower-bounded by*

$$1 - \exp\left[-gd^\alpha(\beta_f + \beta_r) - \lambda\pi C(\alpha,\varphi)\left(s_f^\delta + s_r^\delta\right)\right],$$

and

$$1 - \exp\left[-gd^\alpha(\beta_f + \beta_r) - c\left(s_f^\delta + s_r^\delta\right)\right],$$

respectively, where $C(\alpha,\varphi) = \Gamma(1+\delta)\Gamma(1-\delta,\varphi)$, $c = \lambda\pi\left(\frac{1}{2}+\frac{1}{\alpha}\right)C(\alpha,\varphi)$ *and* $\Gamma(s,x) = \int_x^{+\infty}t^{s-1}e^{-t}dt$ *is the upper incomplete gamma function.*

Theorem 2 *Two-way transmission capacity with guard zone is lower and upper-bounded by*

$$\tau \geq (1-\epsilon)\frac{\ln\left(\frac{1}{1-\epsilon}\right) - gd^\alpha(\beta_f + \beta_r)}{\pi C(\alpha,\varphi)\left(s_f^\delta + s_r^\delta\right)}\left(\frac{R_f + R_r}{W_{total}}\right)$$

$$\tau \leq (1-\epsilon)\frac{\ln\left(\frac{1}{1-\epsilon}\right) - gd^\alpha(\beta_f + \beta_r)}{\pi\left(\frac{1}{2}+\frac{1}{\alpha}\right)C(\alpha,\varphi)\left(s_f^\delta + s_r^\delta\right)}\left(\frac{R_f + R_r}{W_{total}}\right),$$

respectively.

Corollary 1 *The condition for a positive two-way transmission capacity is given by*

$$g \leq \frac{1}{d^\alpha(\beta_f + \beta_r)} \cdot \ln\left(\frac{1}{1-\epsilon}\right).$$

Lemma 3 *If φ satisfies the following inequality, the outage probability of forward transmission is at most σ,*

$$\Gamma(1-\delta,\varphi) \geq g\beta_f d^\alpha - \frac{\ln\left(\frac{1}{1-\sigma}\right)}{\lambda\pi d^2 \beta_f^\delta \Gamma(1+\delta)}.$$

In theory, outage probability can never be 0 due to the existence of channel fading. Using guard zone, we consider a transmission as successful if its success probability is greater than $1 - \sigma$; corresponding size of guard zone is denoted by φ_σ.

7 Two-way transmission capacity with cooperative communication and guard zone

Let S_{relay} denote the set of all relay nodes. We consider a two-phase cooperative protocol. During the broadcasting phase, the sender first broadcasts *probing* message with some fraction of transmission power; the relays which can successfully decode the transmitted signal form a decoding set $C_{dec} \subset S_{relay}$, then the sender, its receiver, and selected relay (selecting process will be given as follows) broadcast *stop* message to interrupt interfering transmissions within φ_σ. In the transmission phase, the sender sends data to the selected relay, and then, the latter transmits towards the receiver. By using the guard zone, interference outside φ_σ can be safely ignored (by using Lemma 3), and there are no interfering transmissions by message broadcasting, then the received SNR at selected relay and receiver are $\frac{Ph_{SR}\cdot d_{SR}^{-\alpha}}{gP}$ and $\frac{Ph_{RD}\cdot d_{RD}^{-\alpha}}{gP}$, respectively.

The received SINR of decode-and-forward (DF) with relay R and guard zone of size φ_σ can be written as [10]

$$\gamma_{SD} = \min\left\{\frac{Ph_{SR}\cdot d_{SR}^{-\alpha}}{gP}, \frac{Ph_{RD}\cdot d_{RD}^{-\alpha}}{gP}\right\}$$

$$= \min\left\{\gamma_{S,R(S)}, \gamma_{R(S),D}\right\},$$

and the forward transmission rate R_f is

$$R_f^{DF} = W_f \min\left\{\log_2\left(1 + \gamma_{S,R(S)}\right), \log_2\left(1 + \gamma_{R(S),D}\right)\right\}.$$

To maximize R_f, the optimal relay for forward transmission, denoted by R^*, must satisfy the following condition

$$R^* : \max\left\{R \in C_{dec}| \min\left\{\gamma_{S,R(S)}, \gamma_{R(S),D}\right\}\right\}.$$

Using R^* as the optimal relay for reverse communication for simplification of analysis.

Proposition 1 *If X_1 is an independent and identically distributed (i.i.d.) exponential random variable with parameter λ_1, the probability density function (PDF) of the new random variable $X = \frac{X_1}{a}$ for constant a is given by*

$$p_X(x) = \begin{cases} \lambda_1 a e^{-\lambda_1 ax}, & x \geq 0 \\ 0, & otherwise \end{cases}.$$

(13)

Proposition 2 *If X_1 and X_2 are two i.i.d. exponential random variables with parameters λ_1 and λ_2, the PDF of the new random variable $X = \min\{X_1, X_2\}$ is given by*

$$p_X(x) = \begin{cases} (\lambda_1 + \lambda_2)e^{-(\lambda_1+\lambda_2)x}, & x \geq 0 \\ 0, & otherwise. \end{cases}$$

Proof Due to independence of random variables X_1 and X_2, we have

$$\begin{aligned} F_X(x) &= \Pr[X \leq x] = 1 - \Pr[X > x] \\ &= 1 - \Pr[X_1 > x, X_2 > x] \\ &= 1 - \Pr[X_1 > x] \cdot \Pr[X_2 > x] \\ &= 1 - (1 - F_{X_1}(x))(1 - F_{X_2}(x)), \end{aligned}$$

where $F_X(\cdot)$ is the cumulative distribution function of random variable X. □

Applying Propositions 1 and 2, the outage probability, considering high SNR, can be written as

$$\begin{aligned} p_{out} &= \Pr\left[\min\{\gamma_{SR}, \gamma_{RD}\} < R_f, \min\{\gamma_{DR}, \gamma_{RS}\} < R_r\right] \\ &= \Pr[\min\{\gamma_{SR}, \gamma_{RD}\} < R_f] \cdot \Pr[\min\{\gamma_{DR}, \gamma_{RS}\} < R_r] \\ &= (\lambda_1 + \lambda_2)^2 \int_0^{R_f} e^{-(\lambda_1+\lambda_2)x} dx \cdot \int_0^{R_r} e^{-(\lambda_1+\lambda_2)x} dx \\ &= \left[1 - e^{-(\lambda_1+\lambda_2)R_f}\right] \cdot \left[1 - e^{-(\lambda_1+\lambda_2)R_r}\right] \\ &= \left[1 - e^{-g(d_{SR}^{-\alpha}+d_{RD}^{-\alpha})R_f}\right] \cdot \left[1 - e^{-g(d_{SR}^{-\alpha}+d_{RD}^{-\alpha})R_r}\right] \\ &\leq \left[1 - e^{-g(d_{SR}^{-\alpha}+d_{RD}^{-\alpha})R_{\max}}\right]^2, \end{aligned}$$

where $R_{\max} = \max\{R_f, R_r\}$.

Therefore, to satisfy the given outage probability ϵ and data rate (i.e., quality of service (QoS)), relay node must be properly selected with the following constraint

$$\frac{1}{d_{SR}^{\alpha}} + \frac{1}{d_{RD}^{\alpha}} \leq \frac{1}{gR_{\max}} \ln\left(\frac{1}{1 - \epsilon^{\frac{1}{2}}}\right), \quad R \in C_{dec}. \qquad (14)$$

Theorem 3 *Two-way transmission capacity with cooperative communication is lower bounded by*

$$\tau \geq (1 - \epsilon)\lambda \left(\frac{R_f + R_r}{W_{total}}\right) bits/sec/Hz/m^2$$

if the selected relay satisfies Inequality (14).

8 Results and discussion

8.1 Results

Simulations are carried out on networks constructed by randomly placing nodes on $100 \times 100\text{m}^2$. The distance between a sender and its corresponding receiver is 10 m; related SIR parameters are set to $P = 1$ mW, $\beta_f = 1$ dB, $\beta_r = 1$ dB, $W_f = 0.99$ MHz, $W_r = 0.01$ MHz, $\alpha = 4$, and $g = 0$, as shown in Table 1. Based on Slivnyak's theorem, we place an additional sender on the origin and the coordinate of its

Table 1 Parameters and meanings

Symbol	Meaning
Φ_T	PPP of sender set
Φ_R	PPP of receiver set
p_a	Access probability
d	Fixing distance between the source and the destination
h_{ii}	Channel fading gain between sender s_i and receiver r_i
τ	Two-way transmission capacity
R_f	Forward transmission rate
R_r	Reverse transmission rate
ϵ	Outage probability constraint
φ	Radius of guard zone

receiver is (10, 0). The density of relay nodes is set to 0.1, $\epsilon = 0.1$ and guard zone size is set to 10 m. Each reported result in the following parts is the average of 1000 runs, unless otherwise specified.

We first consider the impact of node density on two-way outage probability, as shown in Fig. 1. Outage probability increases over node density increasing, since cumulative interference gets greater at the typical receiver. Moreover, outage probability by using cooperative communication decreases largely compared with results in [12, 13] and that of using guard zone, which means that the theoretical analysis for cooperative communication is effective. On average, the decrements are 91.36% and 74.96%, respectively.

Next, we consider the influence of the transmission distance. As shown in Fig. 2, over d increasing, two-way outage probability increases. This is because, on the one

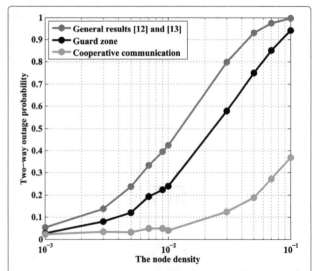

Fig. 1 Two-way outage probability vs. node density. Legends: blue solid line denotes results in [12, 13]; black and red solid lines are results with guard zone and cooperative communication, respectively

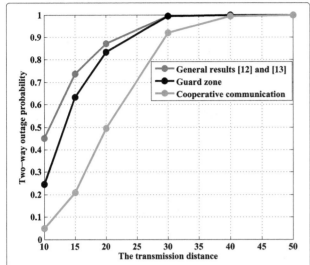

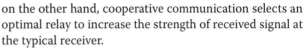

Fig. 2 Two-way outage probability vs. transmission distance. Legends: blue solid line denotes results in [12, 13]; black and red solid lines are results with guard zone and cooperative communication, respectively

hand, the strength of received signal at the typical receiver increases due to longer transmission distance; on the other hand, guard zone size is set to 10 m; when transmission distance is greater than 10 m, there may exist more interferers within d and outside φ, and the typical receiver suffers from more interference. Setting d =10 m, the influence of guard zone size is shown in Fig. 3. We can see that a greater guard zone leads to a smaller two-way outage, since, on the one hand, guard zone interrupts interfering transmissions within φ around the typical receiver;

on the other hand, cooperative communication selects an optimal relay to increase the strength of received signal at the typical receiver.

Finally, we consider the impact of two-way outage probability two-way transmission capacity. As shown in Figs. 4 and 5, two-way transmission capacity first increases and then decreases. The reason is that capacity expressions are proportional to $(1-\epsilon)\ln\left(\frac{1}{1-\epsilon}\right)$. Intuitively, as the outage probability ϵ approaches towards 1, a high density of links is allowed in a unit area; however, most of the links fail; therefore, the amount of successfully received information actually decreases.

Compared with results in [12, 13], the proposed methods can decrease two-way outage probability obviously, then two-way transmission capacity is increased according to Theorems 1, 2, and 6.

8.2 Discussion
To simplify deriving process of outage probability, we assume that the distance between the source and the destination is fixing. In our future research, we will tackle the following two limitations that exist in almost all existing research: first, our theoretical analysis considers a more practical case where the distance between an arbitrary source-destination pair is random; second, our analysis ignores the impacts of noise power and node mobility.

9 Conclusion
In this paper, we study two-way transmission capacity of a wireless ad hoc network by using the tools of stochastic geometry. Furthermore, to increase it, we introduce

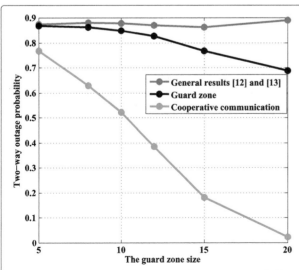

Fig. 3 Two-way outage probability vs. guard zone size. Legends: blue solid line denotes results in [12, 13]; black and red solid lines are results with guard zone and cooperative communication, respectively

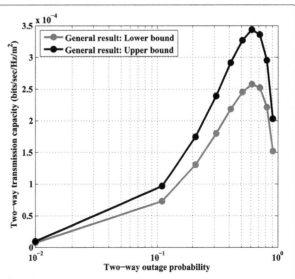

Fig. 4 Two-way transmission capacity vs. outage constraint ϵ. Legends: blue and black solid lines denote lower and upper bounds of results in [12, 13], respectively

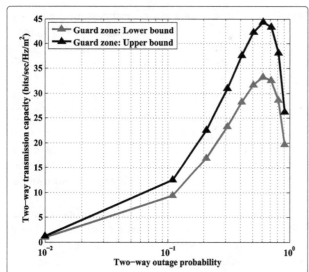

Fig. 5 Two-way transmission capacity vs. outage constraint ϵ.
Legends: blue and black solid lines denote lower and upper bounds
of results with guard zone, respectively

the concepts of guard zone and cooperative communication; theoretical analysis and simulation results show that the proposed scheme can effectively decrease two-way outage and increase two-way transmission capacity. Remarkably, we give a method of how to select an optimal relay.

Acknowledgements
"Not applicable"

Funding
This work is supported by the National Natural Science Foundation of China (61672321, 61771289), the Shandong Provincial Postgraduate Education Innovation Program (SDYY14052, SDYY15049), the Shandong Provincial Postgraduate Education Quality Curriculum Construction Program, and the Qufu Normal University Science and Technology Plan Project (xkj201525), GIF of Shandong University of Science and Technology (SDKDYC180109), and Shandong Provincial College of Science and Technology Plan Project (J15LN05).

Authors' contributions
KY and JY designed and analyzed the strategies. GL and KY designed the simulations. JY, GL, KY, and LN wrote the paper. All authors read and approved the final manuscript.

Competing interests
The authors declare that they have no competing interests.

Author details
[1]College of Computer Science and Engineering, Shandong University of Science and Technology, Qingdao, 266000 Shandong, China. [2]School of Information Science and Engineering, Qufu Normal University, Rizhao, 276826 Shandong, China.

References
1. M. Farooq, H. ElSawy, Q. Zhu, M. Alouini, Optimizing mission critical data dissemination in massive iot networks (2017). WiOpt. https://doi.org/10.23919/WIOPT.2017.7959930
2. T. Song, R. Li, B. Mei, J. Yu, X. Xing, X. Cheng, A privacy preserving communication protocol for iot applications in smart homes. IEEE Internet Things J. **4**(6), 1844–1852 (2017)
3. H. Shi, R. Prasad, E. Onur, I. Niemegeers, Fairness in wireless networks: issues, measures and challenges. IEEE Commun. Surv. Tutor. **16**(1), 5–24 (2014)
4. S. Weber, J. G. Andrews, X. Yang, G. Veciana, Transmission capacity of wireless ad hoc networks with successive interference cancellation. IEEE Trans. Inf. Theory. **53**(8), 2799–2814 (2007)
5. A. M. Hunter, J. G. Andrews, S. Weber, Capacity scaling of ad hoc networks with spatial diversity. IEEE Trans. Wirel. Commun. **7**(12), 2799–2814 (2008)
6. N. Jindal, J. Andrews, S. Weber, Rethinking MIMO for wireless networks: linear throughput increases with multiple receive antennas. IEEE Int. Conf. Commun. (ICC)., 1–6 (2009). https://doi.org/10.1109/ICC.2009.5199417
7. K. Huang, J. Andrews, R. Heath, D. Guo, R. Berry, Spatial interference cancellation for multi-antenna mobile ad hoc networks. IEEE Trans. Inf. Theory. **58**(3), 1660–1676 (2012)
8. R. Vaze, R. Heath, Transmission capacity of ad-hoc networks with multiple antennas using transmit stream adaptation and interference cancelation. IEEE Trans. Inf. Theory. **58**(2), 780–792 (2012)
9. A. Hasan, J. G. Andrews, The guard zone in wireless ad hoc networks. IEEE Trans. Wirel. Commun. **6**(3), 897–906 (2007)
10. J. Lee, H. Shin, J. T. Kim, J. Heo, Transmission capacity for dual-hop relaying in wireless ad hoc networks. EURASIP J. Wirel. Commun. Netw. (2012). https://doi.org/10.1186/1687-1499-2012-58
11. T. Jing, W. Li, X. Chen, X. Cheng, X. Xing, Y. Huo, T. Chen, H. Choi, T. Znati, Achievable transmission capacity of cognitive radio networks with cooperative relaying. EURASIP J. Wirel. Commun. Netw. (2015). https://doi.org/10.1186/s13638-015-0311-8
12. K. Truong, S. Weber, R. Heath, Transmission capacity of two-way communication in wireless ad hoc networks. IEEE Int. Conf. Commun. (ICC)., 1637–1641 (2009). https://doi.org/arXiv:1009.1460
13. R. Vaze, K. Truong, S. Weber, R. Heath, Two-way transmission capacity of wireless ad-hoc networks. IEEE Trans. Wirel. Commun. **10**(6), 1966–1975 (2011)
14. S. Weber, X. Yang, J. G. Andrews, G. Veciana, Transmission capacity of wireless ad hoc networks with outage constraints. IEEE Trans. Inf. Theory. **51**(12), 4091–4102 (2005)
15. S. Weber, J. G. Andrews, N. Jindal, An overview of the transmission capacity of wireless networks. IEEE Trans. Commun. **58**(12), 3593–3604 (2010)
16. R. Vaze, Transmission capacity of spectrum sharing ad hoc networks with multiple antennas. IEEE Trans. Wirel. Commun. **10**(7), 2334–2340 (2011)
17. F. Baccelli, B. Blaszczyszyn, P. Muhlethaler, An Aloha protocol for multihop mobile wireless networks. IEEE Trans. Inf. Theory. **52**(2), 421–436 (2006)
18. D. Kim, S. Park, H. Ju, D. Hong, Transmission capacity of full-duplex-based two-way ad hoc networks with ARQ protocol. IEEE Trans. Veh. Technol. **63**(7), 3167–3183 (2014)
19. Y. Chen, J. G. Andrews, An upper bound on multihop transmission capacity with dynamic routing selection. IEEE Trans. Inf. Theory. **58**(6), 3751–3765 (2012)
20. H. Ding, C. Xing, S. Ma, G. Yang, Z. Fei, Transmission capacity of clustered ad hoc networks with virtual antenna array. IEEE Trans. Veh. Technol. **65**(9), 6926–6939 (2016)
21. Z. Tong, M. Haenggi, Throughput analysis for full-duplex wireless networks with imperfect self-interference cancellation. IEEE Trans. Commun. **63**, 4490–4500 (2015)
22. L. Maraševiú, J. Zhou, H. Krishnaswamy, Y. Zhong, G. Zussman, Resource allocation and rate gains in practical full-duplex systems. IEEE Trans. Networking. **25**(1), 292–305 (2017)
23. F. Baccelli, B. Blaszczyszyn, Stochastic geometry and wireless networks, volume i-theory. Found. Trends Netw. **3**(3-4), 249–449 (2009)
24. M. Haenggi, R. Ganti, Interference in large wireless networks. Found. Trends Netw. **3**(2), 127–248 (2009)
25. K. Yu, J. Yu, X. Cheng, T. Song, Theoretical analysis of secrecy transmission capacity in wireless ad hoc networks. IEEE WCNC, 1–6 (2017). https://doi.org/10.1109/WCNC.2017.7925621
26. J. Dams, M. Hoefer, T. Kesselheim, Scheduling in wireless networks with rayleigh-fading interference. IEEE Trans. Mob. Comput. **14**(7), 1503–1514 (2015)
27. G. Grimmett, *Percolation*. (Springer-Verlag, 1980)

9

TIHOO: An enhanced hybrid routing protocol in vehicular ad-hoc networks

Shirin Rahnamaei Yahiabadi[1], Behrang Barekatain[1,2*] (iD) and Kaamran Raahemifar[3,4]

Abstract

Recently, vehicular ad hoc network (VANET) is greatly considered by many service providers in urban environments. These networks can not only improve road safety and prevent accidents, but also provide a means of entertainment for passengers. However, according to recent studies, efficient routing has still remained as a big open issue in VANETs. In other words, broadcast storm can considerably degrade the routing performance. To address this problem, this research proposes TIHOO, an enhanced intelligent hybrid routing protocol based on improved fuzzy and cuckoo approaches to find the most stable path between a source and a destination node. In TIHOO, the route discovery phase is limited intelligently using the fuzzy logic system and, by limiting the route request messages, the imposed extra overhead can be efficiently controlled. Moreover, in figure of an intelligent hybrid approach, the improved cuckoo algorithm, which is one of the most effective meta-algorithms especially in the large search space, intelligently selects the most stable and optimal route among known routes by calculating an enhanced fitness function. The simulation results using NS2 tool demonstrate that TIHOO considerably improves network throughput, routing overhead, packet delivery ratio, packet loss ratio, and end-to-end delay compared to similar routing protocols.

Keywords: Hybrid routing protocol, Cuckoo-fuzzy approach, Stable path, Vehicular ad hoc network

1 Introduction

One group of ad hoc networks are mobile ad hoc networks (MANET)—a collection of wireless mobile nodes which act independent of central management with no or little use of infrastructures. These networks are characterized by dynamic topologies, energy limitations, limited bandwidth, etc. A special category of MANETs is the vehicular ad hoc network (VANET) (Fig. 1) that has a number of unique characteristics such as predictable mobility, free of constraint energy, rapid changes in the network topology, etc. [1, 2]. In contrast to wireless sensor networks [3, 4] or MANETs, long-life batteries of vehicles have prevented energy-induced constraints in these networks. In VANETs, communications occur between adjacent vehicles and between vehicles and fixed units placed at special locations such as intersections and parking lots. Accordingly, there are two types of

communication in these networks: vehicle-to-vehicle (V2V) and vehicle-to-infrastructure (V2I). In V2V communications, vehicles that are in adjacent to each other engage in data sharing via dedicated short-range communications (DSRC) and wireless access in vehicular environment (WAVE) [5] among others while in V2I, the vehicles are connected to roadside infrastructures to engage in information exchanges [6]. Inter-vehicular networks improve the traffic efficiency and promote safe driving as well as welfare and comfort of the passengers. However, there are numerous challenges that impact the achievements of vehicular ad hoc networks. These challenges include signal losses, limited bandwidth, security and privacy, connection, and routing. Timely delivery of warning messages at the time of collisions and accidents could prevent more accidents, an objective fulfilled through reliable data packets routing. Designing a routing protocol which could deliver data packets in the shortest possible time and with the lowest number of packet loss is essential to improving the security of vehicles and winning the users' satisfaction. High mobility of vehicles, limitations on wireless sources, and the

* Correspondence: Behrang_Barekatain@iaun.ac.ir

[1]Faculty of Computer Engineering, Najafabad Branch, Islamic Azad University, Najafabad, Iran

[2]Big Data Research Center, Najafabad Branch, Islamic Azad University, Najafabad, Iran

Full list of author information is available at the end of the article

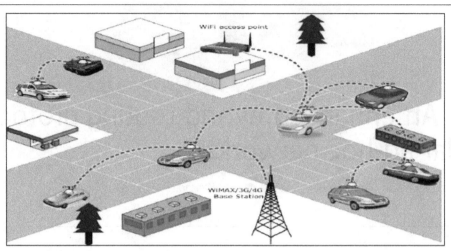

Fig. 1 Vehicular ad hoc network. A special category of MANETs is the vehicular ad hoc network (VANET). In VANETs, communications occur between adjacent vehicles and between vehicles and fixed units placed at special locations such as intersections and parking lots

characteristic wireless channel losses present a serious challenge to carve a route from the source node to the destination node through middle ones. An effective route depends on all the wireless links which make up the route [5, 7, 8].

Routing is faced with serious challenges in these networks. The movement of vehicles constantly changes network topology[9]. In addition, network expansion on a large scale significantly increases routing overheads [2, 10, 11]. Another challenge is to prevent VANET's routing from being trapped in a local optimum [12].

Like other networks such as wireless mesh networks, peer to peer networks use a variety of techniques such as network coding for data sharing in order to enhance the routing effectiveness [13–15]. Various methods have been used to improve routing in VANETs. Some routing protocols draw on the topology of the middle links [16] (occurring between the source and destination nodes) to find the optimal route while some other use the geographical position of vehicles to design a routing protocol [17]. The latter focuses on the location of connections and predicting the next locations of the vehicles to find the appropriate route [18–20]. Fuzzy logic is very effective as well for the introduction of routing protocols [21, 22]. A number of researches apply bio methods for appropriate routing. These methods are very effective for large-scale VANETs and at the same time offer low-complex solutions to computation problems [23–25]. However, and as it was mentioned in research [26], there is no routing protocol in VANET that can satisfactorily perform in every scenario and completely fulfill all routing protocols objectives. Thus, a hybrid method could prove very effective. In order to

address the existing problem in previous studies, this paper proposes TIHOO, an enhanced hybrid routing protocol. This protocol intelligently employed fuzzy [27, 28] and cuckoo [29] approaches by introducing novel fitness functions in which many important parameters for finding the most stable path between the source and the destination node are considered in them.

In other words, the route discovery phase is limited using the fuzzy logic system and, by limiting the route request messages, it somewhat controls the created extra overhead. After identifying the routes in the source node, the cuckoo search algorithm selects the most stable and optimal route among known routes by calculating the fitness function based on criteria such as route lifetime, route reliability, and average available buffer. Simulation results using NS2 (network simulator 2) simulator provides 55.52, 65.85, 79.27, 31.67, and 78.19% in the most important network parameters including packet delivery ratio, throughput, end-to-end delay, routing overhead, and packet loss ratio, respectively.

The rest of this paper is organized as follows. Section 2 is a review of the previous research in this area. In section 3, we provide a brief review over some basic issues. Section 4 explains the method and experimental approach. Section 5 summarizes existing open issues in the related works. Section 6 discusses the proposed routing protocol (TIHOO) in detail and how it addresses the open issues mentioned in Section 5. Performance evaluation parameters and simulation results are discussed in Section 7. Finally, the paper is concluded in Section 8.

2 Related work

As mentioned before, routing is a much challenging issue in VANETs. The existing protocols cannot sufficiently satisfy the routing needs in these networks due to the dynamic nature of VANETs, which is a result of traffic conditions and limitations. This section provides a review of the most important attempts in developing a routing algorithm for VANETs. Adaptive fuzzy multiple attribute decision routing in VANETs (AFMADR) [17] is a location-based protocol in which data packet carriers are employed for selection of the next step. In this protocol, every candid vehicle is specified through distance, direction, road density, and location where each characteristic receives a fuzzy point. The vehicle receiving the highest points is selected for the transference of data packets. This protocol has a desirable profile in the rate of delivery of data packets, transfer delays. However, a more accurate selection of carriers could significantly improve the delivery rate of data packets in this protocol. For example, using city buses which are different locations based on a schedule is a way of selecting suitable carriers. In contrast to AFMADR, TIHOO draws on fuzzy logic for selection of appropriate route request (RREQ) reception links. Stable routes are determined through designing effective fitness functions based on the life of the links, etc. Choosing a stable route improves packet delivery rates and packets lost. Moridi and Barati [22] have proposed a multi-level routing protocol which is based on trust in forbidden search in MANET. In the first step, in order to improve the Ad hoc On-demand Distance Vector routing (AODV) routing protocol, fuzzy logic has been used among cluster members. At this level, the reliability and sustainability of a link are used as the input criterion of the fuzzy system; the ultimate goal is to choose the most reliable link. Afterwards, the best links of cluster heads are presented in a tabular list. The best links are selected based on distance, speed, and direction. This protocol is credited for decreased failures of links and lower losses of packets. However, the overhead has not been considered in this protocol for the purpose of comparison with other protocols. Dharani Kumari and Shylaja [11] have proposed a multi-step geographical routing protocol for V2V communications which has improved Greedy Perimeter Stateless Routing (GPRS) protocol. AHP-based Multimetric Geographical Routing Protocol (AMGRP) improves the data transference mechanism of the geographical routing protocol via four routing criteria of mobility, link life, node density, and node position. Routing criteria are hierarchically organized to be compared in the form of a general unit once similar parameters have been grouped and the sub-criteria of each group are tested and compared with each other. This protocol identifies the next step via computing the function unit weight in a predefined range which could guarantee the improved transfer process. Although AMGRP routing protocol has had some success in improving the rate of data packet delivery compared to other protocols, it has failed in controlling the overhead. F-ANT [2] has designed an improved, fuzzy logic-based framework for ant colony optimization (ACO) protocol. Devised for VANETs, ACO

draws on factors such as bandwidth, the power of received signals, and congestion criterion for recognizing the link credibility. Despite its high rate of packet delivery and low end-to-end delays, this protocol is challenged by high overhead due to employing ant colony algorithm along with fuzzy logic while the proposed metaheuristic algorithm produces considerably lower overheads compared to ant colony algorithm. By combining the advantages of proactive and reactive routing, a reliable routing protocol (R2P) for VANETs was developed by [30]. R2P uses a route discovery mechanism to detect available routes to the destination. It then selects the safest route, which may not be necessarily the shortest one. This routing method is more efficient than previous methods and shortens the delays. However, it is not superior to other methods in term of overhead while its packet delivery is low in some of its default simulation speeds. PFQ-AODV which has been presented in [21] learns the optimal route using Q-learning algorithms and fuzzy constraints on AODV protocol. This protocol employs fuzzy logic and considers a number of criteria such as available bandwidth, the quality of the link, and the direction of the vehicle to assess the wireless link. It then learns the best route through Hello and RREQ messages. The flexible and practical routing that this protocol offers is due to its independence from lower layers. However, it does not have a desirable performance in term of overhead. The intelligent proposed protocol (TIHOO) in this article limits the route discovery through fuzzy logic and decreases the available control packets of the network to keep the overhead in check.

In the method introduced by [31], there is an area-based routing protocol that utilizes fuzzy logic and bacterial foraging optimization algorithm (BFOA) to detect unsafe situation associated with rapid topological changes and the best possible route to the destination node. This protocol applies these three techniques to enhance the sustainability of the selected route. First, the areas are created based on the mobility of vehicles and the characteristic of dispersion while nodes are assigned into different areas. Once a node generates the Hello message from the data transfer phase, it attempts to find the route between the source and the destination points. The fuzzy logic is implemented in the next stage to determine the status of the links based on the quality of the link as well as bandwidth and mobility. BFOA has been applied for finding the best route based on the results of implementing fuzzy logic and the credit of the link. Although this method produces lower delays and faster delivery rates, the computations drastically increase in complexity as a result of increases in population. This is in contrast to the low computation complexity offered by our metaheuristic method [31].

AODV [32] is a Bellman-Ford-based dynamic protocol that creates routes when a node needs to send data packets [33]. AODV comprises three phases of discovery, data transfer, and route maintenance. What primarily distinguishes AODV from other protocols is its use of sequence number. When there is a need for information transfer in this protocol, the source node broadcast RREQ control packet to all neighboring nodes. These nodes keep performing this process until the destination node is confirmed as the receiver node. In the next stage, the destination node unicasts the route reply (RREP) packet to the source node. Data transfer begins once the route has been established. If links are broken in this stage, the maintenance phase will be recalled to help find a new route. AODV suits the networks characterized with high mobility. However, this protocol consumes large amounts of bandwidth due to periodical baking. In [34], this method that attempts to improve the GPRS routing performance controls the network congestion via buffering the nodes and as a result displays higher efficiency in terms of time delays and packet losses. Two factors of the distance between the transfer node and the destination node and the length of the buffers left from the transfer nodes are considered in the proposed method to calculate the possibility of transference by a given node. Those nodes which are assigned higher possibility will be selected as the transference nodes. However, this protocol is a failure in resolving the issue of local optimum and provision of recovery strategy [34]. In [35], the fuzzy control-based AODV routing (FCAR) protocol that takes place on the traditional platform of AODV employs both fuzzy logic and fuzzy control to decide the route. The length of the routes and the percent of vehicles taking the same direction are the two criteria of route assessments. Regarding packet losses and end-to-end delays, the proposed protocol outperforms AODV. However, it produces increased extra overheads compared to AODV if there are few network nodes. In contrast, TIHOO has managed to control the overhead in a relatively satisfactory way via using fuzzy logic for selection of links receiving RREQ and limiting the phase of route discovery.

Using the transfer quality criteria, which is a combination of connections and the rate of data delivery packets, the functionality of each route part is evaluated to be able to select each route part dynamically so that it contains the best path [36]. This method improves the data packet transfer and delivery rates; however, it will go through local maximization and the strategy to deal with it will lead to end-to-end delay as it uses greedy forwarding.

In [37], it is an algorithm predicting the variability of the receiver queue length, the intermediary signal rate, and the noise. The results of this prediction determine the usefulness of the relay vehicles in the candidate set. The results of the utility of the vehicles are based on the

weighting algorithm and weights are the variance of these two criteria. The relay priority of each relay device is determined by its usefulness. This algorithm did not consider the effects of parameters such as packet length and transfer steps in the results obtained.

In [38], there is a unique protocol based on attractive selection, which is an opportunistic protocol and can adapt routing feedback packets to the dynamics and complexity of the environment. A multi-attribute decision-making strategy is applied to reduce the number of candidates to be chosen at the next step to improve the effectiveness of the attractive selection mechanism. When a route is detected, URAS either keeps the current path or finds another better route based on the current pathway efficiency; this self-evaluation is continued until finding the best route. Some of the latest research studies are summarized in Table 1.

3 Preliminary

In this section, we provide a brief review over some basic issues such as the fuzzy logic system and cuckoo search algorithm.

3.1 The fuzzy logic system

The fuzzy logic variables have a true value that ranges from 0 (false) to 1 (true).

Fuzzy logic control is comprised of four main phases of fuzzifying, a base of rules, inferences, and defuzzifying. Linguistic variables and pre-defined membership functions that are used to indicate the digital value of the linguistic variables are converted into fuzzy values. The second phase presents a number of a limited set of rules which are applied in the decision-making process and are intended to guide the final fuzzy value.

In the third inference phase, fuzzy decisions are generated through the rules available in the rule base. The rules of this database are a set of if-then rules which relate fuzzy input variables to the output variables. In the last phase, the membership output functions are defined in advance and non-fuzzifying methods are converted into digital values. This is a very useful method for sophisticated processes where there is no chance of defining simple mathematical methods as well as non-linear models. Since fuzzy logic can control inaccurate and variable information, it is used in solving the routing-associated problems without applying the mathematical models. Fuzzy logic could process approximate data through linguistic, non-digital variables to express the facts [8].

3.2 Cuckoo search algorithm

Cuckoo search (CS) algorithm [29] presented by Young in 2009 has been inspired by the way of life of the bird cuckoo. The way of life of this bird has supplied ideas

Table 1 Latest research studies

Article title/year of publication	Methodology	Advantages	Disadvantages
F-ANT: an effective routing protocol for ant colony optimization based on fuzzy logic in the vehicular ad hoc network (2016)	The improved ant algorithms and fuzzy logic findings regarding the bandwidth, received signal strength, and congestion to find the most efficient route	High data packet delivery rates, low end-to-end delay	The high overhead of the ant colony algorithm and the lack of development in the highway scenario
An adaptive geographic routing protocol based on quality of transmission in urban VANETs (2018)	The use of quality standards that combines connections and packet delivery rates will monitor the performance of each section of the road so that each section can be dynamically selected to encompass the best route	High rate of transfer and delivery rate of data packets	Occurrence of local maximizations. Due to the use of the greedy forward, the strategy used to counteract local maximizations increases the end-to-end delay
Probability prediction-based reliable and effective opportunistic routing algorithm for VANETs (2018)	A probability prediction algorithm based on the reliability and opportunistic function that predicts the variability of the receiver queue length, the intermediary signal, and noise. The predicted results are applied to determine the usefulness of the relay vehicles in the candidate set. The relay priority of each relay device is determined by its usefulness	Reducing the end-to-end delay and delivery rates of high data packets	Not considering the influence of parameters such as packet length and transition steps in the results obtained
Adaptive fuzzy multiple attribute decision routing in VANETs (2015)	AFMADR is determined in this protocol for any vehicle with these four aspects: distance, direction, road density, and location. A fuzzy score and the characteristics are then allocated to each aspect. Data packets are sent to the vehicles with the highest score.	The rate of the end-to-end delay is acceptable	Not selecting better carrier nodes for improving packet delivery

for many optimization algorithms. To raise its babies, the bird lays its eggs in the nest of other birds. Cuckoos push out one of the eggs of the host bird and lay eggs that better imitate the host eggs. The eggs with the highest resemblance to the host have much more chances of being hatched into mature cuckoos while the eggs discovered by the host are destroyed [39]. In CS, every egg in the host's nest indicates a solution to the effectiveness of which is computed through some parameters of the adaptation function. If the new solution is found to be superior to the old one, a replacement takes place [29].

Yang and Deb described the CS in three ideal rules [29]:

1. Each cuckoo lays one egg at a time and dumps its egg in a randomly chosen nest;
2. The best nests with high-quality eggs will be carried over to the next generation;
3. The number of available host nests is fixed, and the egg laid by a cuckoo is discovered by the host bird with a probability of [0..1].

4 Methods

Recent studies have tried to address the routing issue in VANETs by introducing enhanced methods. However, they have not carefully considered the problem of a large number of required exchanged messages in the routing process. Moreover, like other wireless networks, VANETs suffer from unwanted broadcast storm both in routing and data dissemination. In order to efficiently address

these challenges, in the first phase of TIHOO, contrary to previous methods where fuzzy logic was applied as a tool to find the best path, in this method, fuzzy logic is intelligently used to limit the route discovery phase. By limiting the RREQ control packet, the additional overhead created in the network is reduced and the network traffic is prevented from creating a storm.

On the other hand, by reducing the number of wandering packets in the network, the bandwidth usage of the network is also decreased. This not only considerably reduces the routing overhead, but also decreases the number of discovered routes. This can help the source to select the best path in less time. It is important to note that fuzzy logic is used in a way so that only the paths are removed where link failures are very likely to occur.

Another required step is to find the most stable path among the introduced paths by the first phase. An efficient new fitness function is also introduced and used in this phase. The cuckoo search (CS) algorithm selects the most stable route via intelligent configuration of the fitness function based on parameters such as the link lifetime and the reliability of the route and its available buffer in the second phase.

Adjusting the link's lifetime and reliability in the compatibility function will lead to the selection of more stable paths, resulting in fewer failures in paths and higher delivery rates for data packets. Selecting the buffer parameter available in the compatibility function causes the selection of paths with less traffic load. It is therefore effective in reducing the end-to-end delay.

This algorithm, as one of the most effective metaheuristic algorithms, selects the most stable path with higher execution speed and convergence rate compared to other metaheuristic algorithms. The combination of these two phases introduces an enhanced routing protocol named TIHOO for efficient routing in vehicular ad hoc networks.

Finally, TIHOO is evaluated in the NS2 simulation tool under various conditions and compared with similar works. The obtained results indicate that TIHOO provides significant improvement in routing in VANETs and makes this process more effectively and efficiently especially in high network traffic.

5 Problem statement

VANETs are dynamic networks; the frequent changes in the environment caused by various factors such as traffic conditions, and changes in the topology of the road, necessitate a sufficiently adaptive routing protocol. Multiple routing protocols have been proposed for this purpose, a number of which was presented in the previous section. Although these protocols make great efforts in fulfilling the routing objectives and improving the routing conditions, they are challenged by serious shortcomings.

5.0.0.1 Failures of link The life of the links which are affected by the motion of vehicles is an influential factor in identifying stable routes. Not paying attention to this factor by a number of routing protocols leads to consecutive failures of the route which in turn result in lower rates of data packet deliveries and increased number of packet losses.

The intelligent configuration of the parameters applied in the fitness function of the CS [29] metaheuristic algorithms in TIHOO leads to identification and selection of stable routes with a lower possibility of failure and naturally improved data packet delivery rate as well as fewer cases of packet loss.

5.0.0.2 High overhead Another serious shortcoming of some of routing protocols [21, 35] is their inability in controlling the number of control packets and the increased number of the packets over the network. These factors contribute to the development of higher overhead and increased bandwidth consumption which negatively affects the efficiency of the protocols. Moreover, using some metaheuristic algorithms to improve the routing protocols and enhance the quality of the provided services [2, 31] suffers from a number of issues such as high overhead and overly complex computations. Since fuzzy logic system [27] can control inaccurate and variable information, it is used in solving the

routing-associated problems without applying the mathematical models. Fuzzy logic uses linguistic non-digital variables and, thus, can process approximate data [8, 28].

Although fuzzy-based approach is employed in previous researches for deciding this route, TIHOO attempts to use the fuzzy logic in a new manner in which it intelligently selects appropriate links for sending RREQ packets and at the same time limits broadcasting of the route discovery phase and decreases the number of control packets to lower the overhead imposed on the system. Also, the lower overhead and easier computation demands of the metaheuristic algorithms of this research compared to other metaheuristic algorithms such as ACO and BFOA have made this bio algorithm a desirable candidate for application in the improved protocol of the proposed one.

6 The proposed TIHOO protocol

TIHOO is an intelligent hybrid routing protocol as we mentioned in the previous section. This protocol is discussed in detail in the following paragraphs.

6.1 General description of TIHOO

TIHOO is an optimized hybrid protocol developed for routing in VANETs. It is implemented based on the AODV routing protocol. In contrast to previous methods, this research uses fuzzy logic to limit the route discovery phase. Selection of effective input parameters contributes to the selection of more stable links to receive route request control packets. In addition, CS algorithm selects the most stable route via intelligent configuration of the fitness function based on parameters such as the link life and the reliability of the route and its available buffer. CS algorithm is one of the most effective, metaheuristic algorithms with higher execution speed and convergence rate compared to other metaheuristic algorithms. It also is able to discover the best route in a rational time [40]. TIHOO consists of two main phases.

6.1.1 Phase I: Broadcast limitation phase

The first phase of this protocol starts when a node intends to send a data packet but the route from source to destination is not available. In VANET, the communications between the source and destination node occur through a multi-hop and the main purpose of the communications is routing the messages. At runtime, a route is created through several moving vehicles. If, on a route, a device moves outside the communication range of another device, the route connections are disconnected. To create and maintain the routes, the vehicles send and receive control packets. Increasing the number of these packets will considerably affect the quality of service in

VANETs. The first phase of this protocol, which focuses on the fuzzy logic system, sends packet requests to a number of selected neighboring nodes in a targeted way. This part of the neighboring nodes is selected based on having some features. These features are included as inputs of the fuzzy logic system. The neighbors that are the most appropriate receivers of the control packets are determined by the output of this system. By limiting the route requesting receiver nodes, the number of discovered routes is reduced and the system's overhead is partially under control. This process is repeated in the nodes receiving the control packet so that the destination node receives all route requesting control packets from different routes.

6.1.2 Phase II: Find the best route

After identifying all routes, the second phase of the proposed protocol begins and the cuckoo search algorithm is called in the source node. Levy's flight in this algorithm provides the possibility of random to randomly select the routes. The length of the random steps gets use of Levy's distribution in large steps. For each of the routes, the calculation of the proposed enhanced fitness function is performed based on a number of factors. These factors, which determine the path stability, are a criterion for measuring path optimality. Ultimately, the cuckoo algorithm selects the most optimal route by comparing the fitness function of various routes and converges toward it. Since this algorithm has a high performance and convergence speed, this routing protocol is expected to provide a fairly satisfactory result. Table 2 shows the pseudo-code, and Fig. 2 shows the flowchart of the proposed protocol. The steps of the proposed protocol are shown in these figures in detail. In the following section, we will deal with the careful review of the proposed protocol.

6.2 Details of TIHOO

This section discusses the proposed protocol in details. We will also examine each section of the flowchart separately.

AODV reactive protocol provides the platform for the implementation of this research. In the first step of Fig. 2, the current node which is considered as a_i is the source node in the first run of the algorithm while other surrounding nodes which receive the control packets of route requests are set a_j.

In the second step of Fig. 2, the algorithm starts through sending Hello packets from the source node. Sending these packets to the surrounding nodes helps identify the neighboring nodes and learn their positions. Once the source node has identified its surrounding nodes and their positions using GPS (Global Positioning System), the time arrives for sending the packets of

Table 2 Pseudo of TIHOO. The pseudo-code of the proposed protocol. The steps of the proposed protocol are shown in this figure in detail

1: Start procedure
2: Phase 1: Action of the source node
3: Set N_j current node and N_d destination node
4: While N_d destination Receive Packet or (stop criterion)
5:   Current node send Hello packet to neighbor nodes
6:   Calculate input of fuzzy system
7: Call fuzzy logic system to limit broadcasting RREQ
8: Increment j
9: End while
10: Interrupt for receive RREQ of all paths
11: Phase 2: Action of the destination node
12: Get available population of N paths p_i $i = 1,2,...,n$ for Declared paths of source to destination
13: While $i < n$ or (stop criterion)
14: Calculate reliability factor , lifetime , buffer -available for p_i
15:   Call cuckoo search algorithm
16: Calculate Fitness F(p_i)
17: Increment i
18: End while
19: Choose Optimized path is path with best fitness value
20 : If sending data not finished
21: Repeat Step 4 to Step 18
22: Else if
23: End procedure

route requests. Unlike the other protocols, in TIHOO, the source node does not send the packets of route requests to all neighboring nodes via broadcasting. Instead, it sends them only to those neighboring nodes with more effective characteristics such as smaller motion speed difference with the source node (the current node), smaller divergence with the source node (the current node), and a shorter distance. Placing limits on sending the packets of route requests implies limiting the route request phase and decreasing the number of control packets, in turn resulting in reduced control overheads. Since not every neighbor participates in sending the packets of route requests, fewer control packets are sent over the network.

Fuzzy logic is employed in this method for selection of RREQ reception nodes. The inputs of the fuzzy logic system are calculated and determined in the third step of Fig. 2. These inputs consist of the speed difference between the current node and the neighboring node, the direction taken by the current node and the neighboring node, and the distance between the destination node and the neighboring node. The

Fig. 2 Flowchart of TIHOO. The flowchart of the proposed protocol. The steps of the proposed protocol are shown in this figure in detail

smaller the speed difference of these nodes, the higher the duration of being in mutual communication range which boosts the possibility of successful exchange of data packets. However, large differences between the nodes in term of their speed quickly push these nodes out of their mutual ranges and consequently they have fewer chances of sharing their information. Fast movements of vehicles prevent the development of stable connection in the network. The second input factor is using the fuzzy logic for movements of vehicles. The angle of the movement direction of the receiving node and the previous node is calculated and then is set as the second factor of the

fuzzy logic input. α is the angle of two vehicles is calculated via Eq 1.

$$\arccos \alpha = \frac{dx1.dx2 + dy1.dy2}{\sqrt{dx1 + dy1} + \sqrt{dx2 + dy2}}$$

(1)

In this equation, the movement direction of the vehicle 1 is (d_{y1} and d_{x1}) while (d_{y2} and d_{x2}) is assigned to the movement direction of the vehicle 2. The direction vector on x- and y-axes is denoted by d_x and d_y, respectively.

In general, the movement direction of the vehicles is limited by the road, and according to the conducted researches, it has been proved that the route is more consistent with a greater number of vehicles with the same direction[35]. Therefore, the movement direction in this research has been used as one of the inputs of fuzzy criterion to select routes for sending route request packets that have a higher potential.

The distance is the third factor used to limit the route discovery phase. At this stage, we use the information on the location of the neighboring node obtained by the Hello message and then we calculate the distance between the neighboring node and the destination. It is clear that neighboring nodes that are in a far less distance to the destination node are in a better position and are better candidates for sending RREQ packets. The distance between the neighboring node and the destination is calculated using Eq. (2)—the Euclidean distance equation [22].

$$D = \sqrt{(x1-x2) + (y1-y2)} \tag{2}$$

In the equation above (x_1, y_1) indicates the neighboring node's location and (x_2, y_2) represents the spatial location of the destination node.

In the fourth step of Fig. 2, after determining the input parameters of the fuzzy logic and calculating the distance of the neighboring node to the destination and angular displacement, these three criteria are used as the input of this system so that a proper conclusion is provided using the fuzzy logic based on this uncertain and inaccurate information. Figure 3 indicates the fuzzy system used in this paper.

The main processes of fuzzy logic include fuzzification, fuzzy inference, and defuzzification. Fuzzification takes place in this section whereby numerical values are converted into fuzzy values via fuzzy member function.

For each of the mentioned criteria, the membership function is defined in the linguistic set. The words large, medium, and low are used to determine the rate of input criteria.

The linguistic set of the member function of the fuzzy logic input parameters covers three values (low, medium, and high). The diagram of member functions of the speed difference, direction difference, and distance is presented in Fig. 4. Although we could use other complex member functions as well, this triangular member function suits the proposed protocol, since it only needs to identify the best links.

In the fifth step of Fig. 2, once input parameters had been determined, we embarked upon fuzzy inference and developed a set of rules informed by expert knowledge. Fuzzy-based knowledge has been designed to integrate input and output variables which are in turn informed by accurate perceptions of traffic patterns of urban vehicular ad hoc networks. Fuzzy rules are based on the IF-THEN rule [41].

Every fuzzy rule consists of an IF component, a logical relation, and a THEN component. The "IF" part has been formed by predictions and logical relation connects the input logic to the result while "THEN" defines the degree of member function and the usefulness. These rules have been presented in Table 3. The linguistic variables for determining the desirability of the candidate node for receiving RREQ are very bad, bad, unpredictable, acceptable, good, and very good.

For example, a small speed difference between the current vehicle and the neighboring vehicle indicates a short distance between the neighboring vehicles and the destination vehicle and a small angular divergence between the current vehicle and the neighboring vehicle; we then can say that the neighboring vehicle is a very good candidate for receiving RREQ packets.

Since several rules are implemented simultaneously, we apply the Min-Max methodology to combine the assessment

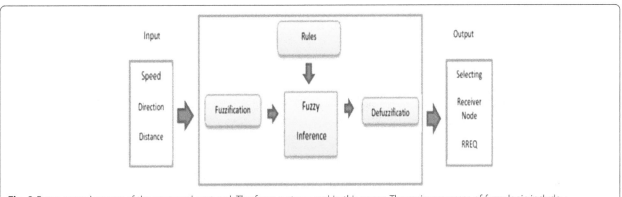

Fig. 3 Fuzzy control system of the proposed protocol. The fuzzy system used in this paper. The main processes of fuzzy logic include fuzzification, fuzzy inference, and defuzzification. Three criteria are used as the input of this system so that a proper conclusion is provided using the fuzzy logic based on this uncertain and inaccurate information

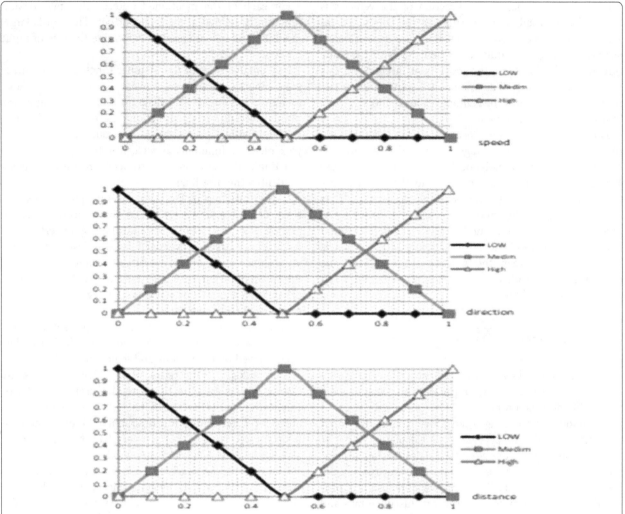

Fig. 4 Fuzzy member function for fuzzy input. For each of the mentioned criteria, the membership function is defined in the linguistic set. The words large, medium, and low are used to determine the rate of input criteria. The linguistic set of the member function of the fuzzy logic input parameters covers three values (low, medium, and high). The diagram of member functions of the speed difference, direction difference, and distance is presented in this figure. Although we could use other complex member functions as well, this triangular member function suits the proposed protocol, since it only needs to identify the best links

results. In this method [42], the minimum prior value is used as the final degree. When different rules are combined, the maximum results are used.

Defuzzification is the process of generating numerical data based on the output member function. The outputs of this system are shown in Fig. 5. In TIHOO, unlike previous protocols, the fuzzy logic system is not used to determine the optimal path but determines the proper neighboring nodes for reception RREQ.

In the sixth step of Fig. 2, after obtaining the output from the fuzzy logic, among the neighboring nodes, the nodes that have a better situation and more stable links between them and the source node (considered in the next iterations of the link between each node and its neighbors) are selected, and the control packets of the

route request are only sent in a limited way on these links.

Sending a limited number of control packets in this part of the algorithm affects the overhead and decreases it in practice.

The seventh step of Fig. 2 shows the control condition. Every neighboring node receiving the route request packet iterates this process and sends the route request packet over a group of neighboring links. This controlled process is repeated for each node until the destination node receives the packets.

The next step is the eighth stage of Fig. 2. Since different routes cover different steps and consequently face varying congestion, the messages sent from some routes are delivered faster, and thus, there is an interruption in

Table 3 Fuzzy rule used in TIHOO

Rules	Movement speed difference	Distance	Direction	Fuzzy priority
1	Low	Low	Low	Very good
2	Low	Low	Medium	Good
3	Low	Low	High	Unpredictable
4	Medium	Low	Low	Good
5	Medium	Low	Medium	Acceptable
6	Medium	Low	High	Bad
7	High	Low	Low	Unpredictable
8	High	Low	Medium	Bad
9	High	Low	High	Very bad
10	Low	Medium	Low	Good
11	Low	Medium	Medium	Acceptable
12	Low	Medium	High	Bad
13	Medium	Medium	Low	Acceptable
14	Medium	Medium	Medium	Unpredictable
15	Medium	Medium	High	Bad
16	High	Medium	Low	Bad
17	High	Medium	Medium	Bad
18	High	Medium	High	Very bad
19	Low	High	Low	Unpredictable
20	Low	High	Medium	Bad
21	Low	High	High	Very bad
22	Medium	High	Low	Bad
23	Medium	High	Medium	Bad
24	Medium	High	High	Very bad
25	High	High	Low	Bad
26	High	High	Medium	Very bad
27	High	High	High	Very bad

this stage to ensure that all route request packets are delivered on time from all available routes.

In the ninth step of Fig. 2, a route reply packet is sent by the destination node to the source node for each identified route. Factors such as reliability, lifetime, and available buffer are calculated for each route. These factors are attached to the reply packet to other usual information. Table 4 presents the structure of the RREP packet.

The reliability of communication links of two vehicles is the ability of transference of information packets with the minimum possibility of the link failure which is a very important parameter for assessing the system stability and effectiveness. The reliability is defined based on Eq. 3 in [22].

Applying this parameter to fitness function helps select a more stable route, resulting in an enhanced data packet delivery rate. More stable routes had the added advantage of decreased losses of data packets.

$$r\left(l\right) = P\left\{\text{to continue to be available until } t + T_{\mathrm{p}} \mid \text{available } t\right\} \tag{3}$$

T_{p} is the predicted time for the availability of link between vehicles c_i, c_j. The reliability will be defined as below:

In case of the availability of the link in time t, this link will be available at time $t + T_p$. Therefore, L_{ij} is the distance between the two vehicles which is obtained through Eq. 4 [22].

$$Lij = \sqrt{\left(X1 - X2\right)^2 + \left(Y1 - Y2\right)^2} \tag{4}$$

Tp can be calculated as follows by Eq. (5) and Eq. (6) [22]; R is radio range in VANET:

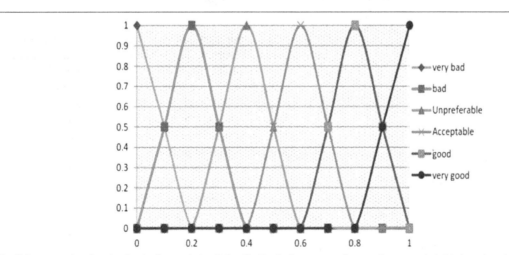

Fig. 5 Fuzzy member function for the fuzzy priority. Defuzzification is the process of generating numerical data based on the output member function. The outputs of this system are shown in this figure. In TIHOO, unlike previous protocols, the fuzzy logic system is not used to determine the optimal path but determines the proper neighboring nodes for reception RREQ

Table 4 Structure of the RREP packet

Type	Dest Ip Address	Num.seq.dest	Nbr_hops	Src IP Address	Lifetime	Reliability	available Buffer	Timestamp

1. If two vehicles have the same direction.

$$Tp \begin{cases} \dfrac{2R\text{–}Lij}{|vi + vj|} & \text{if } vj > vi, \text{ i.e.} Cj \text{ Approaches } Ci \text{ from behind} \\ \dfrac{R\text{–}Lij}{|vi + vj|} & \text{if } vi > vj, \text{ i.e.} Ci \text{ Moves forward in front of } cj \end{cases}$$

$$(5)$$

2. If two vehicles have opposite directions:

$$T = \begin{cases} \dfrac{R + Lij}{vi + vj} & Ci \text{ and } C_i \text{ are moving toward each other} \\ \dfrac{R + Lij}{vi + vj} & Ci \text{ and } C_i \text{ are moving away each other} \end{cases}$$

$$(6)$$

The reliability of the link is determined through Eq. (7):\

$$r_t(1) = \text{Erf}\left(\frac{\frac{2R}{t} - \mu\,\Delta V}{\sigma \Delta V \sqrt{2}}\right) - \text{Erf}\left(\frac{\frac{2R}{t} + Tp\text{-}\mu\,\Delta V}{\sigma \Delta V \sqrt{2}}\right) \quad (7)$$

σ and μ are mean and standard deviation which are included in this equation to figure out the speed differences of vehicles. These parameters could be easily calculated using the maximum and minimum speed limits in urban settings. The speed differences and Erf function are calculated through Eq. 8 [22] and Eq. 9 [22], respectively:

$$\Delta V = |V_2 \text{–} V1| \quad (8)$$

The Erf function is obtained using the following equation:

$$\text{Erf} = \frac{2}{\sqrt{u}} \int e^{-t} dt, -\infty < W < \infty \quad (9)$$

The link lifetime is the shortest time for two vehicles to start communication over a route. Including this parameter in the fitness function helps in finding the most stable route. Selection of the most stable route contributes to the more effective delivery of the packets and lowers the loss rates of data packets during transfers. Suppose that (x_1, y_1) is the last hop position of a given vehicle, Vx_1 and Vy_1 are the components of the speed

vector of the last hop vehicle, (x_2, y_2) is the position of the current vehicle, and Vx_2 and Vy_2 are the components of its speed vector. Therefore, the position of the last hop will be $(y_1 + t.vy_1, x_1 + t.vx_1)$ '$(y_2 + t.vy_2; x_2 + t.vx_2)$.

Where:

$$a = x1\text{–}x2, b = y1\text{–}y2, c = Vx1\text{–}Vx2, d = Vy1\text{–}Vy2$$

Based on Pythagoras' theory, the r or the distance between two vehicles will be calculated using Eq. 10 [35].

$$r^2 = (a + c.t)^2 + (b + d.t)^2 \quad (10)$$

r is the effective communication range between the two vehicles and thus the communication time (t) is estimated using Eq. 11 [35].

$$t = \frac{-(ac + bd) + \sqrt{(ac + bd)^2 - (a^2 + b^2 - r^2)(c^2 + d^2)}}{c^2 + d^2}$$

$$(11)$$

Equation 12 [43] shows the available buffer. Considering available buffer for each node is intended to prevent losses of effective data packets. The application of this parameter helps in the selection of less congested route and prevention of losses of effective data packets. End-to-end delays will be decreased while sending data packets. The $\frac{qi}{bi}$ indicates the available buffer in which qi is the length of the data packet queue in N_i node, while b_i displays the buffer capacity of the node.

$$\text{Avail\_Buffer} = \left(1 - \frac{qi}{bi}\right) \quad (12)$$

The tenth step of Fig. 2 takes the form of call cuckoo search algorithm. The initial cuckoo population and the number of the nests are 30 in this research. CS algorithm demands accurate adjustment of a smaller number of parameters compared to other algorithms such as genetics and optimization of congestion of particles. Three components are included in the cuckoo search algorithm: selection of the best, random discovery through general Levy flight and discovery through local random walks. Selection of the best nest (solution) is done through the protection of the best nest (solution) which guarantees finding the best solution via the next repetition. The best solution has been discovered by random walks. Using levy flights in local random walks causes bigger and effective steps [39] in CS while in F-ANT algorithm, random walks are in a standard format as a result of the application of Gaussian distribution.

Moreover, applying Levy flight to cuckoo algorithm makes it possible to sample the entire search space and find the best solution. Using Levy flight is more effective than random discovery for conducting searches in large spaces [39].

The occurrence of these components in the cuckoo search algorithm makes it one of the best metaheuristic algorithms and gives the edge to our method over its counterparts.

The fast convergence of the algorithm combined with the possibility of searching in a bigger space provides for the optimal route in a shorter time. It also decreases end-to-end delays during sending data packets.

In the 11th to 13th steps of Fig. 2, the source node receives RREPs from different routes and then calculates the fitness function of each route. The fitness function has been configured based on Eq. 13 in TIHOO. In this equation, $r_t(l)$ denotes reliability, t represents the link lifetime, Avail_ Buffer is the length of the available buffer, and α, β, and γ are weighted factors for reliability, link lifetime, and the available buffer, all of which have been set based on experience and repeated tests ($\alpha + \beta + \gamma = 1$).

$$\text{Fitness Function} = \alpha\, r_t\,(l) + \beta\, t \\ + \gamma\, \text{Avail\_Buffer} \qquad (13)$$

Since the factors used for calculation of the fitness function are not measured with the same scales, it is neither possible to compare them nor to include them in this function. Thus, it is necessary to normalize or descale these factors to allow for the application of a single measurement tool. We have used linear descaling in this research. Considering the fact that the indicators are positive and equal, each of the values of data matrices are divided equally to the maximum of the related column.

In the 14th and 15th steps of Fig. 2, once the fitness function of each route has been calculated, the route with the best fitness function is selected as the optimal route for data transfer.

7 Performance evaluation

The performance of TIHOO, F-ANT, and AODV in the challenge of routing is evaluated in this section. TIHOO is compared with two methods, why these protocols?

The F-ANT algorithm is a combination of fuzzy logic and one of the biological algorithms called ant colony. Since our proposed method is also a combination of bio-algorithms and the fuzzy logic, the F-ANT algorithm is an appropriate alternative to our proposed protocol. Based on the articles presented below, the cuckoo search algorithm and the ant colony are also used in combination with other routing

research. Therefore, we have considered an article for comparison which applied an ant colony biology algorithm in order to evaluate the efficiency and effectiveness of the use of the ant colony algorithm in comparison with the use of cuckoo's algorithm.

7.1 Performance metrics

To evaluate the effectiveness and performance of TIHOO, we conduct extensive simulations and compare the results with those of F-ANT and AODV. We evaluate end-to-end delay, packet delivery ratio, packet loss ratio, throughput, and routing overhead.

7.1.1 Packet delivery rate (PDR)

PDR is the number of data packets successfully delivered to destination nodes to the total number of data packets sent from the source node. Accordingly, PDR is calculated via Eq. (14).

$$PDR = \left(\frac{\sum\limits_{j=1}^{n} \text{Packets received}}{\sum\limits_{j=1}^{n} \text{Packets originated}} \right) \qquad (14)$$

7.1.2 End-to-end delay (EED)

The average delay between the moments a packet is sent by the source to the moment it is received in the destination is called end-to-end delay. EED is measured in milliseconds and includes all possible delays that occur on the way, including route discovery,

Table 5 Simulation parameters

Parameter	Value
Simulator version	NS-2.29
Antenna type	Omni Antenna
Simulation time	400 s
Propagation model	Two ray ground
Traffic type	CBR
Network area	$700 \times 700\ \text{m}^2$
Mobility model	Random waypoint
Transmission range	200 m
Data packet size	512 B
Scenario type	Urban
Mobility generator	SUMO
Number of nodes	30, 40, 50, 80, and 120
Speed range	0–30 m/s
MAC layer	IEEE-802.11p
α, β, γ	0.45, 0.3, 0.25

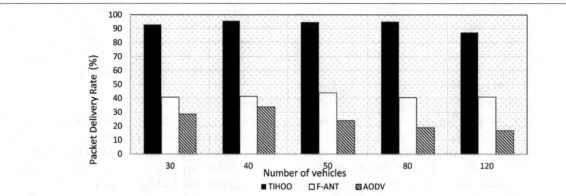

Fig. 6 Packet delivery ratio based on the number of vehicles. Comparison of the performance of TIHOO with that of F-ANT and AODV for packet delivery ratio. This improvement is due to the cuckoo search algorithm in the proposed protocol and the intelligent configuration of the fitness function parameters such as the reliability and the link lifetime which positively affect the data packet delivery rate via the selection of more stable routes as the optimal route. Simulation results show that TIHOO performs better than F-ANT and AODV in terms of PDR

data acquisition, and queuing delays, delay caused by processing at intermediate nodes, propagation time, retransmission delays at the MAC, etc. The lower the value of EED, the better the performance of the protocol. EED is calculated using Eq. (15) below:

$$EED = \sum (PA_i - PS_i) \qquad (15)$$

where PA_i = packet arrival, PS_i = packet start, and ith packet

7.1.3 Throughput
This parameter denotes the number of data packets delivered to the destination node per unit of time. Throughput is calculated as received throughput in

bit per second at the traffic destination. Throughput is calculated via Eq. (16) as follows:

$$Throughput = \frac{\sum_{j=1}^{n} Packets\ received}{t} \qquad (16)$$

7.1.4 Routing overhead
The number of generated control packets is denoted by control overheads. Lower overhead positively affects the assessment of protocol efficiency. The overhead is calculated via Eq. (17) where N_c is the number of control packets and N_t is the total number of packets.

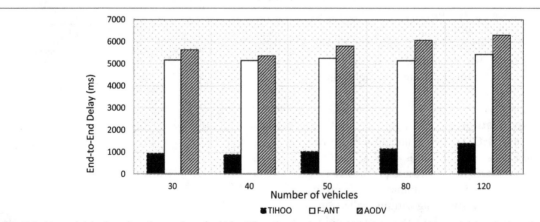

Fig. 7 End-to-end delay based on the number of vehicles. When the number of vehicles increases, end-to-end delay also increases. Applying the cuckoo metaheuristic algorithm to TIHOO has helped in the selection of the best route in a shorter time than the other two protocols. In contrast to F-ANT which is very slow to converge as a result of the ant colony optimization algorithm (ACO), TIHOO has a fast convergence rate and could identify the optimal route in a shorter time. Moreover, the intelligent configuration of the fitness function of this protocol and considering parameters such as available buffer have proved to be very useful in the deselection of congested routes. The selection of less busy routes has decreased end-to-end delays

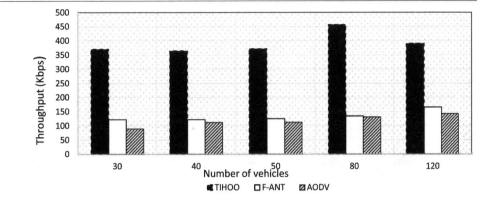

Fig. 8 Throughput based on number of vehicles. Comparison of the performance of TIHOO with that of F-ANT and AODV for throughput. Based on simulation figures, TIHOO has outperformed the other two protocols. Selecting the best route takes more time in F-ANT as it employs ACO and is slow to converge while the application of CS algorithm to TIHOO has brought about faster convergence and less parametric configuration demands during the implementation phase compared to other metaheuristic algorithms which have accelerated the selection of the optimal route and improved the throughput of the proposed protocol

$$\text{Routing Overhead} = \frac{Nc}{Nt} \qquad (17)$$

7.1.5 Packet loss rate (PLR)

PLR is the fraction of the total transmitted packets that were not received at the destination. The packet loss rate is calculated in Eq. (18) as follows:

$$\text{PLR} = \left(\sum_{j=1}^{n} Number\ of\ sent\ packets - \sum_{j=1}^{n} Number\ of\ recieved\ packets \right) * 100 \qquad (18)$$

7.2 Simulation results and analysis

We have implemented TIHOO, F-ANT, and AODV approaches in the NS-2 [7] on Fedora 10. Table 5 shows the simulation parameters.

7.2.1 An assessment of the simulation parameters based on the number of vehicles

This section assesses the parameters of end-to-end delays, throughput, packet delivery rate, routing overhead, and the loss rate of packets as a function of the number of vehicles. The results of this test and comparisons have been satisfactory in this protocol between 30, 40, 50, 80, and 120 nodes. Figures 6, 7, 8, 9, and 10 show that the proposed protocol is a scalable one as the desired results are still produced in face of the growing number of vehicles.

Figure 6 compares the performance of TIHOO with that of F-ANT and AODV for packet delivery ratio.

This improvement is due to the cuckoo search algorithm in the proposed protocol and the intelligent configuration of the fitness function parameters such as the reliability and the link lifetime which positively affect the

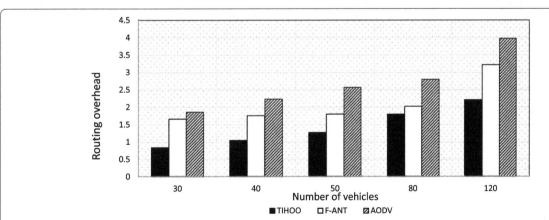

Fig. 9 Routing overhead based on number of vehicles. The routing overhead values of TIHOO, F-ANT, and AODV for various numbers of vehicles. In TIHOO, broadcasting of route request control packets are prevented and the discovery is conducted in a controlled manner while running this phase in AODV algorithm causes increased system overhead. Also, F-ANT protocol imposes increased overhead on the system as a result of employing the ACO metaheuristic algorithm and storing the whole information of the entire colony. Therefore, TIHOO outperforms other protocols in terms of overhead produced

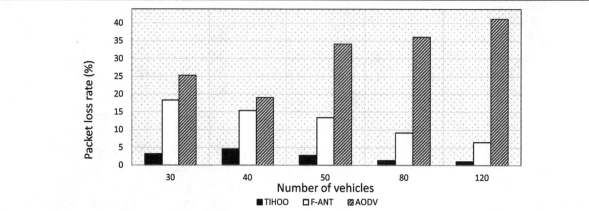

Fig. 10 Packet loss rate based on number of vehicles. The packet loss rate values of TIHOO, F-ANT and AODV protocols for various numbers of vehicles. Loss of data packets was partly due to packet collision, link failure, insufficient bandwidth, overhead of buffer, and etc. In TIHOO, the fitness function of the CS algorithm is configured based on such parameters as route lifetime and its reliability. Employing these factors contributes to the selection of more stable with fewer cases of link failures which affects the packet losses as well. Moreover, accounting for the available buffer for selection of a less congested route is very effective. Less congestion is synonymous with fewer packet losses

data packet delivery rate via the selection of more stable routes as the optimal route. Simulation results show that TIHOO performs better than F-ANT and AODV in terms of PDR (as shown in Fig. 6).

Figure 7 shows the end-to-end delay based on the number of vehicles. In Fig. 7, when the number of vehicles increases, the end-to-end delay also increases. Applying the cuckoo metaheuristic algorithm to TIHOO has helped in the selection of the best route in a shorter time than the other two protocols. In contrast to F-ANT which is very slow to converge as a result of the ant colony optimization algorithm (ACO), TIHOO has a fast convergence rate and could identify the optimal route in a shorter time. Moreover, the intelligent configuration of the fitness function of this protocol and considering parameters such as available buffer have proved to be very useful in the deselection of congested routes. The selection of less busy routes has decreased end-to-end delays.

Figure 8 compares the performance of TIHOO with that of F-ANT and AODV for throughput.

Based on simulation figures, TIHOO has outperformed the other two protocols. Selecting the best route takes more time in F-ANT as it employs ACO and is slow to converge while the application of CS algorithm to TIHOO has brought about faster convergence and less parametric configuration demands during the implementation phase compared to other metaheuristic algorithms which have accelerated the selection of the optimal route and improved the throughput of the proposed protocol.

Figure 9 shows the routing overhead values of TIHOO, F-ANT, and AODV for various numbers of vehicles. In TIHOO, broadcasting of route request control packets are prevented and the discovery is conducted in a controlled manner while running this phase in AODV algorithm causes increased system

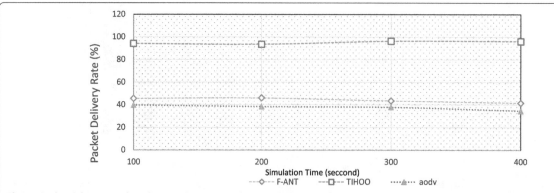

Fig. 11 Packet delivery ratio based on simulation time. The simulation results for packet delivery rate are assessed in this section based on the simulation time. The bigger rate of packet delivery in TIHOO is associated with the movement of packets in the most stable route which in turn is determined thanks to the application of stability factors in the fitness function of TIHOO. In addition, accounting for the available buffer for the selection of a less congested route is very effective in enhancing the number of delivered data packets

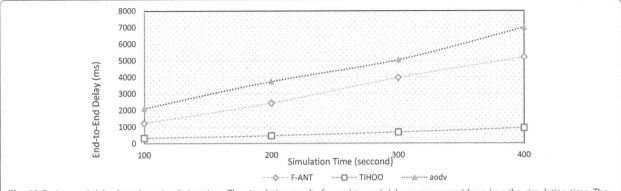

Fig. 12 End-to-end delay based on simulation time. The simulation results for end-to-end delays are assessed based on the simulation time. The CS algorithm which has a fast convergence rate has improved end-to-end delays

overhead. Also, F-ANT protocol imposes increased overhead on the system as a result of employing the ACO metaheuristic algorithm and storing the whole information of the entire colony. Therefore, TIHOO outperforms other protocols in terms of overhead produced.

Figure 10 depicts the packet loss rate values of TIHOO, F-ANT, and AODV protocols for various numbers of vehicles. Loss of data packets was partly due to packet collision, link failure, insufficient bandwidth, overhead of buffer, etc. In TIHOO, the fitness function of the CS algorithm is configured based on such parameters as route lifetime and its reliability. Employing these factors contributes to the selection of more stable with fewer cases of link failures which affect the packet losses as well. Moreover, accounting for the available buffer for selection of a less congested route is very effective. Less congestion is synonymous with fewer packet losses.

7.2.2 Assessment of simulation parameters based on simulation duration

In this section, the packet delivery rate and end-to-end delays are assessed based on the simulation time. Figures 11 and 12 display the simulation results for the packet delivery rate and end-to-end delays which are assessed in this section based on the simulation time. These figures show that TIHOO is superior over the other two protocols. The bigger rate of packet delivery in TIHOO is associated with the movement of packets in the most stable route which in turn is determined thanks to the application of stability factors in the fitness function of TIHOO. In addition, accounting for the available buffer for the selection of a less congested route is very effective in enhancing the number of delivered data packets. CS algorithm which has a fast convergence rate has improved end-to-end delays.

8 Conclusion

VANETs are a special type of ad hoc networks that consist of nodes connected via wireless links without any

fixed infrastructure. This and the lack of a centralized administration have made efficient routing a significant challenge in these networks. In this study, we proposed a new routing protocol called TIHOO for VANETs. This protocol intelligently employed fuzzy and cuckoo approaches. In TIHOO, the fuzzy logic system is used to limit the route discovery phase and, by limiting the route request messages, it somewhat controls the created extra overhead. The inputs in the fuzzy logic system include three factors of vehicle speed, direction of movement, and neighbor node-to-destination distance. After identifying the routes in the source node, the cuckoo search algorithm is called. This algorithm is one of the most effective metaheuristic algorithms, especially in the large search space. It selects the most stable and optimal route among known routes by calculating the fitness function based on criteria such as route lifetime, route reliability, and available buffer.

Simulation results show that TIHOO outperformed F-ANT and AODV in terms of throughput, routing overhead, packet delivery ratio, packet loss ratio, and end-to-end delay. The reliability parameter and the link lifetime in the compatibility function make it possible to select a path that is more stable. These parameters are effective along with the buffers available to increase the delivery rate of data packets. TIHOO is proposed to be better than the F-ANT protocol used by the ant colony algorithm, because it uses the cuckoo algorithm and its unique features, such as the run-rate and convergence resulting from the use of Levy's distribution in this algorithm.

Abbreviations
ACO: Ant colony optimization algorithm; AFMADR: Adaptive fuzzy multiple attribute decision routing in VANETs; AMGRP: AHP-based Multimetric Geographical Routing Protocol; AODV: Ad hoc On-demand Distance Vector routing; BFOA: Bacterial foraging optimization algorithm; CS: Cuckoo search; DSRC: Dedicated short-range communications; EED: End-to-end delay; FCAR: Fuzzy control based AODV routing; GPRS: Greedy Perimeter Stateless Routing; MANETs: Mobile ad hoc networks; NS2: Network simulator 2; PA: Packet arrival; PDR: Packet delivery rate; PLR: Packet loss rate; PS: Packet

start; R2P: Reliable routing protocol; RREP: Route reply; RREQ: Route request; V2I: Vehicle-to-infrastructure; V2V: Vehicle-to-vehicle; VANETs: Vehicular ad hoc networks; WAVE: Wireless access in vehicular environment

Authors' contributions
SRY and BB carried out all research steps including finding the problem, analyzing the related works, and finding, implementing, and simulating the proposed method as well as writing the manuscript. KR has contributed in analyzing the data and editing the paper. All authors read and approved the final manuscript.

Authors' information
Shirin Rahnamaei Yahiabadi received her bachelor in computer science from the University of Yazd, Yazd, Iran, in 2013 and now is a master student in Najafabad Branch, Islamic Azad University, Iran. Her research interest encompasses the fuzzy logic system and metaheuristic base on bio-inspired algorithm, as well as wireless communication.
Behrang Barekatain earned his BSC and MSC in computer software engineering in 1996 and 2001, respectively. He has more than 22 years of experience in computer networking and security. He is as a faculty member in Najafabad Branch, Islamic Azad University, Iran, for 18 years. He received his PhD and post-doc in computer networks from Ryerson University, Canada. His research interests encompass wire and wireless systems, VANETs, FANETs, SDN, NDN, IoT, peer-to-peer networking, network coding, video streaming, network security, and wireless mesh networks using network coding.
Kaamran Raahemifar received his B.Sc. degree (1985-1988) in Electrical Engineering from Sharif University of Technology, Tehran, Iran, his MASc. degree (1991-1993) from Electrical and Computer Engineering Dept., Waterloo University, Waterloo, Ontario, Canada, and his PhD degree (1996–1999) from Windsor University, Ontario, Canada. He was Chief Scientist (1999–2000), in Electronic Workbench, Toronto, Ontario, Canada. He joined Ryerson University in September 1999 and was tenured in 2001. Since 2011, he has been a Professor with the Department of Electrical and Computer Engineering, Ryerson University. He is the recipient of ELCE-GSA Professor of the Year Award (Elected by Graduate Student's body, 2010), Faculty of Engineering, Architecture, and Science Best Teaching Award (April 2011), and Department of Electrical and Computer Engineering Best Teaching Award (December 2011), and Research Award (December 2014). He has been awarded more than $6 M external research fund during his time at Ryerson. His research interests include (1) Optimization in Engineering: Theory and Application, which includes grid optimization and net-zero communities, as well as biomedical signal and image processing techniques, (2) Big Data Analysis (Dictionary/Sparse Representations, Interpolation, Predictions), (3) Modelling, Simulation, Design, and Testing, and (4) Time-Based Operational Circuit designs.

Funding
The authors received no financial support for the research, authorship, and/or publication of this article.

Competing interests
The authors declare that they have no competing interests.

Author details
¹Faculty of Computer Engineering, Najafabad Branch, Islamic Azad University, Najafabad, Iran. ²Big Data Research Center, Najafabad Branch, Islamic Azad University, Najafabad, Iran. ³Electrical and Computer Engineering, Sultan Qaboos University, P.O. Box: 31, Al-Khoud 123, Sultanate of Oman. ⁴Chemical Engineering Department, University of Waterloo, 200 University Avenue West, Waterloo, Toronto, Ontario N2L 3G1, Canada.

References
1. Singh, S. and S. Agrawal, VANET routing protocols: Issues and challenges, 2014 Recent Advances in Engineering and Computational Sciences (RAECS), 1-5,6-8 March 2014,1-5.
2. H. Fatemidokht, M. Kuchaki Rafsanjani, F-Ant: an effective routing protocol for ant colony optimization based on fuzzy logic in vehicular ad hoc networks. Neural Computing and Applications, 1–11 (2016)
3. S. Dehghani, M. Pourzaferani, B. Barekatain, Comparison on energy-efficient cluster based routing algorithms in wireless sensor network. Procedia Computer Science 72, 535–542 (2015)
4. S. Dehghani, B. Barekatain, M. Pourzaferani, An enhanced energy-aware cluster-based routing algorithm in wireless personal networks (Wireless Personal Communications, 2017), pp. 1605–1635
5. S. Bitam, A. Mellouk, S. Zeadally, Bio-inspired routing algorithms survey for vehicular ad hoc networks. IEEE Communications Surveys & Tutorials 17, 843–867 (2015)
6. H. Rana, P. Thulasiraman, R.K. Thulasiram, MAZACORNET: mobility aware zone based ant colony optimization routing for VANET (2013 IEEE Congress on Evolutionary Computation, 2013), pp. 2948–2955
7. S. Al-Sultan et al., A comprehensive survey on vehicular ad hoc network. Journal of Network and Computer Applications 37, 380–392 (2014)
8. C. Wu, S. Ohzahata, T. Kato, Routing in VANETs: a fuzzy constraint Q-learning approach (2012 IEEE Global Communications Conference (GLOBECOM), 2012), pp. 195–200
9. K.N. Qureshi, A.H. Abdullah, A. Altameem, Road aware geographical routing protocol coupled with distance, direction and traffic density metrics for urban vehicular ad hoc networks. Wireless Personal Communications 92, 1251–1270 (2017)
10. B. Barekatain et al., GAZELLE: An enhanced random network coding based framework for efficient P2P live video streaming over hybrid WMNs. Wireless Personal Communications 95, 2485–2505 (2017)
11. N.V. Dharani Kumari, B.S. Shylaja, AMGRP: AHP-based multimetric geographical routing protocol for urban environment of VANETs. Journal of King Saud University - Computer and Information Sciences, 849–857 (2017)
12. T.-Y. Wu, Y.-B. Wang, W.-T. Lee, Mixing greedy and predictive approaches to improve geographic routing for VANET. Wireless Communications, and Mobile Computing 12, 367–378 (2012)
13. B. Barekatain et al., efficient P2P live video streaming over hybrid WMNs using random network coding. Wireless Personal Communications 80, 1761–1789 (2015)
14. Barekatain, B., et al., MATIN: A Random network coding based framework for high quality peer-to-peer live video streaming, PLOS ONE, 8, e69844 (2013), 165-172.
15. B. Barekatain et al., Performance evaluation of routing protocols in live video streaming over wireless mesh networks. Jurnal Teknologi 10, 85–94 (2013)
16. B. Barekatain et al., GREENIE: a novel hybrid routing protocol for efficient video streaming over wireless mesh networks. EURASIP Journal on Wireless Communications and Networking 168(2013), 1–22 (2013)
17. G. Li et al., Adaptive fuzzy multiple attribute decision routing in VANETs. International Journal of Communication Systems 30, 1543–1563 (2017)
18. P. Sermpezis, G. Koltsidas, F.N. Pavlidou, Investigating a junction-based multipath source routing algorithm for VANETs. IEEE Communications Letters 17, 600–603 (2013)
19. L.N. Balico et al., A prediction-based routing algorithm for vehicular ad hoc networks (2015 IEEE Symposium on Computers and Communication (ISCC), 2015), pp. 365–370
20. J.-M. Chang et al., An energy-efficient geographic routing protocol design in vehicular ad-hoc network. Computing 96, 119–131 (2014)
21. C. Wu, S. Ohzahata, T. Kato, Flexible, portable, and practicable solution for routing in VANETs: a fuzzy constraint Q-learning approach. IEEE Transactions on Vehicular Technology 62, 4251–4263 (2013)
22. E. Moridi, H. Barati, RMRPTS: a reliable multi-level routing protocol with tabu search in VANET. Telecommunication Systems, 1–11 (2016)
23. S. Bitam, A. Mellouk, S. Zeadally, HyBR: A hybrid bio-inspired bee swarm routing protocol for safety applications in vehicular ad hoc networks (VANETs). Journal of Systems Architecture 59, 953–967 (2013)
24. H. Dong et al., Multi-hop routing optimization method based on improved ant algorithm for vehicle to roadside network. Journal of Bionic Engineering 11, 490–496 (2014)
25. B. Barekatain, S. Dehghani, M. Pourzaferani, An energy-aware routing protocol for wireless sensor networks based on new combination of genetic algorithm & k-means. Procedia Computer Science 72, 552–560 (2015)

26. M. Al-Rabayah, R. Malaney, A new scalable hybrid routing protocol for VANETs. IEEE Transactions on Vehicular Technology **61**, 2625–2635 (2012)
27. L.A. Zadeh, Fuzzy sets. Information and Control **8**, 338–353 (1965)
28. L. Altoaimy, I. Mahgoub, *Fuzzy logic based localization for vehicular ad hoc networks, 2014 IEEE Symposium on Computational Intelligence in Vehicles and Transportation Systems (CIVTS)* (2014), pp. 121–128
29. X.S. Yang, D. Suash, *Cuckoo Search via Levy flights, 2009 World Congress on Nature & Biologically Inspired Computing (NaBIC)* (2009), pp. 210–214
30. A.I. Saleh, S.A. Gamel, K.M. Abo-Al-Ez, A reliable routing protocol for vehicular ad hoc networks. Computers & Electrical Engineering, 473–495 (2016)
31. K. Mehta, P.R. Bajaj, L.G. Malik, *Fuzzy bacterial foraging optimization zone based routing (FBFOZBR) protocol for VANET, 2016 International Conference on ICT in Business Industry & Government (ICTBIG)* (2016), pp. 1–10
32. C.E. Perkins, E.M. Royer, *Ad-hoc on-demand distance vector routing, Mobile Computing Systems and Applications* (Proceedings. WMCSA '99. Second IEEE Workshop on, New Orleans, 1999), pp. 90–100
33. A. Feyzi, V. Sattari-Naeini, Application of fuzzy logic for selecting the route in AODV routing protocol for vehicular ad hoc networks,2015 23rd Iranian Conference on. Electrical Engineering **684-687**, 684–687 (2015)
34. T. Hu et al., *An enhanced GPSR routing protocol based on the buffer length of nodes for the congestion problem in VANETs, 2015 10th International Conference on Computer Science & Education (ICCSE)* (2015), pp. 416–419
35. X.B. Wang, Y.L. Yang, J.W. An, *An, Multi-metric routing decisions in VANET, 2009 Eighth IEEE International Conference on Dependable, Autonomic and Secure Computing* (2009), pp. 551–556
36. Li, X., et al., An adaptive geographic routing protocol based on quality of transmission in urban VANETs, 2018 IEEE International Conference on Smart Internet of Things (SmartIoT), 17-19 2018, 52-57.
37. N. Li et al., Probability prediction-based reliable and efficient opportunistic routing algorithm for VANETs %J. IEEE/ACM Trans. Netw **26**, 1933–1947 (2018)
38. D. Tian et al., A microbial inspired routing protocol for VANETs. IEEE Internet of Things Journal **5**, 2293–2303 (2018)
39. A.H. Gandomi, X.-S. Yang, A.H. Alavi, Cuckoo search algorithm: a metaheuristic approach to solve structural optimization problems. Engineering with Computers **29**, 17–35 (2013)
40. A. Kout et al., AODVCS, a new bio-inspired routing protocol based on cuckoo search algorithm for mobile ad hoc networks. Wireless Networks, 2509–2519 (2017)
41. Z. Qin, M. Bai, D. Ralescu, A fuzzy control system with application to production planning problems. Information Sciences **181**, 1018–1027 (2011)
42. C. Wu, S. Ohzahata, T. Kato, *VANET broadcast protocol based on fuzzy logic and lightweight retransmission mechanism* (IEICE Transactions on Communications, E95.B, 2012), pp. 415–425
43. Hui, L., et al., An adaptive genetic fuzzy multi-path routing protocol for wireless ad hoc networks, Sixth International Conference on Software Engineering, Artificial Intelligence, Networking, and Parallel/Distributed Computing and First ACIS International Workshop on Self-Assembling Wireless Network,468-475, 2005,468-475.

A reliable path selection and packet forwarding routing protocol for vehicular ad hoc networks

Irshad Ahmed Abbasi[1,2]* (iD), Adnan Shahid Khan[1] and Shahzad Ali[3]

Abstract

Vehicular ad hoc networks (VANETs) have earned a gigantic consideration in the recent era. Wide deployment of VANETs for enhancing traffic safety, traffic management, and assisting drivers through elegant transportation system is facing several research challenges that need to be addressed. One of the crucial issues consists of the design of scalable routing algorithms that are robust to rapid topology changes and frequent link disconnections caused by the high mobility of vehicles. In this article, first of all, we give a detailed technical analysis, comparison, and drawbacks of the existing state-of-the-art routing protocols. Then, we propose a novel routing scheme called a Reliable Path Selection and Packet Forwarding Routing Protocol (RPSPF). The novelty of our protocol comes from the fact that firstly it establishes an optimal route for vehicles to send packets towards their respective destinations by considering connectivity and the shortest optimal distance based on multiple intersections. Secondly, it uses a novel reliable packet forwarding technique in-between intersections that avoids packet loss while forwarding packet due to the occurrence of sudden link ruptures. The performance of the protocol is assessed through computer simulations. Simulation outcomes specify the gains of the proposed routing scheme as compared to the earlier significant protocols like GSR (Geographic Source Routing), GPSR (Greedy Perimeter Stateless Routing), E-GyTAR (Enhanced Greedy Traffic Aware Routing), and TFOR (Traffic Flow-Oriented Routing) in terms of routing metrics such as delivery ratio, end-to-end delay, and routing overhead.

Keywords: Multiple intersections, Position-based routing, Optimal route, Forwarding

1 Introduction

The immense growth of automobiles and irregular behavior of drivers on the road cause traffic congestion, accidents, wastage of fuel, and loss of precious lives, which makes the existing transportation system inefficient. To direct these challenges, a new research field called as Intelligent Transportation System (ITS) has been proposed. It applies a combination of multiple promising technologies of automobiles and transportation system in order to enhance security, safety, effectiveness of transportation systems, vehicle control, and provision of latest mobile services and applications to the on-road public by advancing traffic management system. In ITS, developing vehicle to vehicle and vehicle to infrastructure communication is an

outstanding challenge to ITS industry. Thus, the US Federal Communication Commission has approved 75-MHz spectrum at 5.9 GHz for dedicated short-range communications (DSRC) [1–3] for the successful deployments of WLAN technologies for making vehicular ad hoc networks (VANETs) a reality.

From the last few years, inter-networking over VANETs has been achieving a massive momentum. Realizing its intensifying significance, the academic research society, major car manufacturers, and governmental institutes are making efforts to develop VANETs. Various significant projects are initiated by different countries and famous industrial firms such as Daimler-Chrysler, Toyota, and BMW for inter-vehicular communications. Some of these prominent projects include CarTALK2000 [4], Car-to-Car Communication ConsortiumC2CCC [5], Advanced Driver Assistance Systems (ADASE2), California Partners for Advanced Transit and Highways (California PATH) [6], FleetNet [7], DEMO 2000 by Japan Automobile Research

* Correspondence: irshad_upesh@yahoo.com
[1]Department of Computer Science and Information Technology, Universiti Malaysia Sarawak (UNIMAS), 9300 Kota Samarahan, Malaysia
[2]Department of Computer Science, Faculty of Science and Arts at Balgarn, University of Bisha, P.O. Box 60, Sabt Al-Alaya 61985, Kingdom of Saudi Arabia
Full list of author information is available at the end of the article

Institute (JSK) [8], Chauffeur in EU [9], and Crash Avoidance Metrics Partnership (CAMP) [10]. These developments are the key steps towards the recognition of intelligent transportation services.

VANETs are a particular offshoot of MANETs. Due to the fast motion of the vehicles, they have more rapidly and dynamically changing topology as compared to MANETs. But the mobility of the vehicular nodes is restricted by pre-defined roads layout. Speed limits, congestion level, and traffic control systems like traffic lights and stop symbols also restrict the vehicular node velocities. Additionally, future vehicular nodes can be provided with larger transmission ranges, broad onboard storage and sensing capabilities, and rechargeable energy sources. Unlike MANETs, VANETs are rich in storage and processing power capabilities which make them flexible and make them more compatible of doing computationally intensive tasks [11–15].

There are various technical challenges for the design of efficient vehicular communications. One of the most critical challenges of the vehicular ad hoc network is to develop a scalable and reliable multi-hop routing protocol that is capable of providing an optimal route for forwarding packets towards the destination. Guaranteeing a stable, robust, optimal, and reliable multi-hop routing mechanism over VANETs is a fundamental move towards the realization of efficient vehicular ad hoc communications. Many emerging applications in vehicular communication require the assistance of multi-hop communications. One of them is onboard active safety mechanisms which help drivers to avoid collisions and provide coordination at crucial positions like highway entries and city intersection/junctions [16]. With the help of safety system information about roads like real-time traffic congestion, traffic accidents, road surface conditions, or high-speed levy can be smartly disseminated. This helps in avoiding the road congestion. It also largely reduces vehicle accidents and helps in saving many precious lives. In addition, inter-vehicular communication can provide comfort and infotainment applications. These consists of information about weather conditions, locations of gas stations and restaurants, e-commerce, and infotainment applications like accessing the internet, downloading music, and content delivery [14, 17–20].

The characteristics like non-uniform distribution of vehicles on the road, large size network, high mobility due to the high speed of the vehicles, frequently changing topology, and disruption of communication due to obstacle hindrance make routing of data in the vehicular ad hoc network more challenging. Majority of existing routing techniques are incapable of deciding optimal routes because of inefficiently incorporating aforementioned characteristics of VANETs. One of the problems with these routing protocols is that they often relay the packet towards destination using those streets that do not contain enough vehicular density. As a result, the packet meets a local optimum situation. Local optimum is a situation when a forwarding node is unable to locate the next neighbor because of lack of traffic density on the street. The node keeps the packet in its buffer for a longer time. If packet stays a longer time in the buffer, its time to live field expires and is eventually discarded. This degrades the network performance in terms of end-to-end delay and packet delivery ratio. Designing a routing protocol capable of solving such issues is critical and our proposed protocol intends to overcome this issue by selecting multiple connected streets that contain high vehicular density based on multiple intersections. Our findings focus on those multiple intersections or streets that provide optimal routing path based on closest distance to destination and contain high traffic through which packet can be easily relayed towards the destination. Secondly, besides optimal path based on multiple streets, we also propose a reliable packet forwarding strategy based on link stability and predicted packet propagation time between packet carrier node and next candidate forwarding neighbor to overcome packet loss due to unstable links.

The major contributions of this research are as follows:

1. Presented technical analysis, comparison, and drawbacks of earlier significant position-based routing schemes in VANETs.
2. Provided the significance of dynamic multiple intersection selection mechanism and limitations of earlier dynamic intersection selection mechanism pertaining to the city environment. Presented a novel routing scheme for city environment that is based on multiple intersections selection mechanism, which decides the best multiple streets to forward the packet towards the destination based on optimal distance and traffic density. To the best of our knowledge, this is the first time multiple intersection selection-based routing issue being thoroughly studied.
3. Along with multiple intersections selection mechanism, we introduced a novel reliable forwarding technique that considers link duration time and expected packet delivery time to overcome the packet loss issue due to high mobility and intermittent connectivity of vehicular nodes.
4. We provided an analysis and compared the performance of our routing scheme with existing approaches (GSR, GPSR, E-GyTAR, and TFOR) using ONE simulator. Simulation results indicate the benefits of the proposed routing strategy as compared to the existing protocols like GSR, GPSR, E-GyTAR, and TFOR in terms of packet delivery ratio, end-to-end delay, and routing overhead.

The remaining paper is structured as follows. In Section 2, we briefly present methods used in our study. In Section 3, we summarize the earlier routing protocols by highlighting their drawbacks which act as a source of persuasion for our research. Proposed routing strategy is presented in Section 4. Then, we present the performance evaluation based on extensive simulations in Section 5. In Section 5, we study and recapitulate the evaluation results. Finally, Section 6 concludes the paper.

2 Methods

For the performance evaluation of the proposed protocol, extensive simulations were used to compare the performance of our protocol with other state-of-the-art routing protocols. For the performance evaluation, it is vital to use a state of the art simulation environment capable of performing the simulations reliably and efficiently. In the research community for VANETs, the most versatile and well-trusted simulation environment is provided by the ONE simulator [21]. In this study, we used ONE simulator for all the simulations. Another important aspect regarding the performance evaluation of VANETs is the mobility model to be used. We utilized SUMO (simulation of urban mobility) [22] for generating the realistic vehicular mobility patterns to be used for the performance evaluations. SUMO is a microscopic road traffic simulation package, and the mobility traces generated by SUMO incorporates all essential characteristics of the city environment. More detail about the simulation environment and vehicular mobility patterns is provided in the simulation setup section. The proposed protocol was implemented in SUMO, and similarly, all the other considered routing protocols were also implemented. The performance metrics used for the comparison of the proposed protocol with the existing protocol were packet delivery ratio, end-to-end delay, and routing overhead. The performance was evaluated based on the considered performance metrics, and a variety of results were obtained and presented in this study.

3 Related work

The existing routing protocols (like Dynamic Source Routing (DSR) [23], Ad hoc On-Demand Distance Vector (AODV) [24], and Optimized Link State Routing (OLSR) [25]) that were originally proposed for mobile ad hoc networks are ineffective for VANETs [1, 11, 12, 21, 26–28]. These routing techniques consider the address of mobile nodes while discovering and maintaining end-to-end routing path in between the source node and destination. In vehicular ad hoc communication, the irregular distribution and high mobility of vehicular nodes frequently break the routing paths which make it hard to maintain and find routes. Consequently, these routing techniques generate high control overhead which degrades network performance

[1, 4, 6, 11, 12, 14, 18, 29–34]. The other class of routing which considers a geographical position of mobile nodes instead of address is suitable for vehicular ad hoc communication [11, 26, 27, 31, 32, 34, 35]. This class is more enviable for VANETs because of the following facts. Firstly, in the near future, vehicles will be implanted with navigation systems and Global Positioning Systems (GPS); thus, position-based routing class accomplishes colossal achievement in vehicular communication. Secondly, this class of routing is stateless; there is no need to maintain accomplished routing paths in between the source and target nodes; hence, this class is exceedingly scalable and very robust against high mobility which frequently changes network topology [11, 21, 27, 31, 34].

VANETs can be deployed into two different environments: (1) city/urban and (2) highway. City environment consists of intersections. Intersections are the points where two or more roads meet each other. On the other hand, a highway environment contains no intersection [1, 21, 33, 36]. In a city environment, the different sequence of intersections can play an important role in providing the shortest distance from source to destination. In the existing literature, some protocols are dynamic intersection selection based, some are static intersection selection based, and some having no intersection selection mechanism at all. Figure 1 provides the classification of the protocols. Below, we provide a brief technical description of existing significant position-based routing protocols that are designed for the city environment.

Greedy Perimeter Stateless Routing (GPSR) [28] is designed for handling routing issues in the highway environment. In a highly dense scenario, it performs well. It has two modules that are greedy module and perimeter module. In the greedy module, a vehicular node forwards the packet to one of its one-hop neighbors that is the closest among its one-hop neighbors and itself to the destination. The greedy module meets the local maximum if the packet carrier node has no one-hop neighbor that is close to the destination than itself. Perimeter module is used to handle the local maximum situation. The perimeter module includes two mechanisms, the graph planarization, and the right hand rule. The perimeter module induces long delays in dispatching packets from the source to the target. It also creates routing loops in the network, and unaware of obstacles [37], which makes it difficult to work in city scenarios. Furthermore, the graph planarization partitions the network in the city scenarios due to obstacles and degrades its performance further [11, 12].

Geographic source routing (GSR) [37] is a position-based routing protocol designed for urban scenarios. It uses position awareness with network topological awareness. It accomplishes the shortest route between the

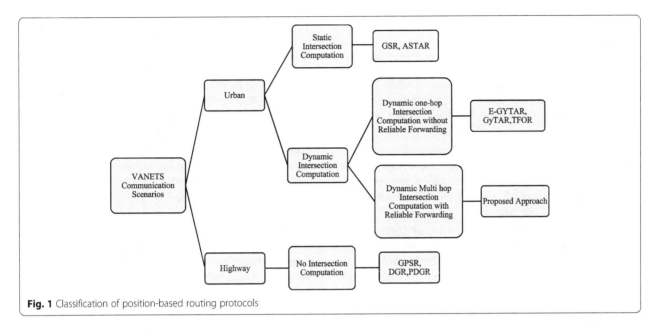

Fig. 1 Classification of position-based routing protocols

source and target node using Dijkstra's algorithm. The shortest route accomplished by GSR with the help of digital map consists of a sequence of intersections through which packet moves towards the destination. The packets are sent to the destination based on greedy forwarding. Greedy forwarding causes local optimum. GSR recovers from local optimum using carry and forward approach. While accomplishing the shortest path, GSR does not consider traffic density between intersections. In low-traffic density scenarios along a preselected route, establishing an end-to-end connection is very hard which degrade the performance of network [14]. Also, in GSR, greedy forwarding suffers from sudden link ruptures.

The Greedy Perimeter Coordinator Routing (GPCR) [38] accomplishes routing path without considering a map. It is composed of restricted greedy forwarding and a repair strategy. In restricted greedy forwarding, a packet carrier node prefers to forward the packet to a node that is located nearest to the intersection or at the intersection. The coordinator node is responsible to choose next-street for relaying packet towards the destination. GPCR uses repair strategy to overcome the local optimum problem. The repair strategy includes perimeter mode without considering graph planarization. It assumes that topological graphs are naturally planner in city environment. Computing graph planarization induces network disconnections which degrade network performance. GPSR does not consider low vehicular density cases while routing [1]. It is not a traffic aware routing protocol [11, 14], and it does not consider the status of link while forwarding the packets that may result in excessive packet loss.

The Anchor-Based Street and Traffic Aware Routing (A-STAR) [30] is a traffic aware routing protocol as compared to GPSR and GPCR. It has two main distinctive characteristics. Firstly, for traffic awareness, it uses statically or dynamically rated maps which assist it to discover routes that have a higher number of vehicles. Secondly, to overcome local optimum problem, it has a novel local recovery strategy which is better than those of GSR and GPSR [14]. The routing path accomplished by A-STAR on the basis of anchors may not be the optimal path, which may induce higher end-to-end delay [1, 11, 14]. It has no mechanism to overcome link ruptures as well that may result in packet loss.

Greedy Traffic Aware Routing Protocol (GyTAR) [14] was designed specifically for city environments. GSR and A-STAR select intersection statically while GyTAR selects intersection dynamically. It has three mechanisms: (a) intersection selection mechanism, (b) Infrastructure Free Traffic Information System (IFTIS), and (c) an improved greedy forwarding in-between intersection. It assigns weights to each neighboring intersection on the basis of the distance to destination and traffic density. The intersection that has highest traffic density and shortest distance to the destination is decided as next intersection, and thereby, packets are routed towards the destination. At low traffic density, its intersection selection mechanism sometime meets local optimum in city environment which degrades its performance [1]. Moreover, GyTAR has no mechanism to minimize packet loss due to link ruptures that may result in packet loss.

Enhanced Greedy Traffic Aware Routing (E-GyTAR) [1] is an enhanced version of GyTAR. It chooses junction on the basis of directional density on multilane streets. Non-directional density is also helpful in relaying packet from source to destination. Hence, it misses some shortest path while routing data [11]. Its forwarding strategy also suffers from sudden links rupture problem due to high mobility of the vehicles.

Directional geographic source routing (DGSR) [36] is an enhanced version of geographic source routing (GSR) with directional forwarding strategy. In this routing scheme, the source vehicular node uses location services [39, 40] to acquire the position of destination vehicle. It establishes the shortest path from source to destination using Dijkstra Algorithm. The shortest path consists of intersections which are ordered sequentially. The packets from source vehicular node follow the sequence of intersections to reach destination. If packet meets local optimum, DGSR uses carry and forward approach to overcome local optimum problem. However, this protocol does not consider the status of link while forwarding. Therefore, in case of high mobility of vehicular nodes, it suffers from packet loss due to link raptures.

Enhanced Greedy Traffic Aware Routing Protocol-Directional (E-GyTAR-D) [36] is an enhanced version of E-GyTAR [1] with directional forwarding. It consists of two mechanisms: (i) intersection selection and (ii) directional greedy forwarding strategy. It uses location services to get the position of destination node. It selects intersections on the basis of directional traffic density and shortest distance to the destination. It forwards packets in between intersection using directional greedy forwarding. Simulation outcomes in realistic urban scenarios show that E-GyTAR-D outperformed GSR and DGSR in terms of packet delivery ratio and end-to-end delay.

Traffic flow-oriented routing protocol (TFOR) [11] is a recently proposed technique which consists of two modules: (a) an intersection selection mechanism based on traffic flows and the shortest routing path and (b) a forwarding strategy based on two-hop neighbor information. It accomplishes shortest optimal path based on shortest distance to the recipient node and vehicular traffic density. Simulation results show that TFOR outperforms E-GyTAR and GyTAR in terms of packet delivery ratio and end-to-end delay. Table 1 shows the comparative characteristic of all the aforementioned routing protocols.

A majority of the aforementioned protocols (like GSR, DGSR, GPSR, DGR, PDGR, PDVR, and GPCR) do not consider traffic density while accomplishing routing path. Traffic density is a major source of providing connectivity. Consequently, these routing protocols relay the packets towards destination through those city streets which have low traffic density or connectivity. As a result, packets meet frequently local optimum and this leads to a decrease in packet delivery ratio. This drawback can be overcome by having a mechanism that is capable of giving timely information about city street traffic density. Although, few of the aforementioned routing schemes such as E-GyTAR, GyTAR, E-GYTAR-PD, and TFOR

are traffic-aware, all of these approaches prove to be incompetent in making full use of the real-time traffic density. The intersection selection mechanism of these routing protocols decides sometime those streets that lead to a local optimum problem regardless of the availability of effective streets which may overcome such problem. In such a situation, the probability of packet loss increases. It also increases end-to-end delay because the packets are carried in the buffer for a long time. Consequently, they prove to be ineffective in relaying packets from source to destination. Also, in between successive intersections, the aforementioned routing protocols use different types of forwarding strategies like simple greedy forwarding, restricted greedy forwarding, directional greedy forwarding, and improved greedy forwarding. However, their forwarding approaches do not consider link status and link lifetime between packet carrier node and next candidate neighbor forwarding node. Therefore, these strategies miss some appropriate candidate neighbor nodes which are stable and reliable for packet forwarding. Hence, all aforementioned routing suffer from sudden link rupture problem due to the high mobility of the vehicular nodes.

In this research work, we devise a routing scheme that addresses the aforementioned drawbacks of existing protocols. The proposed routing scheme is capable of accomplishing optimal routing paths in city scenarios. The scheme is envisaged to function well for various kinds of vehicular communication applications by assuring user connectivity. These appliances consist of road safety services such as coordinated communication of two vehicles, managing flows of traffic, triggering driving-related alerts such as traffic congestion alerts, road situation alerts, and accident warnings. The other appliances include finding locations of petrol stations and restaurants, accessing internet, downloading music, and playing games.

4 Reliable Path Selection Packet Forwarding (RPSPF) Routing Protocol
4.1 Problem formulation

In this work, we considered the protocols that are based on one-hop dynamic intersection selection mechanism and forwarding strategy in between intersections for forwarding packets. These protocols include GyTAR, E-GyTAR, and TFOR. All these routing protocols choose dynamically one intersection at a time on the basis of vehicular traffic density and shortest distance to the destination while relaying a packet towards the destination. However, their one-hop dynamic intersection selection mechanisms have some limitations. As an example, we imagine one of the possible situations revealed in Fig. 2.

Assume S is the source vehicular node which is at intersection I_1. It intends to forward the packet to

Table 1 Comparative features of position-based routing protocols

Routing protocol	Traffic-aware	Link ruptures	GPS required	Environment	Traffic data required	Location service	Carry and forward	Nature of the network (sparse/dense)	Forwarding strategy	Multiple intersection selection	Link stability	Digital map
1- GPSR [28]	No	Yes	Yes	Highway	No	Yes	No	Dense	Greedy forwarding	No	No	Yes
2- GSR [37]	No	Yes	Yes	City	No	Yes	No	Sparse	Greedy forwarding	No	No	Yes
3- GPCR [38]	No	Yes	Yes	City	No	Yes	No	Sparse	Restricted greedy forwarding	No	No	No
4- A-STAR [30]	Yes	Yes	Yes	City	No	Yes	Yes	Sparse	Greedy forwarding	No	No	Yes
5- GyTAR [14]	Yes	Yes	Yes	City	Yes	Yes	Yes	Both	Improve greedy forwarding	No	No	Yes
6- DGR [41]	No	Yes	Yes	Highway	No	Yes	Yes	Both	Greedy forwarding	No	No	Yes
7- PDGR [41]	No	Yes	Yes	Highway	No	Yes	Yes	Both	Greedy forwarding	No	No	No
9- PDVR [29]	No	Yes	Yes	Highway	No	Yes	Yes	Sparse	Directional forwarding	No	No	No
10- E-GyTAR [1]	Yes	Yes	Yes	City	Yes	Yes	Yes	Both	Improve greedy forwarding	No	No	Yes
11- TFOR [11]	Yes	Yes	Yes	City	Yes	Yes	Yes	Both	Improved greedy forwarding based on 2-hops	No	No	Yes
12-DGSR [36]	No	Yes	Yes	City	No	Yes	Yes	Sparse	Directional greedy forwarding	No	No	Yes
13-E-GyTARD [36]	Yes	Yes	Yes	City	Yes	Yes	Yes	Both	Directional greedy forwarding	No	No	Yes

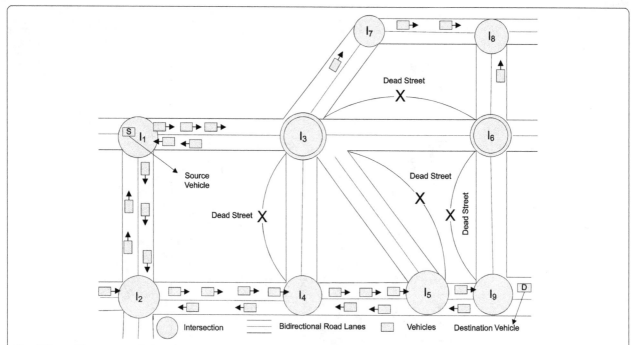

Fig. 2 The problem scenario. Each circle represents a junction. Double lines represent bidirectional two-lane roads, and the small rectangular box with arrow represents vehicular node moving in the direction of the arrow, S represents source vehicle, and D indicates the destination

destination vehicular node D. Current dynamic intersection selection-based routing protocols like E-GyTAR, GyTAR, and TFOR select next intersection based on vehicular traffic density and shortest distance to the destination; therefore, their intersection selection mechanism will bound them to choose I_3 as the next intersection. At I_3, TFOR, GyTAR, and E-GyTAR will be inept to choose next appropriate intersection through which packet can be relayed towards the destination. This is because the next shortest path offering roads/streets that direct towards destination have no traffic density for dispatching packet towards destination. When the packet reaches I_3, all the aforementioned protocols are unable to choose next streets because from I_3 to I_6, there is no vehicular density. Likewise, I_3 to I_5 and I_3 to I_4, there is no vehicular traffic density. Each of these streets is out of traffic, and there is no vehicular traffic density for routing packet further. This improper selection of intersections degrades the network performance as it decreases the packet delivery ratio and increases end-to-end delay regardless of the availability of the optimal path (I_1-I_2-I_4-I_5-I_9) with rich vehicular traffic density. Consequently, all these protocols can prove to be inefficient in intersection selection mechanism because of consideration of just immediate intersection while selection of next intersection. Furthermore, their current intersection selection mechanism is very limited to move the packet progressively closer to the destination due to the consideration of just the immediate

intersection from the current intersection. This increases the probability of incurring a dead street (street without packet carriers/vehicles) along a selected path to the destination due to consideration of just the immediate intersection.

The second problem with these routing protocols is that their forwarding strategies suffer from sudden link rupture problem due the high mobility of the vehicular nodes which cause packet loss. Existing forwarding strategies include greedy forwarding [28], restricted greedy forwarding [38], improved greedy forwarding [14], directional greedy forwarding [41], and predictive directional greedy forwarding [41]. None of these forwarding strategies have mechanism which is capable of handling sudden link rupture. Figure 4 in Section 4.2.3 describes the limitation of existing forwarding strategies without considering link stability mechanism.

We need a routing protocol that provide an efficient intersection selection mechanism which selects intersections by considering connectivity and a reliable forwarding mechanism which overcomes packet loss due to sudden link ruptures so that packet delivery ratio is maximized and routing overhead and end-to-end delay are minimized. We propose a novel routing strategy called Reliable Path Selection and Packet Forwarding Routing Protocol (RPSPF), which selects the next junction by considering multiple intersections based on the shortest path and traffic density and forwards packet using reliable forwarding strategy for tackling aforementioned circumstances.

4.2 RPSPF protocol

In this section, we explain the basic work of the proposed protocol.

4.2.1 Protocol assumptions

RPSPF is an intersection-based geographic routing protocol. It has certain assumptions similar to the assumptions made in [1, 11, 14, 36]. A vehicle locates its position using GPS. Location service like GLS (grid location service) [42] can be used to locate the location of the destination vehicle. Each vehicle is equipped with an onboard navigation system that gives the position of neighboring intersections and valuable city street level awareness with the assistance of preloaded digital maps. Furthermore, it is also assumed that each vehicle is aware of its speed and direction. We also assume that each vehicle is aware of vehicular traffic density in between intersections which can be accomplished by deploying traffic sensors besides intersections or a distributed mechanism for road traffic density estimation apprehended by all the vehicles [43].

RPSPF comprises of two phases: (i) dynamic multiple intersection selection mechanism and (ii) reliable greedy forwarding mechanism between the intersections. The detailed description of both of these phases is given below.

4.2.2 The dynamic multiple intersections selection mechanism

RPSPF routing scheme employs anchor-based routing approach with city streets awareness like other routing routing schemes such as E-GyTAR [1], GyTAR [4], and TFOR [11]. It applies street map topology to route data packets between vehicular nodes. The foremost difference between our routing scheme and the existing routing schemes is its intersection selection mechanism and reliable forwarding mechanism in between intersections. RPSPF choose the next suitable intersection by taking into consideration the next two of the immediate intersections dynamically from the current intersection on the basis of vehicular density and shortest curve metric distance to the destination. Now, this raises a very important question about our proposed routing scheme that is, why two intersections are significant to consider instead of three, four, and so on? This query is addressed in detail in Section 5.2.4. Consideration of two intersections reduces the likelihood of incurring connectivity problem that prevailed in earlier intersection selection techniques. It also minimizes the possibility of facing those streets that contain no packet carriers or vehicular

nodes along a chosen route to destination. Such streets are called dead streets. Hence, RPSPF uses those streets that are rich in connectivity and moves packets progressively closer to the destination in comparison to earlier routing schemes. While selecting the next two-hop neighbor intersection, the sender or intermediate vehicle uses digital city street map and finds the locations of two-hop neighboring intersections. It assigns weight to each of candidate two-hop neighbor intersections based on traffic density and curve metric distance of candidate intersections to the destination using Eq. 1. The two-hop neighbor intersection with the highest weight is chosen as the next destination intersection. Algorithm 1 is used to allocate weight to each of candidate intersection.

$$\text{Weight} = H_1.(W_j) + H_2.(W_k),.... \tag{1}$$

where

$$W_j = \alpha \times (1 - \text{Dp}_1) + \beta \times \text{TD}_1$$

and

$$W_k = \alpha \times (1 - \text{Dp}_2) + \beta \times \text{TD}_2$$

W_j is the weight for candidate one-hop neighbor intersection of the current intersection; Dp_1 is closeness of candidate neighbor intersection with respect to destination. TD_1 is the traffic density between current intersection and next neighbor intersection. W_K is the weight for candidate neighbor intersection of the one-hop neighbor intersection; Dp_2 is closeness of candidate two-hop neighbor intersection w.r.t destination. TD_2 is the traffic density between the one-hop neighbor intersection and its next neighbor intersection. Alpha (α) and beta (β) are the weighting factors for distance and traffic density respectively between the intersections. By adjusting the value of α and β, we can make tradeoff between distance and traffic density when selecting next intersection. Traffic density is an important source of giving connectivity for dispatching packet towards the destination. H_1 and H_2 are the weighting factors for candidate one-hop neighbor intersection and two-hop neighbor intersection, respectively. An adjustment in the value of H_1 and H_2 can make tradeoff between the importance of candidate one-hop neighbor intersection and two-hop neighbor intersection.

In the aforementioned algorithm 1, lines 1 to 10 state the parameters that are utilized in our algorithm to allocate weights to neighbor intersections of the candidate intersection. Line 12 checks if candidate one-hop neighbor has next neighbor intersection that leads to the

---

Algorithm 1: The Dynamic Multiple Intersection Selection Mechanism

Input: Area, α, H_2

Output: The next destination intersection NDj

1. begin
2.   set weight $\leftarrow 0$
3.   set $\beta \leftarrow 1 - \alpha$
4.   set $H_1 \leftarrow 1 - H_2$
5.   for each candidate intersection j do
6.   set $D_n \leftarrow$ the curvemetric distance between NC_j and destination
        /* NC_j is the next candidate intersection (one-hop) */
7.   set $Dc \leftarrow$ the curvemetric distance between C_j and destination
        /* C_j is the current intersection */
8.   set $Dp_1 \leftarrow D_n/D_c$ /*Closeness of candidate intersection to destination*/
9.   set $TD_1 \leftarrow$ no. of vehicles between NC_j and C_j in both directions
10. $W_j = (\alpha \times (1 - Dp_1) + \beta \times TD_1)$
11.   if weight $< W_j$
12.   if NC_j contains next candidate neighbour intersection NC_k //2-hop
13.   GetnextneighbourjuctionKofJ(ND_k, Weight)
14.   set $W_k =$ Weight
15.   set $NC_j \leftarrow ND_k$
16.   set $ND_j \leftarrow NC_j$
17.   set Weight $\leftarrow H_1.(W_j) + H_2.(W_k)$
18.   else
19.   set $ND_j \leftarrow NC_j$
20.   set weight $\leftarrow W_j$
21.   end
22.   else
23. $NDj = D$// Where D is the destination vehicle
24.   end
25.   end
26.   GetnextneighbourintersectionKofJ(NC_k, Weight)
27.   begin
28.   for each candidate intersection K of J do
29.   set $Dn_k \leftarrow$ the curvemetric distance between NC_k and destination
            /* NC_k is the next candidate intersection */
30.   set $Dc_k \leftarrow$ the curvemetric distance between C_k and destination
            /* C_k is the current intersection */
31.   set $Dp_2 \leftarrow Dn_k /Dc_k$ //closeness of second hop w.r.t destination
32.   set $TD_2 \leftarrow$ no. of vehicles between NC_k and C_k in both directions
33. $W_k = (\alpha \times (1 - Dp_2) + \beta \times TD_2)$
34.   if weight $< W_k$ then
35.   set $ND_k \leftarrow NC_k$
36.   set weight $\leftarrow W_k$
37.   end
38.   end
39.   return (ND_k, weight)
40.   end
41.   return ND_j
42. end

---

destination then algorithm call procedure mentioned in line 13 which uses line 26 to 40 to allocate weight to each of two-hop neighbor intersections. Line number 17 calculates the weight of the two-hop intersection, and the intersection having the maximum weight will be chosen as the next two-hop candidate neighbor intersection through which packet is relayed towards the destination. Any candidate two-hop neighbor intersection that is the closest to the destination and provides higher traffic density will be selected as next destination intersection through which packet moves towards the destination.

In order to understand the working of RPSPF, let us consider the scenario presented in Fig. 3. S is the source vehicle which is present at intersection I_1. It uses RPSPF for dispatching packet towards the destination D. In this scenario, current intersection I_1 contains four candidate two-hop neighbor intersections through which packets can be relayed towards the destination. These intersections are I_4, I_5, I_6, and I_8. The intersection selection mechanism in RPSPF will allocate weights to each of two-hop neighbor intersections of I_1 on the basis of vehicular traffic density and the shortest curve metric distance to the destination. There is a higher concentration of vehicular traffic along the streets that are connecting I_1 to I_4 as compared to I_1–I_5, I_1–I_8, and I_1–I_6. Accordingly, RPSPF will allocate more weight to I_4 as compared to I_5, I_6, and I_8 and it will be selected as the

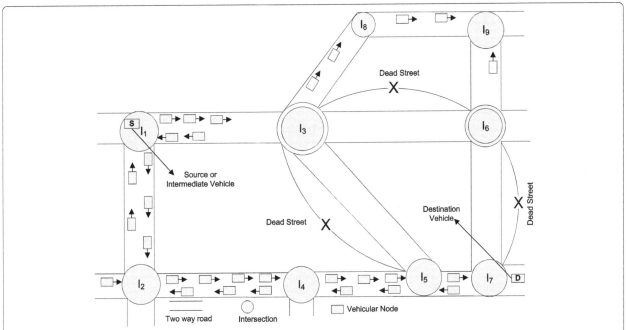

Fig. 3 Working of RPSPF protocol. Each circle represents a junction. Double lines represent bidirectional two-lane roads, and the small rectangular box with arrow represents vehicular node moving in the direction of the arrow, S represents source vehicle, and D indicates the destination

next destination intersection. In this manner, it will overcome the precincts of earlier routing approaches like TFOR, GyTAR, and E-GyTAR. Since in earlier routing schemes, I_3 would have been chosen. This is because all these routing schemes are based on one-hop intersection selection mechanism, which is the sub-optimal option in this scenario. The packet would have been trapped in local optimum because after choosing I_3, intersections I_5 and I_6 have lack of traffic density as these are along dead streets which contain no vehicular node for carrying packet towards the destination. Our routing scheme route the packet from current intersection I_1 to I_4 through I_2, and in this manner, it will dispatch the packet towards the destination. Consequently, RPSPF will move packet successively closer towards the destination beside the city streets where there are plenty of vehicles to give connectivity.

4.2.3 Reliable forwarding mechanism between intersections
The forwarding strategies that are used by TFOR, GyTAR, E-GyTAR, GPSR, and GSR suffer from a sudden link rupture problem due to high mobility in VANETs. Figure 4 shows a scenario where GyTAR and E-GyTAR prove to be inefficient. If the forwarding vehicle F is using GyTAR, E-GyTAR, GPSR, or GSR, the forwarding strategies used in these routing protocols will compel F to select B as the next hop for forwarding packet as B is the closest to the destination and moving in the direction of the

destination. But if F forwards packet at time t_1 to B and at the same time B leaves the range of F due to high speed, then the packet will not be delivered to B and it will be lost. In general, we can say that if packet delivery time is greater than the link duration time between forwarding node and the next neighbor node, then packet cannot be delivered to the next neighbor node. This will decrease the packet delivery ratio and thus reduces the throughput of the network. We try to overcome this problem by taking into account link duration time between two mobiles nodes and the expected packet delivery time.

4.2.4 Link duration time
If the parameters like speed, position, radio propagation range, and direction that are related to the motion of two neighboring vehicular nodes are known, then the time duration for which these two nodes will remain in contact can be determined [44]. Consider two vehicular nodes j (packet carrier or forwarding node) and k (next candidate packet receiving neighbor node) inside the transmission range of each other. Let us assume the (X_j, Y_j) be the coordinates of the location for vehicular node j and (X_k, Y_k) be the coordinates of vehicular node k. Let V_j and V_k be the velocities, (θ_j, θ_k) be the directions of movement for vehicular nodes j and k, respectively. The time duration for which vehicular node j and vehicular node k will stay in range of each other is given by the Eq. 2.

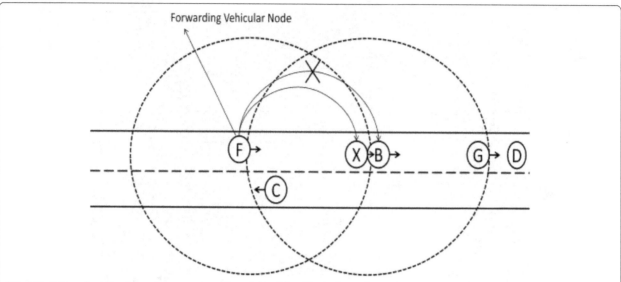

Fig. 4 Limitation of existing forwarding strategies without link stability. F is the forwarding node, X and B are the neighbors that are moving in the direction of destination, and C is a vehicle traveling in opposite direction to the destination. D is the destination node. The big dotted circles represent the transmission ranges of the nodes

$$LDT = \frac{-(ab + cd) + \sqrt{(a^2 + c^2)r^2 - (ad-bc)^2}}{a^2 + c^2} \quad (2)$$

where

$$
\begin{aligned}
a &= V_k\cos\theta_k - V_j\cos\theta_j \\
b &= X_k - X_j \\
c &= V_k\sin\theta_k - V_j\sin\theta_j \\
d &= Y_k - Y_j.
\end{aligned}
$$

In the aforementioned formula, LDT represents the link duration time between the nodes; V_j and V_k are the velocities of the packet carrier or forwarding node and next candidate receiver node, respectively. The transmission range of the wireless vehicular node is given by r. Theta (θ) represents the direction of movement of the nodes with respect to each other. (X_j, Y_j) and (X_k, Y_k) are the coordinates of the packet sender and receiver, respectively. Factor a represents the relative velocity of the candidate receiver node with respect to packet sender node along the Y-axis. Factor b represents the distance of the receiver node from the packet sender node along the X-axis. The relative velocity of the receiver node with respect to the packet sender node along the Y-axis is represented by factor c. The distance of the receiver node from the sender node along with the Y-axis is represented by factor d.

In order to discard selection of a node that is about to leave the range of the sender until the packet is delivered to that node, as a first simple step, let us simplify the communication model to get a rough estimate about the

time required for transmission and receiving of a packet of size "s" kilobytes. There are two different things to consider:

1) The transmission time

This is the amount of time from the beginning until the end of a message transmission. In the case of a digital message, it is the time from the first bit until the last bit of a message has left the transmitting node. The packet transmission time in seconds can be obtained from the packet size in bits and the bit rate in bits/s as packet transmission time = packet size/bit-rate, and both of these units are in bits [45].

2) Propagation delay

It is the time it takes for the first bit to travel from the sender to the receiver. The distance here is the distance between the sender and the receiver in meters and the propagation speed is the speed of light, i.e., 3×10^8 m/s.

Propagation time = distance/propagation speed

3) Packet delivery time (PDT)

The total time it will take for the packet to be completely transmitted from sender node to the receiver node is given by:

Packet delivery time (PDT) = transmission time (TT) +propagation delay (PD)

Any neighbor whose expected link duration time is less than the expected packet delivery time will not be considered as the next node. Algorithm 2 decides the neighbor that is chosen as the next forwarding node.

4) Estimation of closeness and direction of motion

The packet carrier vehicle uses the following equation to choose a neighbor vehicle that moves towards the destination and is closest to the destination.

$$\text{Score}_i = \sigma \times (1 - D_{n_i}/D_{cv}) + \rho \times \cos(V_{n_i}, \text{posn}_i, dv) \quad (3)$$

In the aforementioned Eq. 3, the first factor $(1 - D_{n_i}/D_{cv})$ represents the closeness of a neighbor vehicle to the destination. The second factor represents the moving direction of the vehicular node using cosine values of two vectors (velocity and position). Here, σ and ρ are the weighting factors for closest position and direction of the vehicular node with respect to the destination. We can adjust the value of σ and ρ to make a tradeoff between position and direction when forwarding. If the value of ρ is set to 0, then the protocol uses greedy forwarding. On the other hand, if σ is set to 0, then protocol uses directional forwarding which will be unable to select a neighbor node that is the closest to the destination. Therefore, our approach will consider both by setting the value of $\sigma = 0.5$ and $\rho = 0.5$.

Below, in our reliable packet forwarding algorithm 2, the packet carrier or forwarding vehicle considers the position, direction, expected packet delivery time, and link duration time for assigning the score to each of candidate neighboring nodes. It assigns the higher weighting score to those neighbors that are the closest to the destination and moving towards the destination that are capable of successfully receiving the message based on link duration time. In our algorithm, the current packet carrier vehicular node uses line 8–17 for determining its next candidate neighbor. In line 15, the forwarding vehicle compares each neighbor link duration time with expected packet delivery time. For any neighbor, if link duration time is less then packet delivery

---

Algorithm 2: The Reliable Packet Forwarding Mechanism

Algorithm Notations: cv current vehicle, nv: next candidate neighbor vehicle
Poscv: position of current vehicle Vcv: speed of current vehicle
Posdv: position of destination, Dcv: Distance between current vehicle and destination
$\overrightarrow{\text{Poscv,dv}}$: vector from the position of current vehicle to the position of destination
Posn$_i$: position of neighbor vehicle, Vn$_i$: speed of neighbor vehicle i
Dn$_i$: Distance between neighbor vehicle i and destination
LETcv,nv$_i$: Link duration time between cv and nv
PDTcv,nv$_i$=Packet delivery time between cv and nv

Input: α, β, link duration time (LET), Packet delivery time (PDT). **Output:** Next forwarding hop

```
1.   Begin
2.       Poscv←getPosition(currentvehicle) // position of current vehicle
3.       Vcv←getSpeed(currentvehicle)        // speed of current vehicle
4.       Posdv←getPosition (destinationvehicle) // position of destination
5.       Dcv = distance (Poscv,Posdest) // Distance between current vehicle and destination
6.       Poscv,dv =Posdv − Poscv//vector from the position of current vehicle i to the position of dest
7.       Score = β×cos(Vcv, poscv,dv)  // the cosine value for the angel made by these two vectors.
8.       NextHop = currentVehicle
9.       for all neighbors of the currentVehicule do
10.          Posn_i←getPosition(neighVehicle_i) // position of neighbor vehicle
11.          Vn_i←getSpeed(neighVehicle_i) // speed of neighbor vehicle
12.          Dn_i = distance(Posn_i, Posdv) // Distance between neighbor vehicle and destination
13.          posn_i,dv = Posdv − Posn_i
14.          LETcv,nv_i= getLinkDurationTime () // Link duration time between cv and nv
15.          PDTcv,nv_i= getPacketDeliveryTime () // Packet delivery time between cv and nv
16.          if PDTcv,nv_i < (LETcv,nv_i − PDTcv,nv_i) do
17.              Score_i = σ×(1 − Dn_i/Dcv) + ρ×cos(Vn_i, posn_i,dv)
18.              if Scoren_i > Score then
19.                  Score = Scoren_i
20.                  NextHop = neighVehicle_i
21.              else
22.                  drop neighVehicle_i
23.              end
24.          end
25.      end
26.      if NextHop! = currentVehicle then
27.          forward packet to NextHop
28.      else
29.          keep the packet with Cv(Currentvehicle)
30.      end
31.  end
```

time or not satisfying the condition mentioned in line 15 will be not be considered. On the other hand, if link duration time is greater than expected packet delivery time and satisfying the equation mentioned in line 13, then a score is assigned according to the equation mentioned in line 13. This equation considers the position as well as the direction of motion of the neighbors. Among all the neighbors, the neighbor that is closest to the destination and moving in the direction of destination is assigned highest weighting score and is selected as the next forwarding vehicle. σ and ρ are the weighting factors for the closeness of the next hop neighbor and its movement direction with respect to destination respectively. Their value is set to $\sigma = \rho = 0.5$. Sometimes a packet carrier vehicle is unable to locate a next candidate forwarding vehicle to forward the packet. In this case, the packet carrier vehicle will hold the packet until next intersection or an appropriate forwarder node in its vicinity is found.

5 Simulation and results analysis

In this section, we evaluate the performance of RPSPF. The simulations are carried out in ONE [19] simulator. The mobility model can affect the traffic characteristics which affects the performance of a routing protocol. Therefore, the selection of an appropriate realistic mobility model for generating realistic mobility traces for simulations is a very vital step [46]. We used SUMO (simulation of urban mobility) [22] to generate the realistic vehicular mobility patterns. It is a microscopic road traffic simulation package which is open source. The realistic mobility traces generated by SUMO include all essential characteristics of city environment like multiple lane roads, communication obstacles, and vehicle speed regulations in accordance with traffic signals at intersections.

5.1 Simulation setup

The simulation scenario comprises of 3000×2500 m² city area with 32 multilane bidirectional roads and 32 intersections. At the start of the simulations, all the vehicular nodes were positioned randomly over the multilane bidirectional roads. The movement of the vehicular nodes in both directions on a multilane road is based on the intelligent driving model [22]. The simulation parameters that are used for performance analysis are presented in Table 2. The simulation outcomes are based on an average of 15 simulation runs.

5.2 Results

The metrics that are used for evaluation of the routing protocols include the packet delivery ratio, end-to-end delay, and routing overhead. Packet delivery ratio is the fraction of packets that are effectively dispatched to their destination vehicular nodes. End-to-end delay is the average delay incurred by a packet while moving from its source to destination. While routing overhead is the fraction of total control packets generated to the total data packets delivered at destinations during one complete simulation run. The detailed depiction of the aforementioned metrics is provided in [1, 11, 14, 36, 47]. We provide a comparative study and in-depth analysis of RPSPF with respect to GPSR, GSR, GyTAR, and TFOR considering the aforementioned metrics.

5.2.1 Packet delivery ratio

Figure 5 presents the impact of increasing node density on the packet delivery ratio. In Fig. 5, it is illustrated that as the vehicular density increases, the packet delivery ratio of all the considered routing protocols increases. The figure exhibits that the packet delivery ratio of the RPSPF is the highest as compared to GPSR, GSR, E-GyTAR, and TFOR. This is because of two reasons, firstly, in RPSPF, the routing path is set up based on multiple intersections guaranteeing that the intersections with rich traffic density are chosen. Therefore, the packet will move successively towards destination along the streets which contains enough vehicles ensuring rich network connectivity. Secondly, RPSPF provides a reliable link stability mechanism that ensures that before forwarding the data packet, the link has enough life-time

Table 2 Simulation setup

Simulation/scenario		MAC/routing	
Simulation time	250 min	MAC protocol	802.11 DCF
Map size	3000×2500 m²	Channel capacity	54 Mbps
Mobility model	SUMO	Transmission range	266 m
Number of intersections	32	Traffic model	15 CBR connections
Number of double lane roads	36	Packet sending rate	(1–10 packet(s)/s)
Number of vehicles	100–400	Vehicle speed	35–60 Km/h
Weighting factors	$\alpha = 0.5$, $\beta = 0.5$, $H_1 = 0.5$, $H_2 = 0.5$ $\sigma = 0.5$, $\rho = 0.5$	Beacon interval	1 s

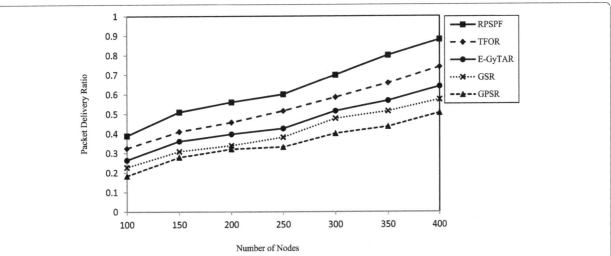

Fig. 5 Packet delivery ratio as a function of the number of nodes (@5 packets/s). RPSPF represents Reliable Path Selection and Packet Forwarding Routing Protocol which is proposed approach and is represented by a solid line with a square. TFOR is traffic flow-oriented routing protocol and is represented by a dashed line having rhombus. E-GyTAR is Enhanced Greedy Traffic Aware Routing protocol and is represented by a solid line with circle. GSR is geographic source routing protocol and is presented by a dotted line having a cross. GPSR is Greedy Perimeter Stateless Routing Protocol and is presented by a dashed line having a triangle

to accomplish successful forwarding. This avoids link ruptures before a packet is completely delivered to the destination. GSR computes a sequence of intersections statically without considering vehicular traffic density. As a result, sometimes it selects routing paths that consist of city streets with low vehicular traffic density. As a result, packets are unable to move towards destination due to lack of connectivity which degrades the packet delivery ratio. E-GyTAR and TFOR accomplish routing path dynamically based on one intersection at a time. Dynamically, considering one intersection at a time might lead to a selection of those intersections whose next streets contain no or very low vehicular traffic. Selection of such streets results increases the probability of encountering local optimum at street level which leads to a reduction in packet delivery ratio. Also, GSR, E-GyTAR, and TFOR protocols during forwarding mechanism may select those neighbor nodes which are the closest to the destination but their link duration time is not enough to get packets successfully transferred from the forwarding nodes. In other words, the links break before successful delivery of the packets and packets are lost which also degrades the packet delivery ratio.

Figure 6 presents the impact of packet sending rate on the packet delivery ratio. Increase in packet sending rate causes network congestion and packets collision which brings down the packet delivery ratio. It affects the performance of all the routing protocols. Some vehicular nodes along the preselected routing paths in GSR transmit more control messages due to static intersection selection mechanism which causes a reduction in the packet delivery ratio as compared to RPSPF, E-GyTAR,

and TFOR. Unreliable forwarding mechanism in GSR, GPSR, E-GyTAR, and TFOR also affects the packet delivery ratio because more packets are re-generated and dropped due to link ruptures.

Figure 7 illustrates the influence of increasing file size (data size) on packet delivery ratio. If we increase the file size, the packet delivery ratio tends to decrease for all the considered protocols. This is because, in VANETs, due to the high mobility of nodes, the topological connections are ephemeral. More time is required for transfer of files having a larger size as compared to smaller

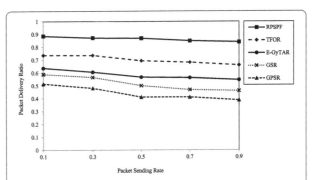

Fig. 6 Packet delivery ratio as a function of the packet sending rate (400 nodes). RPSPF represents Reliable Path Selection and Packet Forwarding Routing Protocol which is proposed approach and is represented by a solid line with a square. TFOR is traffic flow-oriented routing protocol and is represented by a dashed line having rhombus. E-GyTAR is Enhanced Greedy Traffic Aware Routing protocol and is represented by a solid line with circle. GSR is geographic source routing protocol and is presented by a dotted line having a cross. GPSR is Greedy Perimeter Stateless Routing Protocol and is presented by a dashed line having a triangle

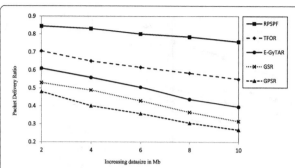

Fig. 7 Influence of increasing data size on packet delivery ratio (400 nodes). RPSPF represents Reliable Path Selection and Packet Forwarding Routing Protocol which is proposed approach and is represented by a solid line with a square. TFOR is traffic flow-oriented routing protocol and is represented by a dashed line having rhombus. E-GyTAR is Enhanced Greedy Traffic Aware Routing protocol and is represented by a solid line with circle. GSR is geographic source routing protocol and is presented by a dotted line having a cross. GPSR is Greedy Perimeter Stateless Routing Protocol and is presented by a dashed line having a triangle

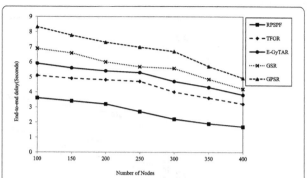

Fig. 8 End-to-end delay as a function of the number of nodes (@5 packets/s). RPSPF represents Reliable Path Selection and Packet Forwarding Routing Protocol which is proposed approach and is represented by a solid line with a square. TFOR is traffic flow-oriented routing protocol and is represented by a dashed line having rhombus. E-GyTAR is Enhanced Greedy Traffic Aware Routing protocol and is represented by a solid line with circle. GSR is geographic source routing protocol and is presented by a dotted line having a cross. GPSR is Greedy Perimeter Stateless Routing Protocol and is presented by a dashed line having a triangle

files. Choosing a suitable node with sufficient contact duration for forwarding a larger file is of immense importance. It can be observed from the figure that as we increase the file size from 2 to 10 MB, the packet delivery ratio degrades for all the protocols. However, our proposed protocol outperforms the other protocols. This is because, firstly, our protocol dispatches the packets towards destination through those streets that maximize connectivity. Secondly and most importantly, the forwarding node always prefers to forward the packet to a neighbor that offers enough link life-time to carry out successful file transfer which decreases the probability of packet loss. The rest of the protocols prove to be ineffective in providing higher packet delivery ratio. The major reason behind this is that their forwarding strategies look to forward the packet to a node without considering link lifetime. The links often break before the successful transfer of the file due to the short link lifetime. As a result, packet loss increases and this leads to decrease in packet delivery ratio.

5.2.2 End-to-end delay

Figure 8 exhibits the performance of RPSPF, E-GyTAR, GPSR, GSR, and TFOR in terms of end-to-end delay with respect to increasing traffic density. RPSPF outperforms GPSR, GSR, E-GyTAR, and TFOR in terms of end-to-end delay as well. This is because, RPSPF progressively accomplishes routing path based on multiple intersection selection mechanism by considering network connectivity when relaying data packets from source to destination. The reliable greedy forwarding mechanism maintaining one-hop information based on link stability accomplishes successful forwarding avoids packet retransmission and helps to reduce the end-to-

end delay as well. While in E-GyTAR and TFOR, their intersection selection mechanisms sometimes compel them to select those routing paths that have negligible connectivity; therefore, a packet stays the longer time in a buffer which results in long delays. Also, E-GyTAR uses only directional density to find the path but in urban scenarios with a two-lane road, there are a lot of streets having the non-directional density for providing the shortest path which avoids end-to-end delay. Pre-determination of end-to-end path routing path in GSR before dispatching data packets without considering connectivity causes delays due to lack of traffic density along some routes. In GPSR, perimeter phase establishes longer routes while relaying the packet towards the destination. It also causes routing loops which results in long delays. The novel combination of multiple intersection selection mechanism with reliable forwarding mechanism in RPSPF leads to a considerable reduction of end-to-end delay in comparison to the other protocols.

Figure 9 provided end-to-end delay with respect to different packet sending rates on the logarithmic scale. It shows that an increase in packet sending rate does not have a considerable impact on the performance of RPSPF as compared to the considered routing techniques. This is because of its new intersection selection mechanism and reliable forwarding which can incorporate the city challenges like high mobility and finding streets which provide better connectivity for routing in a better way as compared to earlier techniques.

5.2.3 Routing overhead

Routing overhead with respect to a different number of the nodes for all the routing protocols is illustrated in

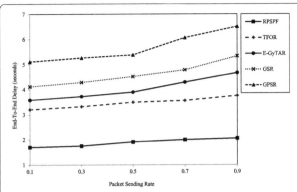

Fig. 9 End-to-end delay as a function of the packet sending rate (400 nodes). RPSPF represents Reliable Path Selection and Packet Forwarding Routing Protocol which is proposed approach and is represented by a solid line with a square. TFOR is traffic flow-oriented routing protocol and is represented by a dashed line having rhombus. E-GyTAR is Enhanced Greedy Traffic Aware Routing protocol and is represented by a solid line with circle. GSR is geographic source routing protocol and is presented by a dotted line having a cross. GPSR is Greedy Perimeter Stateless Routing Protocol and is presented by a dashed line having a triangle

Fig. 10. An increase in traffic density increases routing overhead for all the protocols. This is because the number of control messages produced during simulation is directly proportional to the number of vehicles. The routing overhead incurred by RPSPF is least as compared to the other routing protocol. For acquiring neighbor locations, GSR produces more beacon messages which increase its routing overhead. The number of beacon messages generated by GSR is three times more as compared to E-GyTAR [1]. GPSR recovery strategy incurs more packet transmission which induces higher

routing overhead. While both E-GyTAR and TFOR also incur more routing overhead due to inappropriate intersection selection mechanism and unreliable forwarding strategies. Packets are lost due to sudden link ruptures while forwarding which results in retransmission of the packets resulting in an increase in routing overhead. Also, maintenance of two-hop neighbor information in TFOR causes more routing overhead in high-traffic density scenarios like traffic jams. As shown in Figs. 5 and 6, the packet delivery ratio for the proposed protocol is better as compared to all the other considered protocols; therefore, it means that lesser packets need to be transmitted in case of the proposed protocol resulting in decreased routing overhead. Forwarding mechanism without link reliability causes packet loss and results in increased routing overhead for all the routing protocols.

Figure 11 shows the routing overhead with respect to the different packet sending rates. The increase in packet sending rate increases the routing overhead for all the considered protocols. The proposed protocol is least affected with respect to an increase in packet sending rate. The major reason behind this is that due to the enhancement that the proposed protocol brings, the packet delivery ratio is higher and the number of packets lost is lower. As a result, increase in packet sending rate does not substantially degrade the performance of the proposed routing protocol as compared to the other protocols.

5.2.4 Influence on performance by increasing the number of considered intersections dynamically

One essential query that needs to be explored is that what is the optimal number of intersections that should

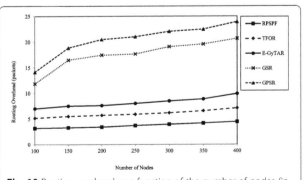

Fig. 10 Routing overhead as a function of the number of nodes (in 5 packet/s). RPSPF represents Reliable Path Selection and Packet Forwarding Routing Protocol which is proposed approach and is represented by a solid line with a square. TFOR is traffic flow-oriented routing protocol and is represented by a dashed line having rhombus. E-GyTAR is Enhanced Greedy Traffic Aware Routing protocol and is represented by a solid line with circle. GSR is geographic source routing protocol and is presented by a dotted line having a cross. GPSR is Greedy Perimeter Stateless Routing Protocol and is presented by a dashed line having a triangle

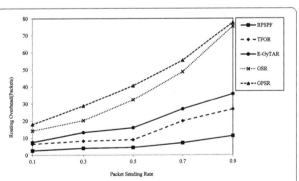

Fig. 11 Routing overhead as function of packet sending rate (400 nodes). RPSPF represents Reliable Path Selection and Packet Forwarding Routing Protocol which is proposed approach and is represented by a solid line with a square. TFOR is traffic flow-oriented routing protocol and is represented by a dashed line having rhombus. E-GyTAR is Enhanced Greedy Traffic Aware Routing protocol and is represented by a solid line with circle. GSR is geographic source routing protocol and is presented by a dotted line having a cross. GPSR is Greedy Perimeter Stateless Routing Protocol and is presented by a dashed line having a triangle

be considered for accomplishing the best performance? For responding such query, we have compared the performance in terms of packet delivery ratio with respect to increasing the number of considered intersections. The outcome of simulation depicts that consideration of two intersections gives us the better performance as compared to increasing intersection beyond two.

According to Fig. 12, increasing the number of considered intersections from 1 to 2 gives us an increase in packet delivery ratio. On the other hand, as we begin to increase the number of intersection beyond 2, RPSPF performance begins to decline in terms of packet delivery ratio. The primary reason behind this study lies in one of the very fundamental characteristics of VANETs, i.e., the VANETs have a very dynamic network in nature and its topology alters very quickly. For this reason, considering more than two intersections while deciding for packet forwarding degrades the performance instead of accomplishing an enhanced performance. Furthermore, maintaining all the information that lies between multiple intersections, such as traffic density and vehicle direction, in a highly dynamic topology also direct to an increased overhead in terms of processing and storage.

The simulation outcomes show that our protocol RPSPF that contains dynamic multiple intersections selection mechanism with reliable forwarding brings the considerable improvement over other routing approaches. It enhances the performance of the network by providing paths with better connectivity and forwarding packets through stable links. The novel mechanisms introduced in RPSPF lead to performance enhancement in terms of end-to-end delay, packet delivery ratio, and routing overhead.

6 Conclusions

In this paper, we have presented a new reliable path selection packet forwarding protocol for VANETs called

as Reliable Path Selection and Packet Forwarding Routing Protocol (RPSPF). At the beginning, we have discussed a detailed technical analysis and comparison of the existing state-of-the-art routing strategies by highlighting the major limitations countenanced by these approaches. After that, we provided details of our proposed routing technique and explained how it overcomes the limitations of existing techniques. RPSPF chooses a couple of intersections at a time on the basis of the shortest curve metric distance to the target node and vehicular traffic density. RPSPF make use of a new multiple intersections selection mechanism, a novel reliable and stable greedy forwarding approach to relay packets in between intersections, and a recovery technique, which is capable of incorporating city environments more efficiently. Simulation results have revealed that RPSPF surpasses TFOR, E-GyTAR, GPSR, and GSR in terms of various metrics like packet delivery ratio, end-to-end delay, and routing overhead. This is due to its ability of incorporating city surroundings' main challenges to routing in a better way as compared to earlier routing techniques.

Abbreviations
ADASE2: Advanced driver-assistance systems; AODV: Ad hoc On-Demand Distance Vector; A-STAR: The Anchor Based Street and Traffic Aware; CAMP: Crash Avoidance Metrics Partnership; DGR: Directional Greedy Routing Protocol; DGSR: Directional Greedy Source Routing; DSR: Dynamic Source Routing; DSRC: Dedicated short-range communication; E-GyTAR: Enhanced Greedy Traffic Aware Routing; E-GyTARD: Enhanced Greedy Traffic Aware Routing Directional; E-GyTAR-PD: Enhanced Greedy Traffic Aware Predictive Directional; GLS: Grid Location Service; GPCR: Greedy Perimeter Coordinator Routing; GPS: Global Positioning System; GPSR: Greedy Perimeter Stateless Routing; GSR: Geographic source routing; GyTAR: Greedy Traffic Aware Routing; IFTIS: Infrastructure Free Traffic Information System; ITS: Intelligent Transportation System; LDT: Link duration time; MANET: Mobile ad hoc network; OLSR: Optimized Link State Routing; PD: Propagation delay; PDGR: Predictive Directional Greedy Routing; PDT: Packet delivery time; PDVR: Predictive Directional Vehicular Routing; RPSPF: Reliable Path Selection Packet Forwarding; SUMO: Simulation of urban mobility; TFOR: Traffic flow-oriented routing protocol; TT: Transmission time; VANET: Vehicular ad hoc networks

Acknowledgements
The first author (the main author) acknowledges the useful feedback provided by the reviewers for improving this article. The main author also acknowledges the financial assistance and support provided during this research work by the Universiti of Malaysia Sarawak.

Funding
The work is fully funded by the Research and Innovation Management Center University of Malaysia Sarawak (RIMC-UNIMAS) under the grant number F08/SpSG/1403/16/4.

Authors' contributions
IAA is the main author of this manuscript. He conceived the novel ideas, designed the algorithms and experiments, and performed the analysis. He wrote the entire manuscript. He accomplished all the revisions provided during entire peer review process until publication. He conducted the final proof reading as well. This manuscript is the outcomes of the reseach activities carried out only by the main author. IAA, ASK and SA checked and reviewed the manuscript. ASK supervised the work. All authors have read and approved the final manuscript.

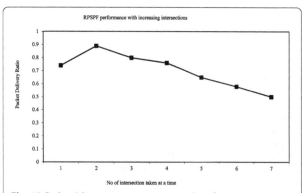

Fig. 12 Packet delivery ratio vs increasing number of intersections dynamically at a time in 400 nodes. RPSPF indicates Reliable Path Selection and Packet Forwarding Routing Protocol which is the proposed approach. It is presented in the figure with a solid line having a square

Authors' information

Irshad Ahmed Abbasi is working as a researcher at the Department of Computer Science and Information Technology, Universiti Malaysia Sarawak, Malaysia. He served as a senior lecturer at King Khalid University, Kingdom of Saudi Arabia. He is also a senior lecturer at the Department of Computer Science, University of Bisha, Kingdom of Saudi Arabia. He has over ten years of research, and teaching experience. Abbasi has received multiple awards, scholarships, and research grants. His research interests include routing in VANETs and MANETs, Mobile Computing, IoT, and Cloud Computing. He is an author of many articles including Traffic Flow Oriented Routing Protocol for VANETs, published in Springer Eurasip Journal on Wireless Communication and Networking. He is also acting as a reviewer of many well reputed peer reviewed international journals and conferences.

Adnan Shahid Khan is currently a senior lecturer at Faculty of Computer Science and Information Technology, Universiti Malaysia Sarawak. He has completed his Postdoctoral, PhD, and Masters in Networks in 2013, 2012, and 2008, respectively, from Universiti Teknologi Malaysia, Johor Bahru, Malaysia, and BSc (Hons) Computer Sciences in 2005 from University of Punjab, Lahore Pakistan. His research interest includes Network and Network security in Wireless communication. He has published more than 60 papers in peer reviewed international conferences and journals.

Shahzad Ali received his M.Sc. in telematics engineering in 2011 and his Ph.D. in telematics engineering from University Carlos III of Madrid in 2014. Currently, he is working as an assistant professor at COMSATS Institute of Information Technology, Abbottabad, Pakistan. His research interests include performance analysis of context-aware applications, wireless sensor networks, vehicular ad hoc networks, and opportunistic networks. He has published many articles in peer reviewed international conferences and journals.

Competing interests

The authors declare that they have no competing interests.

Author details

[1]Department of Computer Science and Information Technology, Universiti Malaysia Sarawak (UNIMAS), 9300 Kota Samarahan, Malaysia. [2]Department of Computer Science, Faculty of Science and Arts at Balgarn, University of Bisha, P.O. Box 60, Sabt Al-Alaya 61985, Kingdom of Saudi Arabia. [3]Department of Computer Science, COMSATS Institute of Information Technology, Abbottabad, Pakistan.

References

1. S.M. Bilal, S.A. Madani, I.A. Khan, Enhanced junction selection mechanism for routing protocol in VANETs. Arab J. Inf. Technol. **8**(4), 422–429 (2011)
2. G.M.T. Abdalla, M.A.A. Rgheff, S.M. Senouci, Current trends in vehicular ad hoc networks. UbiCC J. Special Issue UbiRoads **10**(21), 1–14 (2008)
3. K.C. Lee, U. Lee, M. Gerla, *Advances in Vehicular Ad-Hoc Networks: Developments and Challenges, chapter Survey of Routing Protocols in Vehicular Ad Hoc Networks* (IGI Global, Hershey, 2009)
4. Car-talk 2000. http://cartalk2000.net.outerstats.com/. Accessed 10 May 2013
5. CAR 2 CAR Communication Consortiums. https://www.car-2-car.org/index.php?id=5. Accessed 2 June 2011
6. California Partners for Advanced Transit and Highways. https://path.berkeley.edu. Accessed 15 May 2011
7. FleetNet. https://fleetnetamerica.com. Accessed 17 Oct 2009
8. N. Gupta, A. Prakash, R. Tripathi, Adaptive beaconing in mobility aware clustering based MAC protocol for safety message dissemination in VANET, Hindawi Publishing Corporation. Wireless Commun. Mob. Comput. **5**(4), 1–15 (2017)
9. Chauffeur in EU. https://www.autoeurope.com/chauffeur-services/. Accessed 7 Sept 2015
10. Crash Avoidance Metrics Partnership. https://www.mentor.com. Accessed 5 Feb 2005
11. I.A. Abbasi, B. Nazir, A. Abbasi, S.M. Bilal, S.A. Madani, A traffic flow oriented routing protocol for VANET. Springer EURASIP J Wireless Commun Netw **2014**(121), 1–14 (2014)
12. B.T. Sharef, R.A. Alsaqour, M. Ismail, Review: vehicular communication ad hoc routing protocols: a survey. J. Netw. Comput. Appl. **5**(4), 363–396 (2014)
13. A. Dahiya, R.K. Chauhan, A comparative study of MANET and VANET environment. J. Comput. Secur. **2**(7), 87–91 (2010)
14. M. Jerbi, S.M. Senouci, R. Meraihi, Y.G. Doudane, in *Communications, 2007. ICC '07. IEEE International Conference.* An improved vehicular ad hoc routing protocol for city environments (2007), pp. 3972–3979
15. J. Kumar, V. Mutneja, I.S. Gill, Behavior of position based routing in VANET. Int. J. Comput. Appl. **145**(1), 49–52 (2016)
16. Y.W. Lin, Y.S. Chen, S.L. Lee, Routing protocols in vehicular ad hoc networks: a survey and future perspectives. J. Inf. Sci. Eng. **26**(3), 913–932 (2010)
17. H. Hartenstein, K.P. Laberteaux, *VANET Vehicular Applications and Inter-networking Technologies*, 1st edn. (Wiley Online Library, United Kingdom, UK, 2010)
18. K. Prasanth, K. Duraiswamy, K. Jayasudha, C. Chandrasekar, K. Prasanth, K. Duraiswamy, K. Jayasudha, C. Chandrasekar, Improved packet forwarding approach in vehicular ad hoc networks using RDGR algorithm. Int. J. Next Gene. Netw. (IJNGN) **2**(1), 64–77 (2010)
19. A. Keranen, J. Ott, T. Karkkainen, in *Proceedings of the 2nd International Conference on Simulation Tools and Techniques.* The ONE simulator for DTN protocol evaluation (2009)
20. S Dhankhar, S AgrawalA survey on routing protocols and issues. International Journal of Innovative Research in Science, Eng. Technol. 3(6), 1–14(2014)
21. J. Cheng, J. Cheng, M.C. Zhou, F.Q. Liu, S.C. Gao, C. Liu, Routing in internet of vehicles: a review. IEEE Transac. Intel. Transport. Syst. **16**(5), 1–15 (2015)
22. M. Behrisch, L. Bieker, J. Erdmann, D. Krajzewicz, in *International Conference on Advances in System Simulation.* SUMO–simulation of urban mobility (2011), pp. 63–68
23. D.B. Johnson, D.A. Maltz, in *Mobile Computing.* Dynamic source routing in ad hoc wireless networks (Kluwer Academic Publishers, Alphen aan den Rijn, 1996), pp. 153–181
24. CE Perkins, EM Royer, Ad-hoc on-demand distance vector routing, in Mobile Computing Systems and Applications, 1999. Proceedings. WMCSA '99. Second IEEE Workshop, 1999, pp. 90–100
25. T. Clausen, P. Jacquet, Optimized Link State Routing Protocol (OLSR), RFC Editor, United States, 2003. https://doi.org/10.17487/RFC3626
26. S. Jaap, M. Bechler, L. Wolf, *Evaluation of Routing Protocols for Vehicular Ad Hoc Networks in City Traffic Scenarios, in Proceedings of the 5th International Conference on Intelligent Transportation Systems Telecommunications (ITST)* (Brest, France, 2005)
27. F. Li, Y. Wang, Routing in vehicular ad hoc networks: a survey. Vehic Technol Magaz IEEE **2**(2), 12–22 (2007)
28. B Karp, HT Kung, GPSR: greedy perimeter stateless routing for wireless networks, in Proceedings of the 6th Annual International Conference on Mobile Computing and Networking, MobiCom '00 (ACM, New York, NY, USA, 2000), pp. 243–254
29. D Tian, K Shafiee, VCM Leung Position-based directional vehicular routing, in Proceeding of the Global Telecommunications Conference, GLOBECOM (IEEE, Hoboken, 2009), pp.1–6
30. B.C. Seet, G. Liu, B.S. Lee, C.H. Foh, K.J. Wong, K.K. Lee, in *Lecture Notes in Computer Science: NETWORKING 2004, Networking Technologies, Services, and Protocols; Performance of Computer and Communication Networks; Mobile and Wireless Communications.* A-STAR: a mobile ad hoc routing strategy for metropolis vehicular communications (2004), pp. 989–999
31. R. Sharma, A. Choudhry, An extensive survey on different routing protocols and issue in VANETs. Int. J. Comput. Appl. **106**(10), 1–16 (2014)
32. Shah et al., Unicast routing protocols for urban vehicular networks: review, taxonomy, and research issues. J Zhejiang Univ-SCIENCE (Computers & Electronics). **15**(7) 489–513 (2014)
33. I.A. Abbasi, A.S. Khan, A Review of Vehicle to Vehicle Communication Protocols for VANETs in the Urban Environment. MDPI J. Future Internet **10**(2), 1–14 (2018)
34. A. Srivastava, B.P. Chaurasia, Survey of routing protocol used in vehicular ad hoc networks. Int. J. Curr. Eng. Technol. **7**(3), 1–12 (2017)
35. A. Husain, R. Shringar, B. Kumar, A. Doegar, Performance comparison topology based and position based routing protocols in vehicular network environments. Int. J. Wireless Mob. Netw. (IJWMN) **3**(4), 1–13 (2011)
36. S.M. Bilal, A.R. Khan, S. Ali, Review and performance analysis of position based routing protocols. Springer Wireless Person. Area Commun. **16**(73), 1–13 (2016)
37. C Lochert, H Hartenstein, J Tian, H Fu¨ßler, D Hermann, M Mauve, A routing strategy for vehicular ad hoc networks in city environments, in Proceedings of the Intelligent Vehicles Symposium, 2003. IEEE, 2003, pp. 156–161
38. C Lochert, M Mauve, H Fu¨ßler, H Hartenstein, Geographic routing in city scenarios. SIGMOBILE Mob. Comput. Commun. Rev. 9, 69–72 (2005)
39. W. Kieb, H. Fubler, J. Widmer, M. Mauve, Hierarchical location service for mobile ad-hoc networks: SIGMOBILE Mobile. Comput. Commun. Rev. 1(2), 47–58 (2004)

40. M. Kasemann, H. Fubler, H. Hartenstein, M. Mauve, A reactive location service for mobile ad hoc networks, Technical Report, TR-14-2002. Dept. of Computer Science, University of Mannheim; 2002. http://citeseerx.ist.psu.edu/viewdoc/download;jsessionid=64A84973DF600096DE16D8801C37B86E?doi=10.1.1.12.8778&rep=rep1&type=pdf

41. J Gong, CZ Xu, J Holle, Predictive directional greedy routing in vehicular ad hoc networks, in *Distributed Computing Systems Workshops, 2007. ICDCSW '07.* 27th International Conference on June 2007

42. J. Li, J. Jannotti, D.S.J. Decouto, D.R. Karger, R. Morris, in *Proceedings of the 6th Annual International Conference on Mobile Computing and Networking, MobiCom '00.* A scalable location service for geographic ad hoc routing (ACM, New York, NY, USA, 2000), pp. 120–130

43. M. Jerbi, S.M. Senouci, T. Rasheed, Y. Ghamri, in *Proceedings of the First IEEE International Symposium on Wireless Vehicular Communication WiVec '07.* An infrastructure free intervehicular communication based traffic information (Baltimore, USA, 2007), pp. 1–10

44. W. Su, S.J. Lee, M. Gerla, Mobility prediction and routing in ad hoc wireless networks. Int. J. Netw. Manag. **11**(1), 3–30 (2001)

45. K.R. Kumar, VANETs parameters and applications: a review. Glob. J. Sci. Technol. (GJCST) **10**(7), 1–6 (2010)

46. M.F. Politecnico, C.D.D. Abruzzi, J. Harri, F. Filali, *Vehicular mobility simulation for VANETs, Christian Bonnet Institut Eurecom, Department of Mobile Communications 06904, Sophia, France* (2011), pp. 1–14

47. S. Ali, G. Rizzo, V. Mancuso, M.A. Marsan, in *IEEE Conference on Computer Communications (INFOCOM).* Persistence and availability of floating content in a campus environment (2015), pp. 2326–2334

11

A cross-layer optimization framework for congestion and power control in cognitive radio ad hoc networks under predictable contact

Long Zhang[1]* ⓘ, Fan Zhuo[1] and Haitao Xu[2]

Abstract

In this paper, we investigate the cross-layer optimization problem of congestion and power control in cognitive radio ad hoc networks (CRANETs) under predictable contact constraint. To measure the uncertainty of contact between any pair of secondary users (SUs), we construct the predictable contact model by attaining the probability distribution of contact. In particular, we propose a distributed cross-layer optimization framework achieving the joint design of hop-by-hop congestion control (HHCC) in the transport layer and per-link power control (PLPC) in the physical layer for upstream SUs. The PLPC and the HHCC problems are further formulated as two noncooperative differential game models by taking into account the utility function maximization problem and the linear differential equation constraint with regard to the aggregate power interference to primary users (PUs) and the congestion bid for a bottleneck SU. In addition, we obtain the optimal transmit power and the optimal data rate of upstream SUs by taking advantage of dynamic programming and maximum principle, respectively. The proposed framework can balance transmit power and data rate among upstream SUs while protecting active PUs from excessive interference. Finally, simulation results are presented to demonstrate the effectiveness of the proposed framework for congestion and power control by jointly optimizing the PLPC-HHCC problem simultaneously.

Keywords: Cognitive radio, Cross-layer optimization, Congestion control, Power control, Predictable contact

1 Introduction

1.1 Background and motivation

Cognitive radio (CR) [1] has been widely recognized as a critical technique to mitigate the spectrum scarcity problem and enhance the overall efficiency of spectrum usage, aiming to accommodate for the evolution of wireless systems towards 5G [2]. In a CR network (CRN), unlicensed secondary users (SUs) are allowed to opportunistically access the spectrum allocated to licensed or primary users (PUs) without interfering with the coexisting PUs. That is, the SUs do not violate the quality of service (QoS) requirements of the PUs. Most of the existing research efforts in CRNs mainly focus on the issues of the physical and media access control (MAC) layers for an infrastructure-based single-hop scenario, including spectrum sensing, spectrum access, and sharing techniques [3–5]. In addition, SUs can also form a multi-hop decentralized ad hoc network without the support of infrastructure. In a multi-hop cognitive radio ad hoc network (CRANET) [6], SU can access the licensed spectrum by seeking to overlay, underlay, or interweave its signal with the existing PUs' signals [7]. For the underlay approach, SUs are permitted to concurrently share the licensed spectrum with PUs while guaranteeing the power of interference and noise at the PU not beyond the interference temperature limit. In this context, the interference caused by the SUs should be controlled and mitigated through effective power control strategies. Many studies on power control for CRNs have been reported from different perspectives, such as imperfect channel knowledge [8], arbitrary input distributions [9], and social utility maximization [10].

* Correspondence: zhanglong@hebeu.edu.cn
[1]School of Information and Electrical Engineering, Hebei University of Engineering, Handan 056038, China
Full list of author information is available at the end of the article

In comparison with the lower layer solutions as stated before, recent work indicates that there are many new challenges towards routing problem at the network layer in multi-hop CRANETs, aiming to give more insights into the impact of spectrum uncertainty on routing strategies [11, 12]. However, the constraints and challenges with regard to SUs including random mobility, low deployment density, and limited resource along with discontinuous spectrum availability will give rise to intermittent connectivity of links among SUs in a decentralized CRANET [13]. Clearly, stochastic link outage further has a bearing on the successful transmission of data packets between a pair of SUs. To describe effective continuous transmission of SUs, the paradigm contact has been presented from different types [14], e.g., persistent contact, on-demand contact, and scheduled contact. Conceptually, a contact can be defined as a communication opportunity during which two adjacent SUs can communicate with each other. In a scheduled contact-based CRANET, multiple contacts or the set of communication opportunities can be easily derived from the statistical data of a priori available contact. In this case, the scheduled contact can be predicted and calculated accurately.

Similar to a wireline Internet or most other traditional wireless networks, network congestion in CRNs will also occur when the offered data load exceeds the available capacity of SU due to buffer overflow caused by accumulated data packets injected from upstream SUs. This therefore leads to aggressive retransmission, queuing delay, and blocking of new flows from upstream SUs. Indubitably, congestion control policy in the transport layer is essential to balance resource load and to avoid excessive congestion. However, the conventional Transmission Control Protocol (TCP) as a congestion control mechanism via window-based or acknowledgement-triggered methods is initially designed and optimized to perform in reliable wired links with constrained bit error rates (BERs) and round trip times (RTTs) [15]. Recent study by [16] has reported that the performance of HTTP download deteriorates as much as 40% under the TCP window control in an IEEE P1900.4-based cognitive wireless system using User Datagram Protocol (UDP) and TCP transport protocols. Alternatively, to accommodate for the challenging multi-hop wireless environments, some other research efforts about congestion control techniques have been conducted from the perspective of finding methods to modify TCP protocol [17]. Unfortunately, it has been also shown that these schemes of TCP modifications and extensions cannot be applied into CRANETs because of sudden large-scale bandwidth fluctuation and periodic interruption caused by spectrum sensing and channel switching [18, 19].

It is also noted that the TCP congestion control is targeted to regulate the data rate of upstream SUs so that the total accumulated data load does not exceed the available capacity of SU. In principle, the link capacity between a pair of SUs depend strongly on transmit power of SU coupled with wireless channel conditions [20]. By leveraging the congestion control technique, on the one hand, the attainable data rate on a wireless link between a pair of SUs depends on the interference level, which in turn rests on power control policy. On the other hand, each SU is expected to increase its transmit power in order to obtain as much link capacity that each flow requires [21]. However, increasing the link capacity on one link may reduce the link capacities on other links owing to mutual interference of SUs. From the above discussions, we can see that jointly optimizing transmit power in the physical layer and data rate in the transport layer for attaining the optimal link capacity becomes highly valuable. With a joint cross-layer design, the physical layer is able to share its information and configuration about optimal transmit power with the transport layer without breaking the hierarchical structure of the traditional layered architecture [22]. This motivates us to reinvestigate the cross-layer coupling between capacity supply by power control and rate demand by rate control.

1.2 Related works

Congestion control in wireless multi-hop networks has been widely discussed via the NUM optimization problem maximizing the total utility, subject to some different constraints including the efficiency and fairness of resource allocation [22], heterogeneous traffic [22], lossy link [23], and multipath transmission [24]. Under the condition of outage probability caused by lossy links, another work [25] investigates the rate-effective network utility maximization problem to meet with delay-constrained data traffic requirement. However, although all of the aforementioned studies consider some realistic constraints, they apply to the traditional wireless multi-hop networks only and do not consider the spectrum uncertainty in CRNs. To the best of our knowledge, some studies on congestion control for CRNs have been reported recently, although the mainstream research effort is aimed at the problems of the physical and MAC layers. In [26], Xiao et al. developed a robust active queue management scheme to stabilize the TCP queue length at base station in an infrastructure-based CRN. By using the multiple model predictive control, the proposed scheme absorbed the disturbances caused by busty background traffic and capacity variation. It is found that [26] is not suitable for decentralized CRANET scenario due to a lack of centralized control and global information. Unlike the condition of infrastructure-based CRN, other studies in [18, 27, 28] have been undertaken in multi-hop CRN scenarios. In [27], Song et al. proposed an end-to-end congestion control framework without the aid of common control channel by taking into account the non-uniform channel availability. The explicit feedback mechanism without timeouts and the timeout mechanism were also investigated. In [28], Zhong et al. presented a TCP network coding dynamic generation size adjust scheme by jointly considering network coding gain and delay. The proposed scheme can significantly reduce the retransmissions and guarantee the QoS and enhance the TCP performance. In [18], Al-Ali et al. proposed an end-to-end equation-based TCP friendly rate control mechanism, which achieves rate adjustment by identifying network congestion. However, the end-to-end control policy in [18, 27] is ill suited for operation over wireless transmission links characterized by higher RTTs, particularly if the links present the feature of intermittent connectivity in CRANET under predictable contact. On the contrary, the hop-by-hop control reacts to congestion faster where the rates are adjusted at upstream nodes by feedback information about the congestion state of intermediate nodes.

Other recent schemes that exploit the cross-layer interaction information try to deal with congestion control problem in decentralized CRNs from a cross-layer design perspective. The objective of these schemes is to improve the overall network utility while protecting active PUs' communications from excessive interference introduced by SUs. In [29], Cammarano et al. presented a distributed cross-layer framework for joint optimization of MAC, scheduling, routing, and congestion control in CRAHNs, by maximizing the throughput of a set of multi-hop end-to-end packet flows. However, similar to [18, 27], it is not clear how good the performance of the end-to-end rate control is compared under a wireless transmission environment with higher RTTs. In [30], Nguyen et al. proposed a cross-layer framework to jointly attain both congestion and power control in OFDM-based CRNs through nonconvex optimization method. By means of the adaptation of dual decomposition technique also used by [20], the distributed algorithm was developed to obtain the global optimization. In [31], Nguyen et al. further devised an optimization framework achieving trade-off between energy efficiency and network utility maximization for CRAHNs. By adjusting transmit power, persistence probability, and data rate simultaneously via the interaction between MAC and other layers, the proposed framework can

jointly balance interference, collision, and congestion among SUs However, both of the frameworks in [30, 31] fail to take into account the impact of predictable contact or priori available contact between any pair of SUs on overall cross-layer performance.

1.3 Our approach and contributions

Our work in this paper mainly focuses on a decentralized CRANET under predictable contact in that it is easy to obtain the set of communication opportunities derived from the statistical available contacts among SUs. Owing to a lack of global information to achieve the centralized schedule, the cross-layer coupling between capacity supply by power control and rate demand by rate control needs to be carried out distributively by each SU via local information. Distributed implementation for power control and rate control depends on interactive processes among competitive SUs to figure out the cross-layer coupling relationships. Moreover, the objectives of SUs to maximize their utility functions are conflicting and their decisions are interactive. Apparently, it will be far more realistic to dynamically adjust transmit power and data rate according to the current instant time in the practical dynamic environment. The reason for adopting a differential game model rather than other decentralized optimization approaches is that the differential game is a continuous time dynamic game to investigate interactive decision making over time. In a differential game, the interactions among individual players are characterized by time dependency. This is in line with the nature of dynamic spectrum environment in practical CRANET scenario. Therefore, motivated by cross-layer coupling between capacity supply by power control and rate demand by rate control, we present a cross-layer optimization framework for CRANET under predictable contact by achieving the joint congestion and power control using a differential game theoretic approach. The main contributions of this paper are summarized as follows:

- To measure the uncertainty of contact between a pair of SUs, a predictable contact model is presented by deriving the probability distribution of contact via a mathematical statistics theory. By using Shannon entropy theory, we further devise an entropy paradigm to characterize quantitatively the probability distribution of contact.
- We propose a distributed cross-layer optimization framework for hop-by-hop congestion control (HHCC) and per-link power control (PLPC) for upstream SUs. The HHCC and the PLPC problems are formulated as two

noncooperative differential game models, by taking into account the utility function maximization and linear differential equation constraint with regard to the aggregate power interference to PUs and congestion bid for bottleneck SU.

- We convert the noncooperative differential game models for the PLPC and the HHCC problems into two dynamic optimization problems. By adopting dynamic programming and maximum principle, we obtain the optimal transmit power and the optimal data rate of upstream SUs, respectively. The cross-layer optimization framework is implemented in a distributed manner through the cross-layer coordination mechanism between capacity supply by power controller and rate demand by rate controller.

1.4 Organization and notation

The rest of this paper is organized as follows. We firstly describe the system model in Section 2. Then, the problem formulation is presented in Section 3. In Section 4, we derive the optimal solutions to the proposed noncooperative differential game models and propose the distributed implementation approach to construct the cross-layer optimization framework. Simulation results are provided in Section 5, followed by the conclusions in Section 6.

Notation: $\mathcal{A}$ denotes a set, and $|\mathcal{A}|$ denotes the cardinality for any set $\mathcal{A}$. We use a boldface capital to denote vector $\mathbf{A}$ to discriminate vectors from scalar quantities. $|\cdot|$ and $\|\cdot\|$ represent the absolute value of a polynomial function and the Euclidean distance between the pair of variables, respectively. $\mathbb{E}[\cdot]$ stands for the statistical expectation operator.

2 System model

2.1 Network model

Consider an underlay multi-hop CRANET coexisting with a cellular primary network as depicted in Fig. 1, wherein PUs can send their data traffic to a primary base station (PBS) via the licensed uplink channels. We denote the set of uplink channels by $\mathcal{H} = \{ch_1, ch_2, \cdots, ch_\phi\}$ where ϕ is the number of uplink channels. The uplink channel is either occupied by PUs or unoccupied. We employ the independent and identically distributed alternating ON-OFF process to model the occupation time length of PUs in uplink channels. Specifically, the OFF state indicates the idle state where the uplink channels can be freely occupied by SUs. By performing spectrum sensing on all the uplink channels periodically, S SUs leverage the OFF state to access the unoccupied uplink channels by PUs. Let $\mathcal{V} = \{v_1, v_2, \cdots, v_S\}$ refer to the set of S SUs. Each SU is equipped with two radio transceivers. One with a cognitive radio is used to opportunistically access the uplink channels for transmissions of data packets. The other is used for exchange of control signaling. Due to the randomness of data traffic and the dynamic behavior of PUs, we assume that the licensed uplink channels are available for usage by SU v_i with a probability of δ_i, for $v_i \in \mathcal{V}$. Based on the aforementioned ON-OFF process, the occupancy probability of uplink channel ch_ξ by PUs is defined as $\alpha_\xi/(\alpha_\xi + \beta_\xi)$ [32], where α_ξ is a probability that uplink channel ξ transits from OFF to ON state, and β_ξ is probability that uplink channel ξ transits from ON to OFF state, for $\xi = 1, 2, \cdots, \phi$. It is assumed that SUs can determine the occupancy probability of uplink channels by PUs through a priori knowledge of PUs' activities and local spectrum sensing. Owing to the mutually independent occupancy probability of uplink channel ch_ξ, the probability δ_i of uplink channels used by SU v_i can be expressed as:

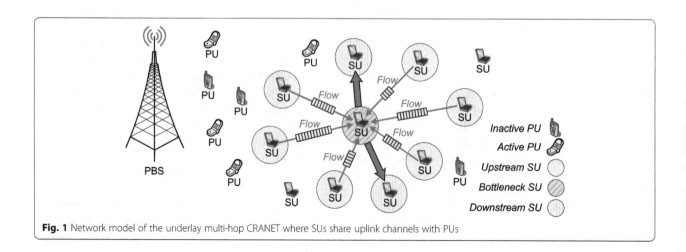

Fig. 1 Network model of the underlay multi-hop CRANET where SUs share uplink channels with PUs

$$\delta_i = \prod_{\xi=1}^{\phi} \left(1 - \frac{\alpha_\xi}{\alpha_\xi + \beta_\xi} \right). \tag{1}$$

Different from the assumption that time is divided into fixed time slots in a discrete way, we exploit the continuous time model to represent the operation duration of the CRANET. The continuous-time operation is confined to a predefined time interval $[t_0, T]$. Use $\vartheta_i(t)$ and $z(t)$ to denote the positions of SU v_i, PU z at time $t \in [t_0, T]$, respectively. Let R_T and R_I stand for the maximum transmission range of SU and the interference range of PU, respectively. Without interference with PUs, a pair of SU v_i and SU v_j can successfully communicate with each other on channel ch_ξ at time t only if the Euclidean distance between SU v_i and SU v_j satisfies $\|\vartheta_i(t) - \vartheta_j(t)\| \le R_T$ and when there is no any PU z on channel ch_ξ, i.e., $\|\vartheta_i(t) - z(t)\| > R_I$ and $\|\vartheta_j(t) - z(t)\| > R_I$, for $v_i, v_j \in \mathcal{V}$ and $\xi = 1, 2, \cdots, \phi$. In this context, there exists a successful transmission link denoted by $l_{(i,j)}$ from SU v_i to SU v_j on channel ch_ξ at time t. For the sake of conciseness, instant time t will be restricted to the time interval $[t_0, T]$ henceforth.

Under the constraint of successful transmission links, we assume that there are multiple different sessions from source SUs to destination SUs. Each session is associated with a route from a source SU to a destination SU. Figure 2 illustrates an example of logical topology of the underlay multi-hop CRANET, where a series of red solid line denote a session along a route from source SU1 to destination SU5, which is one of the different routes. It is assumed that a session consists of several per-link flows with elastic traffic. We use the term *per-link flow* to describe a sequence of data packets with elastic traffic transmitted along a successful transmission link. With

regard to the route from source SU1 to destination SU5, data packets of a flow enter upstream SU2, travel via single hop, then converge at bottleneck SU3 and, finally, move to downstream SU4. We focus on a scenario that multiple different sessions converge at bottleneck SU, aiming to reinvestigate the cross-layer coupling between capacity supply by power control and rate demand by rate control at upstream SUs. From Fig. 2, the convergence of multiple flows from upstream SU2, SU6, and SU8 via single hop may result in a possible congestion at bottleneck SU3 when the offered data load exceeds the available capacity of SU3 due to a buffer overflow, although the amount of data packets has been delivered to downstream SU4, SU7, and SU9. We assume that there are N flows of elastic traffic along the successful transmission links from N upstream SUs to bottleneck SU v_b via single hop, for $v_b \in \mathcal{V}$ and $N < S$. Let $\mathcal{V}_{UP}$ and $\mathcal{N} = \{1, 2, \cdots, N\}$ represent the set of N upstream SUs and the set of flows of elastic traffic from N upstream SUs to bottleneck SU v_b, for $\mathcal{V}_{UP} \subset \mathcal{V}$. For notational simplicity, the flow of elastic traffic along link $l_{(i,b)}$ from upstream SU v_i to bottleneck SU v_b is described by flow i, for $i \in \mathcal{N}$ and $v_i \in \mathcal{V}_{UP}$. Assuming that flow i of elastic traffic along link $l_{(i,b)}$ arrives as a Poisson process of flow arrival intensity λ_i with a size drawn independently from a common distribution of mean $\mathbb{E}[\lambda_i]$ [33]. When $\Psi_{(i,b)} < 1$, the transmission link load, denoted by $\Psi_{(i,b)}$, induced by elastic traffic along link $l_{(i,b)}$ is equal to [33]:

$$\Psi_{(i,b)} = \frac{\lambda_i \times \mathbb{E}[\lambda_i]}{C_{(i,b)}(\mathbf{P})}, \tag{2}$$

where $C_{(i,b)}(\mathbf{P})$ denotes the capacity of link $l_{(i,b)}$, and $\mathbf{P} = \{p_1(t), p_2(t), \cdots, p_N(t)\}$ corresponds to the transmit power vector of N upstream SUs at time t. Here, we use

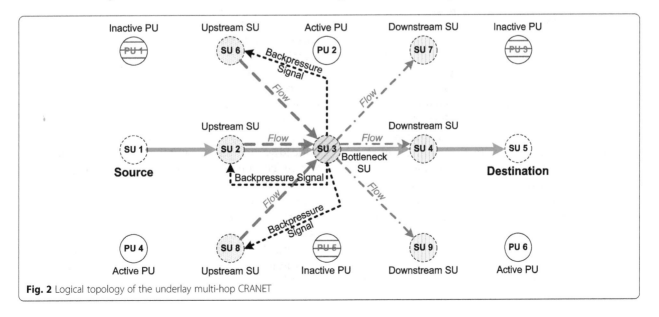

Fig. 2 Logical topology of the underlay multi-hop CRANET

$p_i(t)$ to represent the instant transmit power of upstream SU v_i at time t. Noticing that the transmit power $p_i(t)$ of upstream SU v_i can be adjusted in a continuous way but is also limited by a maximum transmit power threshold denoted by $\bar{p}_i$, i.e., $p_i(t) \in [0, \bar{p}_i]$. Based on elastic traffic model for each flow, the expected duration D_i of flow i with size $\mathbb{E}[\lambda_i]$ is given by [33]:

$$D_i = \frac{\mathbb{E}[\lambda_i]}{C_{(i,b)}(\mathbf{P}) \times \left(1 - \Psi_{(i,b)}\right)}. \qquad (3)$$

Under spectrum underlay scenario, SUs can simultaneously transmit with PUs but have to strictly control their transmit power to avoid interfering with coexisting PUs. Note that the simultaneous transmissions among SUs along a successful transmission link must be undertaken on the same channel, which will further incur the co-channel multiple access interference. We assume that the simultaneous transmissions among N SUs along a successful transmission link on channel ch_ξ can be undertaken under the CDMA-based medium access in the physical layer [34]. The reason for adopting the CDMA-based medium access model is that transmit power of upstream SU can be controlled to induce a different signal-to-interference-plus-noise ratio (SINR) of successful transmission link due to a co-channel multiple-access interference [20, 35, 36]. In principle, link capacity under this scenario cannot remain fixed but depends on SINR of successful transmission link between a pair of SUs. Let $\text{SINR}_{(i,b)}(\mathbf{P})$ be the received SINR of bottleneck SU v_b along link $l_{(i,b)}$ on channel ch_ξ. Therefore, the capacity of link $l_{(i,b)}$ from upstream SU v_i to bottleneck SU v_b can be characterized by a global and non-linear nonconvex function of the transmit power vector and channel conditions as follows [20]:

$$C_{(i,b)}(\mathbf{P}) = \frac{1}{T_s} \log_2 \left(1 + \chi \cdot \text{SINR}_{(i,b)}(\mathbf{P})\right), \qquad (4)$$

where T_s is a symbol period and $\chi = -\phi_1/\log_2(\phi_2 \cdot \text{BER})$ is a constant processing gain factor with ϕ_1 and ϕ_2 depending upon an acceptable BER along with the specific modulation and coding scheme. We assume that a large-scale slow-fading channel model is adopted to describe the line-of-sight wireless transmission environment. In this case, channel gain is subject to distance-dependent power attenuation or log-normal shadowing. As for the practical non-line-of-sight scenario, we use a Rayleigh fading model, in which the channel gain is assumed to be independent exponentially distributed random variables with unit mean [37]. Let $G_{(i,b)}$ and $F_{(i,b)}$ denote the large-scale slow-fading and the Rayleigh fading channel gain of link $l_{(i,b)}$ from upstream SU v_i to bottleneck SU v_b, respectively. Thus, we have

the normalized Rayleigh fading channel gain $\mathbb{E}[F_{(i,b)}] = 1$. By using the certainty-equivalent transmit power and interference power [34, 37], the received SINR of link $l_{(i,b)}$ at bottleneck SU v_b can be expressed as:

$$\text{SINR}_{(i,b)}(\mathbf{P}) = \frac{p_i(t)G_{(i,b)}}{I_i + I_p + n_0}, \qquad (5)$$

where n_0 is the thermal noise power at bottleneck SU v_b, I_p is the interference caused by the PBS, and I_i is the aggregate power interference introduced by other upstream SUs except upstream SU v_i. The aggregate power interference is given by $I_i = \sum_{j \in \mathcal{N} \setminus i} p_j(t)G_{(j,b)}$. In what follows, we are targeted at the line-of-sight wireless transmission environment with the large-scale slow-fading channel gain. How to apply the dynamic fast-fading or Rayleigh fading model under the non-line-of-sight scenario into the cross-layer optimization framework for CRANET will be our further work in the future. Under spectrum underlay scenario, the interference power constraint shall be imposed to protect active PUs' communications from harmful interference caused by all the upstream SUs. We assume that the interference measurement point is located at bottleneck SU v_b for convenience. Hence, the total interference caused by all the upstream SUs should be kept below the interference temperature limit ϖ_{PBS} at the interference measurement point of PBS:

$$\sum_{i \in \mathcal{N}} p_i(s)G_{(i,b)} \le \varpi_{\text{PBS}}. \qquad (6)$$

2.2 Predictable contact model

Considering that a contact is viewed as a communication opportunity during which two adjacent SUs can communicate with each other, we move on to model the predictable contact between a pair of SUs from a priori available contact perspective. Based on the insight into successful transmission link $l_{(i,j)}$ as noticed earlier, an encounter $e^{(i,j)}$ is defined as an effective continuous transmission between SU v_i and SU v_j with a certain duration, for $v_i, v_j \in \mathcal{V}$. It is worth pointing out that an encounter rests on the time of incidence and the duration of an effective continuous transmission between a pair of SUs [38]. Let $t^{0,}_{(i,j)}$ and $\Delta t^{(i,j)}$ represent the time of incidence and the duration of an encounter $e^{(i,j)}$, respectively, for $0 < \Delta t^{(i,j)} < T - t_0$. Therefore, an encounter $e^{(i,j)}$ between SU v_i and SU v_j can be formulated as:

$$e^{(i,j)} = \left\{ v_i, v_j, t^{0,(i,j)}, \Delta t^{(i,j)} \right\}. \tag{7}$$

Suppose that there exist K encounters between SU v_i and SU v_j within a predefined time interval $[t_0, T]$. In particular, the ℓth encounter $e_\ell^{(i,j)}$ between SU v_i and SU v_j with a duration $\Delta t_\ell^{(i,j)}$, for $\ell = 1, 2, \cdots, K$, is given as:

$$e_\ell^{(i,j)} = \left\{ v_i, v_j, t_\ell^{0,(i,j)}, \Delta t_\ell^{(i,j)} \right\}. \tag{8}$$

where $t_\ell^{0,(i,j)}$ refers to the time of incidence of the ℓth encounter $e_\ell^{(i,j)}$. For mathematical tractability, we use the duration $\Delta t_\ell^{(i,j)}$ in (8) to characterize the ℓth encounter, i.e., $e_\ell^{(i,j)} \triangleq \Delta t_\ell^{(i,j)}$. Thus, within time interval $[t_0, T]$, contact $\mathcal{C}^{(i,j)}$ between SU v_i and SU v_j can be rigorously regarded as the set of all encounters, i.e., $\mathcal{C}^{(i,j)} = \{e_1^{(i,j)}, e_2^{(i,j)}, \cdots, e_K^{(i,j)}\}$ and $|\mathcal{C}^{(i,j)}| = K$. Note that the ℓth encounter $e_\ell^{(i,j)}$ in contact $\mathcal{C}^{(i,j)}$ can be referred to a random variable due to the uncertainty of communication opportunity between SU v_i and SU v_j. In this way, we turn to employ a mathematical statistics theory to attain the probability distribution $\Upsilon^{(i,j)} = \{\rho_1^{(i,j)}, \rho_2^{(i,j)}, \cdots, \rho_M^{(i,j)}\}$ of contact $\mathcal{C}^{(i,j)}$, which has been derived from Algorithm 1. In Algorithm 1, we introduce a coefficient M to denote the number of the subintervals, which is obtained by dividing interval $[a, b]$ equally. According to the approximate derivation of the sample distribution in mathematical statistics, coefficient M should be reasonably assigned, depending on the number of encounters of K. That is, when $K \leq 100$, coefficient M can range from 5 to 12. Obviously, it will be possible to measure the uncertainty of contact $\mathcal{C}^{(i,j)}$ between SU v_i and SU v_j by the aid of the probability distribution $\Upsilon^{(i,j)}$. Thus, by analyzing the statistical data of a priori available contact or all encounters between a pair of SUs, it is implicitly understood that the contact can be in a sense predicted very accurately.

Table 1 summarizes the constrained relationship between the number of encounters within subinterval $(d_{l-1}, d_l]$ and contact probability $\rho_l^{(i,j)}$ in Algorithm 1, for $l = 1, 2, \cdots, M$. It is important to emphasize that the probability distribution $\Upsilon^{(i,j)}$ belongs to a complete probability distribution, i.e., $\sum_{l=1}^{M} \rho_l^{(i,j)} = 1$. Technically, the entropy paradigm is widely used for a measure of the uncertainty or randomness associated with a random variable in information theory [39]. In order to characterize quantitatively the probability distribution $\Upsilon^{(i,j)}$, we put forward an entropy paradigm by using Shannon entropy theory to measure the uncertainty of contact $\mathcal{C}^{(i,j)}$. Specifically, the entropy $H(\Upsilon^{(i,j)})$ of the probability distribution $\Upsilon^{(i,j)}$ can be given as:

$$H\left(\Upsilon^{(i,j)}\right) = -\sum_{l=1}^{M} \rho_l^{(i,j)} \log_2 \rho_l^{(i,j)}. \tag{9}$$

Based on Algorithm 1, it is obvious to find that coefficient M impacts the structure of the probability distribution $\Upsilon^{(i,j)}$. Thereby, the entropy $H(\Upsilon^{(i,j)})$ will depend on the selection of coefficient M. Recall that the probability δ_i of uplink channels used by SU v_i determines the stability of successful transmission link $l_{(i,j)}$ or even the contact $\mathcal{C}^{(i,j)}$ between SU v_i and SU v_j due to the impact of PUs' activities on the licensed uplink channels. As such, we formally devise a contact affinity metric to describe the stability of the contact between a pair of SUs. Without losing generality, the contact affinity metric $A_{(i,j)}$ between SU v_i and SU v_j within time interval $[t_0, T]$ is formally expressed as:

Algorithm 1 Probability Distribution Generation Procedure of Contact $\mathcal{C}^{(i,j)}$

1: **Input:** The K encounters $e_1^{(i,j)}, e_2^{(i,j)}, \cdots, e_K^{(i,j)}$.

2: **Output:** The probability distribution $\Upsilon^{(i,j)} = \left\{\rho_1^{(i,j)}, \rho_2^{(i,j)}, \cdots, \rho_M^{(i,j)}\right\}$ of contact $\mathcal{C}^{(i,j)}$.

3: Sort $e_1^{(i,j)}, e_2^{(i,j)}, \cdots, e_K^{(i,j)}$ in ascending order with the minimum value labeled by $e_1^{*(i,j)}$ and maximum value labeled by $e_K^{*(i,j)}$, i.e., $e_1^{*(i,j)}, e_2^{*(i,j)}, \cdots, e_K^{*(i,j)}$.

4: $\forall a, b > 0$, for $a < e_1^{*(i,j)}$ and $b > e_K^{*(i,j)}$.

5: Divide interval $[a, b]$ into M equal subintervals, i.e., $a = d_0 < d_1 < d_2 \cdots < d_{M-1} < d_M = b$.

6: **for** $l = 1 \to M$ **do**

7: $\quad d_l - d_{l-1} = (b-a)/M$.

8: $\quad$ Calculate the number of the K encounters within subinterval $(d_{l-1}, d_l]$, i.e., $q_l^{(i,j)}$.

9: $\quad$ Generate the contact probability $\rho_l^{(i,j)} = q_l^{(i,j)}/K$.

10: **end for**

Table 1 Constrained relationship from algorithm 1

Subinterval	Number of encounters	Contact probability
$(d_0, d_1]$	$q_1^{(i,j)}$	$\rho_1^{(i,j)} = q_1^{(i,j)}/K$
$(d_1, d_2]$	$q_2^{(i,j)}$	$\rho_2^{(i,j)} = q_2^{(i,j)}/K$
$\vdots$	$\vdots$	$\vdots$
$(d_{M-1}, d_M]$	$q_M^{(i,j)}$	$\rho_M^{(i,j)} = q_M^{(i,j)}/K$

$$A_{(i,j)} = H\left(\Upsilon^{(i,j)}\right) \times \delta_i \times \delta_j. \tag{10}$$

3 Problem formulation

In this section, we intend to employ the differential game theoretic approach to formulate the PLPC problem in the physical layer and the HHCC problem in the transport layer. Clearly, the distributed strategy needs to be used to design the cross-layer optimization framework for congestion and power control due to the lack of centralized control and global information under an underlay CRANET scenario. As depicted in Fig. 3, each upstream SU will serve as power and rate controller in charge of joint optimized allocation of transmit power in the physical layer and data rate in the transport layer. Note that the change of power and rate will be continuous in time due to the fact that dynamic congestion and power control will be more realistic in a practical environment.

3.1 Per-link power control in the physical layer

Given the channel conditions, the capacity of successful transmission link between a pair of SUs is a nonconvex function of transmit power vector **P**. In fact, increasing the link capacity on one link may reduce the link capacities on other links because of the mutual interference caused by SUs [20]. Instead, each SU is expected to increase its transmit power to provide as much link capacity that per-link flow requires [21]. However, this adjustment of power will generate extra interference to other SUs. It is necessary to achieve an optimal per-link power allocation in the physical layer for upstream SUs to meet link capacity supply for all the flows. By letting the transmit power of upstream SU v_i equals the maximum transmit power threshold, we can easily obtain the maximum transmit power vector $\overline{\mathbf{P}} = \{\overline{p}_1, \overline{p}_2, \cdots, \overline{p}_N\}$. The transmission loss along link $l_{(i,b)}$ from upstream SU v_i to bottleneck SU v_b on channel ch_ξ is denoted by $\eta_{(i,b)}$. Due to the line-of-sight wireless transmission environment with slow-fading channel model, the transmission loss along link $l_{(i,b)}$ is represented by $\eta_{(i,b)} = [c/(4\pi f_\xi \cdot \|\vartheta_i(t) - \vartheta_b(t)\|)]^2$, where f_ξ is the carrier frequency operating on channel ch_ξ and c is the speed of light. Therefore, the maximum transmit power threshold $\overline{p}_i$ of upstream SU v_i along link $l_{(i,b)}$ from upstream SU v_i to bottleneck SU v_b can be formulated as:

$$\overline{p}_i = \frac{\mathrm{P}_{(i,b)}^{\mathrm{ref}}}{\eta_{(i,b)}}, \tag{11}$$

where $\mathrm{P}_{(i,b)}^{\mathrm{ref}}$ is the received reference power at bottleneck SU v_b along link $l_{(i,b)}$. Given the maximum transmit power threshold $\overline{p}_i$, the capacity of link $l_{(i,b)}$ from upstream SU v_i to bottleneck SU v_b can be denoted by $C_{(i,b)}^*(\overline{\mathbf{P}})$. Owing to maximum transmit power threshold, the value of power reduction for upstream SU v_i is equal to $\overline{p}_i - p_i(t)$. Recall that bits-per-Joule capacity usually serves as a metric to measure the energy

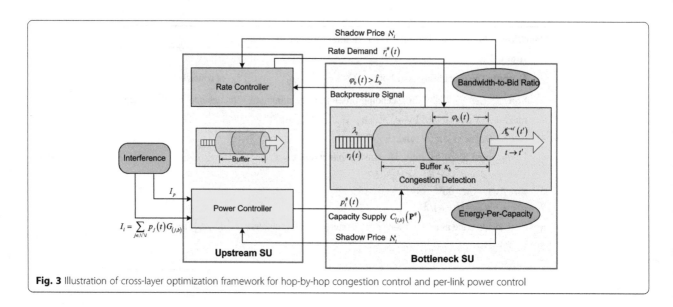

Fig. 3 Illustration of cross-layer optimization framework for hop-by-hop congestion control and per-link power control

efficiency of a communication system [40]. Considering the impact of power reduction on energy efficiency with a link capacity constraint, energy efficiency for power reduction is formally defined as the power reduction achieved per capacity obtained under the maximum transmit power threshold. Thus, we plan to use the energy efficiency for power reduction to characterize a pricing factor of energy-per-capacity, aiming to design the revenue function of power reduction for upstream SU. Given the maximum transmit power vector $\overline{\mathbf{P}}$, the pricing factor $\Phi_i(t)$ of energy-per-capacity of upstream SU v_i at time t can be formally defined as:

$$
\Phi_i(t) = \frac{\overline{p}_i - p_i(t)}{C^*_{(i,b)}(\overline{\mathbf{P}})}
$$

$$
= \frac{\overline{p}_i - p_i(t)}{\dfrac{1}{T_s} \log_2 \left(1 + \chi \cdot \dfrac{\overline{p}_i G_{(i,b)}}{\displaystyle\sum_{j \in \mathcal{N} \setminus i} \overline{p}_j G_{(j,b)} + n_b + n_0}\right)}.
$$

(12)

Revisiting the pricing factor of energy-per-capacity of upstream SU v_i, we define the revenue function of power reduction for upstream SU v_i at time t by attaining the product of the pricing factor together with power reduction value, i.e., $\Phi_i(t)(\overline{p}_i - p_i(t))$. Let ω denote the pricing factor announced by upstream SU v_i to measure the cost of the amount of aggregate power interference to PUs. The amount of aggregate power interference to PUs is denoted by $I(t)$. The cost function of aggregate power interference to PUs for upstream SU v_i at time t is given by $\omega I(t)$. Note that $I(t)$ is a dynamic variable influenced by transmit power $p_i(t)$ of upstream SU v_i and instant level $I(t)$ within time interval $[t_0, T]$. Thus, the aggregate power interference $I(t)$ can be characterized as a linear differential equation given as:

$$
\begin{cases} \dfrac{\mathrm{d}I(t)}{\mathrm{d}t} = \displaystyle\sum_{i \in \mathcal{N}} p_i(t) - \gamma I(t) \\ I(t_0) = I_{t_0} \end{cases},
$$

(13)

where $\gamma > 0$ is a penalty factor of the amount of aggregate power interference and I_{t_0} is an initial aggregate power interference to PUs at time t_0. Therefore, based on both revenue and cost functions as mentioned, the utility function $U_1^i(t)$ of upstream SU v_i at time t is given as:

$$
U_1^i(t) = \frac{\overline{p}_i - p_i(t)}{C^*_{(i,b)}(\overline{\mathbf{P}})}(\overline{p}_i - p_i(t)) - \omega I(t).
$$

(14)

Note that utility function $U_1^i(t)$ is a continuously differentiable function of $p_i(t)$ and $I(t)$. We can find

that utility function in (14) mainly relies on pricing factor of energy-per-capacity in that the marginal effect on utility function stems from aggregate power interference. As shown in Fig. 3, we figuratively define pricing factor $\Phi_i(t)$ as a shadow price $\aleph_i$ which is a function of transmit power of upstream SU v_i. Our optimization objective is to maximize utility function $U_1^i(t)$ by choosing optimal transmit power $p_i^{\#}(t)$ of upstream SU v_i according to $\Phi_i(t)$ and ω:

$$
\underset{p_i(t)}{\text{Maximize}} : \int_{t_0}^{T} \left(\frac{\overline{p}_i - p_i(t)}{C^*_{(i,b)}(\overline{\mathbf{P}})}(\overline{p}_i - p_i(t)) - \omega I(t) \right) e^{-a(t - t_0)} \mathrm{d}t,
$$

(15)

where a is a discount factor, for $0 < a < 1$. Note that discount factor a is an exponential factor between 0 and 1 by which the future utility must be multiplied in order to obtain the present value under the underlying structure of differential game. Hence, utility function $U_1^i(t)$ has to be discounted by the factor $e^{-a(t - t_0)}$. Formally, the PLPC problem in the physical layer is formulated as a differential game model Γ_{PLPC}:

$$
\Gamma_{PLPC} = \left\{ \mathcal{N}, \{p_i(t)\}_{i \in \mathcal{N}}, I(t), \{U_1^i(t)\}_{i \in \mathcal{N}} \right\},
$$

(16)

where

- Player set $\mathcal{N}$: $\mathcal{N} = \{1, 2, \cdots, N\}$ is the set of all the upstream SUs in the PLPC problem as power controllers playing the game. Note that upstream SU v_i stands for the ith player which is a rational policy maker and acts throughout time interval $[t_0, T]$.
- Set of strategies $\{p_i(t)\}_{i \in \mathcal{N}}$: The strategy of the ith player refers to its instant transmit power limited by the maximum transmit power threshold, i.e., $p_i(t) \in [0, \overline{p}_i)$.
- State variable $I(t)$: The state variable of the ith player corresponds to the amount of aggregate power interference to PUs.
- Set of utility functions $\{U_1^i(t)\}_{i \in \mathcal{N}}$: $U_1^i(t)$ is the utility function of the ith player. The objective of the ith player is to maximize its utility function by rationally selecting optimal strategy $p_i^{\#}(t)$ and optimal state $I^{\#}(t)$.

3.2 Hop-by-hop congestion control in the transport layer

Under the scenario that multiple flows from upstream SU2, SU6, and SU8 via single hop converge at bottleneck SU3 in Fig. 2, bottleneck SU3 is a little more inclined to be a congested SU when offered data load exceeds available capacity of SU3 due to buffer overflow. The amount of data packets with elastic traffic in the buffer of bottleneck SU v_b at time t is denoted by $\phi_b(t)$. Given time t, $t' \in [t_0, T]$ and $t < t'$, the amount of data packets $\phi_b(t)$ of bottleneck SU v_b within the time interval $[t, t']$ satisfies the following iterative equation:

$$\phi_b(t') = \max\left\{ \min\left\{\phi_b(t) + \phi_b^{t\to t'}(t'), \kappa_b\right\} - \Lambda_b^{t\to t'}(t'), 0\right\}, \tag{17}$$

where κ_b is the buffer size of bottleneck SU v_b, and $\phi_b^{t\to t'}(t')$ and $\Lambda_b^{t\to t'}(t')$ are the amount of data packets accumulated in the buffer of bottleneck SU v_b and the amount of data packets that could be delivered successfully to downstream SUs, respectively. Considering the constraint of the saturation value $\hat{L}_b$ of the buffer of bottleneck SU v_b, we have the buffer constraint $\phi_b(t) \le \hat{L}_b$ to guarantee that bottleneck SU v_b will not become the real congested SU.

In the end-to-end congestion control, the congestion detection information is piggybacked over data packets to the destination and then sent to the source through the acknowledgement packet from the destination. However, to feed the congestion detection information back to the upstream SU2, SU6, and SU8, bottleneck SU3 will generate a backpressure signal to notify that the congestion occurs as shown in Fig. 2. Compared with the end-to-end mechanism, the backpressure signal is directly sent back to the corresponding upstream SUs from the bottleneck SU3 via single hop. Instead of passing the congestion detection information sent to the source in an end-to-end approach, the idea of hop-by-hop congestion control policy in this paper is that upstream SU v_i directly adjusts the data rate $r_i(t)$ according to the backpressure signal of bottleneck SU v_b when $\phi_b(t) > \hat{L}_b$. Based on the radio transceiver equipped by each SU, the backpressure signal is assumed to be transmitted through a common control channel.

Given the received SINR of link $l_{(i, b)}$, we proceed to derive the required bandwidth of upstream SU v_i for transmissions of data packets with elastic traffic. According to Shannon's capacity formula, the required bandwidth of upstream SU v_i can be expressed by $r_i(t)/\log_2(1 + \chi \cdot \text{SINR}_{(i, b)}(\mathbf{P}))$. We further assume that bottleneck SU v_b acts as a bidder and pays for upstream SU v_i to accommodate the consumption of its network resources while regulating the data rate $r_i(t)$. We use $x(t)$ to represent the congestion bid that bottleneck SU v_b is willing to pay. We then characterize a bandwidth-to-bid ratio

aiming to describe the efficiency of bid with regard to the required bandwidth, i.e., $[r_i(t)/\log_2(1 + \chi \cdot \text{SINR}_{(i, b)}(\mathbf{P}))]/x(t)$. Hence, the cost function $C_i(t)$ of upstream SU v_i at time t can be defined as:

$$C_i(t) = r_i(t) \times \frac{r_i(t)/\log_2(1 + \chi \cdot \text{SINR}_{(i,b)}(\mathbf{P}))}{x(t)} \times D_i. \tag{18}$$

We also use the bandwidth-to-bid ratio to figuratively express the shadow price $\aleph_i$ which is a function of the data rate of upstream SU v_i. Note that the shadow price $\aleph_i$ depends on an optimal per-link power allocation in the PLPC problem due to the constraint of the received SINR of link $l_{(i, b)}$ in bandwidth-to-bid ratio. In this way, we turn our attention to the cross-layer coordination mechanism between capacity supply by power controller and rate demand by rate controller based on shadow prices $\aleph_i$ and $\aleph_i$ as shown in Fig. 3. By taking into account the stability of the contact between a pair of SUs, we conclude that $A_{(i, b)}x(t)$ is the accumulated revenue obtained by upstream SU v_i that bottleneck SU v_b needs to pay. We also remark that the congestion bid $x(t)$ is a dynamic variable influenced by rate $r_i(t)$ as well as by instant level of $x(t)$ within time interval $[t_0, T]$. Accordingly, the congestion bid $x(t)$ can be formulated as a linear differential equation:

$$\begin{cases} \dfrac{\mathrm{d}x(t)}{\mathrm{d}t} = x(t) - \displaystyle\sum_{i \in \mathcal{N}} v r_i(t), \\ x(t_0) = x_{t_0} \end{cases} \tag{19}$$

where x_{t_0} is an initial congestion bid that bottleneck SU v_b needs to pay at time t_0 and v is an average bid per rate, which is assumed to be a unit value for all the upstream SUs, i.e., $v = 1$. Formally, based on both revenue and cost functions as stated before, the utility function $U_2^i(t)$ of upstream SU v_i at time t can be expressed as:

$$U_2^i(t) = A_{(i,b)}x(t) - C_i(t). \tag{20}$$

Noticing that utility function $U_2^i(t)$ is also a continuously differentiable function of $r_i(t)$ and $x(t)$. Our optimization objective is to maximize utility function $U_2^i(t)$ by choosing optimal data rate $r_i^{\#}(t)$ of upstream SU v_i while satisfying the buffer constraint at the same time:

$$\text{Maximize}: \quad \int_{t_0}^{T} \left(A_{(i,b)}x(t) - C_i(t)\right)e^{-\tau(t-t_0)}\mathrm{d}t, \tag{21}$$

where τ is a discount factor, for $0 < \tau < 1$. Similarly, $U_2^i(t)$ will also be discounted by the factor $e^{-\tau(t-t_0)}$. Correspondingly, the HHCC problem in the transport layer can be also defined as a differential game model Γ_{HHCC}:

$$\Gamma_{HHCC} = \left\{ \mathcal{N}, \{r_i(t)\}_{i\in\mathcal{N}}, x(t), \{U_2^i(t)\}_{i\in\mathcal{N}} \right\}, \quad (22)$$

where

- Player set $\mathcal{N}$: $\mathcal{N} = \{1, 2, \cdots, N\}$ is the set of all the upstream SUs in the HHCC problem as rate controllers playing the game. Upstream SU v_i is also known as the ith player which is a rational policy maker and act throughout time interval $[t_0, T]$.
- Set of strategies $\{r_i(t)\}_{i\in\mathcal{N}}$: The strategy of the ith player corresponds to its instant data rate $r_i(t)$.
- State variable $x(t)$: The state variable of the ith player refers to the congestion bid that bottleneck SU v_b is willing to pay.
- Set of utility functions $\{U_2^i(t)\}_{i\in\mathcal{N}}$: $U_2^i(t)$ is the utility function of the ith player. The objective of the ith player is to maximize its utility function by rationally choosing optimal strategy $r_i^\#(t)$ and optimal state $x^\#(t)$.

4 Optimal solution and distributed implementation

Conventionally, upstream SUs as players of the game are expected to act cooperatively and maximize their joint utility functions with fairness for players by constituting the collaborative coalition. As a result, the global optimization of transmit power and data rate will be attained through cooperation among players with group rationality, which has been recently reported in a cooperative bargaining game [41]. However, each upstream SU is unwilling to jointly adjust the power and rate because of the selfish behavior in forwarding data packets. This is a natural idea due to the fact that the transmissions lead to the consumption of network resources of upstream SUs, such as energy and spectrum. Therefore, the cross-layer optimization framework for congestion and power control will be restricted to noncooperation scenario. In the noncooperative differential game models Γ_{PLPC} and Γ_{HHCC}, the ith player competes to maximize the present value of its utility function derived over time interval $[t_0, T]$. For mathematical tractability, we define the starting time of the differential game models Γ_{PLPC} and Γ_{HHCC} as $t_0 = 0$ hereinafter, but the results can be easily extended to more general cases.

4.1 Optimal solution to Γ_{PLPC}

For the noncooperation scenario, we formulate a dynamic optimization problem $\mathbb{P}1$ to derive the optimal solution to the noncooperative differential game model Γ_{PLPC} by taking into account the utility function maximization problem coupled with the linear differential equation constraint in (13):

$$\mathbb{P}1 \quad \underset{p_i(t)}{\text{Maximize}}: \int_0^T \left(\frac{\bar{p}_i - p_i(t)}{C_{(i,b)}^*(\bar{\mathbf{P}})}(\bar{p}_i - p_i(t)) - \omega I(t) \right) e^{-at} \mathrm{d}t$$

$$\text{Subject to}: \frac{\mathrm{d}I(t)}{\mathrm{d}t} = \sum_{i\in\mathcal{N}} p_i(t) - \gamma I(t),$$

$$I(t_0 = 0) = I_0.$$

$$(23)$$

We aim at deriving an optimal solution to $\mathbb{P}1$ by employing the theory of dynamic programming developed by Bellman [42]. Remark that the optimal solution is also viewed as a Nash equilibrium solution to $\mathbb{P}1$ if all the players play noncooperatively. Here, we relax the terminal time of Γ_{PLPC} to explore when T approaches ∞ (i.e., $T \to \infty$) as an infinite time horizon. It is more realistic to obtain the long-term optimal power allocation for upstream SUs due to spectrum underlay strategy with cellular primary network. We use $p_i^\#(t)$ to represent the optimal solution to $\mathbb{P}1$ and assume that there exists a continuously differentiable function $V^i(p_i, I)$ satisfying the following partial differential equation:

$$aV^i(p_i, I) = \underset{p_i(t)}{\text{Maximize}}: \left\{ \frac{(\bar{p}_i - p_i(t))^2}{C_{(i,b)}^*(\bar{\mathbf{P}})} - \omega I(t) + \frac{\partial V^i(p_i, I)}{\partial I}\left(\sum_{j\in\mathcal{N}\backslash i} p_j^\#(t) + p_i(t) - \gamma I(t) \right) \right\}.$$

$$(24)$$

4.1.0.1 Theorem 1 *A vector of optimal transmit power $\mathbf{P}^\# = \{p_1^\#(t), p_2^\#(t), \cdots, p_N^\#(t)\}$ of upstream SUs constitutes a Nash equilibrium solution to $\mathbb{P}1$ if and only if the optimal transmit power $p_i^\#(t)$ of the ith player and the continuously differentiable function $V^i(p_i, I)$ can be formulated as follows:*

$$p_i^\#(t) = \bar{p}_i - \frac{\omega C_{(i,b)}^*(\bar{\mathbf{P}})}{2(a+\gamma)}, \quad (25)$$

$$V^i(p_i^\#, I) = \frac{\omega}{a(a+\gamma)}\left(\frac{\omega C_{(i,b)}^*(\bar{\mathbf{P}})(1+2N)}{4(a+\gamma)} - aI - \sum_{i\in\mathcal{N}} \bar{p}_i \right). \quad (26)$$

Proof: See Appendix 1. ∎

From Theorem 1, we can observe that the existence and uniqueness of the Nash equilibrium point to $\mathbb{P}1$ are guaranteed under the constraint of analytical solution in (25) and (26). It is also revealed that the optimal transmit power $p_i^\#(t)$ has been characterized by a fixed and unique value in (25). Evidently, Theorem 1 mathematically ensures the convergence of $p_i^\#(t)$ to a Nash equilibrium point. The key point to derive the optimal solution to the differential game model Γ_{PLPC} is illustrated with a block diagram shown in Fig. 4a.

4.1.0.2 Proposition 1 *For the given large-scale slow-fading channel model, by letting $G_1 = \varpi_{PBS}/(10^6 g_0)^N$, the*

Algorithm 2 Distributed Optimal Transmit Power Update Algorithm

1: **Input:** $\mathbf{P}$, $\bar{\mathbf{P}}$, $\mathbf{P}^{\#}$.
2: **Output:** Updated instant transmit power vector $\mathbf{P}^{\odot} = \left\{ p_1^{\odot}(t), p_2^{\odot}(t), \cdots, p_N^{\odot}(t) \right\}$.
3: **Initialization:** $C_{(i,b)}^{*}(\bar{\mathbf{P}})$, $\chi = -\varphi_1 / \log_2 (\varphi_2 \cdot \mathrm{BER})$, a, ω, γ.
4: Update shadow price $\aleph_i$ using (28).
5: **for** $i = 1 \to N$ **do**
6:      $p_i^{\odot}(t) \leftarrow p_i^{\#}(t)$ using (25).
7: **end for**
8: **if** $\sum_{i \in \mathcal{N}} p_i(t) G_{(i,b)} > \varpi_{\mathrm{PBS}}$ **then**
9:      **repeat**
10:         **for** $i = 1 \to N$ **do**
11:             **if** $0 < \aleph_i < 1$ **then**
12:                 $p_i^{\odot}(t) \leftarrow \left| p_i^{\odot}(t) - \aleph_i \times \bar{p}_i \right|$.
13:             **else**
14:                 $p_i^{\odot}(t) \leftarrow \left| p_i^{\odot}(t) - \left(\aleph_i \right)^{-1} \times \bar{p}_i \right|$.
15:             **end if**
16:         **end for**
17:     **until** $\sum_{i \in \mathcal{N}} p_i(t) G_{(i,b)} \leq \varpi_{\mathrm{PBS}}$
18: **end if**

---

optimal transmit power $p_i^{\#}(t)$ of the ith player should follow the interference power constraint:

$$\prod_{i \in \mathcal{N}} p_i^{\#}(t) \leq G_1 \prod_{i \in \mathcal{N}} \| \vartheta_i(t) - \vartheta_b(t) \|^4. \tag{27}$$

Proof: See Appendix 2. ∎

Note that the optimal transmit power $p_i^{\#}(t)$ of the *i*th player is fully constrained by the Euclidean distance between upstream SU v_i and bottleneck SU v_b under the given channel model. Substituting for $p_i^{\#}(t)$ with its expression from Theorem 1 and taking into account the previous expression of shadow price $\aleph_b$, we can easily rewrite $\aleph_i$ as:

$$\aleph_i = \frac{\omega}{2(a + \gamma)}. \tag{28}$$

Apparently, shadow price $\aleph_i$ tends to be a constant value for all the upstream SUs. Although (25) and (26) offer an analytical solution to $\mathbb{P}1$, it still remains to design an algorithm to ensure fast convergence of the update of optimal transmit power. Therefore, we devise a distributed optimal transmit power update (OTPU) strategy given in Algorithm 2 to update the optimal transmit power vector $\mathbf{P}^{\#}$ for upstream SUs. Similar to [43], shadow price $\aleph_i$ in Algorithm 2 needs to carefully be chosen to ensure fast convergence of the update of instant transmit power $p_i^{\odot}(t)$. It is

also noted that the update of instant transmit power $p_i^{\odot}(t)$ for upstream SU v_i can be made locally according to its optimal transmit power $p_i^{\#}(t)$ along with interference power constraint.

4.2 Optimal solution to Γ_{HHCC}

For notational simplicity, we begin by defining a notation $B_{(i,b)} \triangleq -D_i / \log_2(1 + \chi \cdot \mathrm{SINR}_{(i,b)}(\mathbf{P}))$. For the noncooperation scenario, we formulate a dynamic optimization problem $\mathbb{P}2$ to derive the optimal solution to the noncooperative differential game model Γ_{HHCC} by taking into account both the utility function maximization problem and the linear differential equation constraint in (19):

$$\mathbb{P}2 \quad \underset{r_i(t)}{\mathrm{Maximize}} : \int_0^T \left(A_{(i,b)} x(t) + B_{(i,b)} \frac{r_i^2(t)}{x(t)} \right) e^{-rt} dt$$
$$\mathrm{Subject\ to} : \frac{dx(t)}{dt} = x(t) - \sum_{i \in \mathcal{N}} r_i(t), \tag{29}$$
$$x(t_0 = 0) = x_0.$$

We turn to take advantage of the theory of maximum principle developed by Pontryagin [42] to derive an optimal solution or a Nash equilibrium solution to $\mathbb{P}2$. We further use $r_i^{\#}(t)$ to represent the optimal solution to $\mathbb{P}2$ and assume that there exists a continuously differentiable function $W^{\ell}(r_b, x)$ satisfying the partial differential equation as follows:

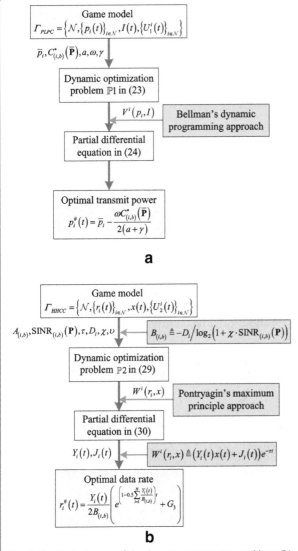

Fig. 4 The block diagram of the dynamic optimization problems $\mathbb{P}1$ and $\mathbb{P}2$ coupled with their optimal solutions: **a** optimal solution to Γ_{PLPC} and **b** optimal solution to Γ_{HHCC}

$$-\frac{\partial W^i(r_i, x)}{\partial s} = \underset{r_i(t)}{\text{Maximize}} : \left\{ \left(A_{(i,b)}x(t) + B_{(i,b)}\frac{r_i^2(t)}{x(t)} \right) e^{-\tau t} \right. \tag{30}$$
$$\left. + \frac{\partial W^i(r_i, x)}{\partial x} \left(x(t) - \sum_{j \in \mathcal{N}\backslash i} r_j^\#(t) - r_i(t) \right) \right\}.$$

For tractability, we introduce two extra introduced auxiliary variables $Y_i(t)$ and $J_i(t)$ to characterize $W^i(r_i, x)$. Specifically, we define $W^i(r_i, x) \triangleq (Y_i(t)x(t) + J_i(t))e^{-\tau t}$.

4.2.0.1 Theorem 2

A vector of optimal data rate $\mathbf{R}^\# = \{ r_1^\#(t), r_2^\#(t), \cdots, r_N^\#(t) \}$ *of upstream SUs constitutes a Nash equilibrium solution to* $\mathbb{P}2$ *if and only if the optimal data rate* $r_i^\#(t)$ *of the ith player can be expressed as:*

$$r_i^\#(t) = \frac{Y_i(t)x(t)}{2B_{(i,b)}}, \tag{31}$$

where $Y_i(t)$ *and* $J_i(t)$ *satisfy the following differential equations:*

$$\frac{dY_i(t)}{dt} = \frac{1}{4B_{(i,b)}} Y_i^2(t) + (\tau-1)Y_i(t) - A_{(i,b)} + \frac{Y_i(t)}{2} \sum_{j \in \mathcal{N}\backslash i} \frac{Y_j(t)}{B_{(j,b)}}, \tag{32}$$

$$\frac{dJ_i(t)}{dt} = \tau \cdot J_i(t). \tag{33}$$

Proof: See Appendix 3. ∎

For notational simplicity, we set $\Omega_{(i,b)} = 1/(4B_{(i,b)}) + 0.5\sum_{j \in \mathcal{N}\backslash i} 1/B_{(j,b)}$ and $\varepsilon = \sqrt{(\tau-1)^2 + 4\Omega_{(i,b)}A_{(i,b)}}$, for $4\Omega_{(i,b)}A_{(i,b)} + (\tau-1)^2 > 0$. We also denote G_2 as a constant number. Substituting $\Omega_{(i,b)}$ into ε, we can rewrite ε as follows:

$$\varepsilon = \sqrt{(\tau-1)^2 + \frac{A_{(i,b)}}{B_{(i,b)}} + 2A_{(i,b)} \sum_{j \in \mathcal{N}\backslash i} \frac{1}{B_{(j,b)}}}. \tag{34}$$

4.2.0.2 Proposition 2

The auxiliary variable $Y_i(t)$ *in the Nash equilibrium solution* $r_i^\#(t)$ *to* $\mathbb{P}2$ *can be further given as:*

$$Y_i(t) = \frac{((\tau-1) + \varepsilon)e^{(t-G_2)\varepsilon} + \varepsilon - \tau + 1}{2\Omega_{(i,b)}(1 - e^{(t-G_2)\varepsilon})}. \tag{35}$$

Proof: See Appendix 4. ∎

Combining $Y_i(t)$ in (35) and ε in (34) yields the expression for $Y_i(t)$ as:

$$Y_i(t) = \frac{\left(\tau-1 + \sqrt{(\tau-1)^2 + \frac{A_{(i,b)}}{B_{(i,b)}} + 2A_{(i,b)} \sum_{j \in \mathcal{N}\backslash i} \frac{1}{B_{(j,b)}}}\right) e^{(t-G_2)\sqrt{(\tau-1)^2 + \frac{A_{(i,b)}}{B_{(i,b)}} + 2A_{(i,b)} \sum_{j \in \mathcal{N}\backslash i} \frac{1}{B_{(j,b)}}}} + \sqrt{(\tau-1)^2 + \frac{A_{(i,b)}}{B_{(i,b)}} + 2A_{(i,b)} \sum_{j \in \mathcal{N}\backslash i} \frac{1}{B_{(j,b)}}} - \tau + 1}{2\left(\frac{1}{4B_{(i,b)}} + 0.5 \sum_{j \in \mathcal{N}\backslash i} \frac{1}{B_{(j,b)}}\right)\left(1 - e^{(t-G_2)\sqrt{(\tau-1)^2 + \frac{A_{(i,b)}}{B_{(i,b)}} + 2A_{(i,b)} \sum_{j \in \mathcal{N}\backslash i} \frac{1}{B_{(j,b)}}}}\right)}. \tag{36}$$

Substituting $B_{(i,b)} \triangleq -D_i/\log_2(1+\chi \cdot \text{SINR}_{(i,b)}(\mathbf{P}))$ into (36), we can rewrite $Y_i(t)$ as follows:

$$Y_i(t) = \frac{\left(\tau-1+\sqrt{(\tau-1)^2-\frac{A_{(i,b)}}{D_i}\log_2(1+\chi \cdot \text{SINR}_{(i,b)}(\mathbf{P}))-2A_{(i,b)}\sum_{j\in\mathcal{N}\backslash i}\frac{\log_2(1+\chi \cdot \text{SINR}_{(j,b)}(\mathbf{P}))}{D_j}}\right)e^{(t-G_2)\sqrt{(\tau-1)^2\frac{A_{(i,b)}\log_2(1+\chi \text{SINR}_{(i,b)}(\mathbf{P}))}{D_i}-2A_{(i,b)}\sum_{j\in\mathcal{N}\backslash i}\frac{\log_2(1+\chi \cdot \text{SINR}_{(j,b)}(\mathbf{P}))}{D_j}}}}{2\left(\frac{\log_2(1+\chi \cdot \text{SINR}_{(i,b)}(\mathbf{P}))}{4D_i}+0.5\sum_{j\in\mathcal{N}\backslash i}\frac{\log_2(1+\chi \cdot \text{SINR}_{(j,b)}(\mathbf{P}))}{D_j}\right)\left(e^{(t-G_2)\sqrt{(\tau-1)^2\frac{A_{(i,b)}\log_2(1+\chi \text{SINR}_{(i,b)}(\mathbf{P}))}{D_i}-2A_{(i,b)}\sum_{j\in\mathcal{N}\backslash i}\frac{\log_2(1+\chi \cdot \text{SINR}_{(j,b)}(\mathbf{P}))}{D_j}}}-1\right)}$$
$$+\frac{\sqrt{(\tau-1)^2-\frac{A_{(i,b)}\log_2(1+\chi \cdot \text{SINR}_{(i,b)}(\mathbf{P}))}{D_i}-2A_{(i,b)}\sum_{j\in\mathcal{N}\backslash i}\frac{\log_2(1+\chi \cdot \text{SINR}_{(j,b)}(\mathbf{P}))}{D_j}}-\tau+1}{2\left(\frac{\log_2(1+\chi \cdot \text{SINR}_{(i,b)}(\mathbf{P}))}{4D_i}+0.5\sum_{j\in\mathcal{N}\backslash i}\frac{\log_2(1+\chi \cdot \text{SINR}_{(j,b)}(\mathbf{P}))}{D_j}\right)\left(e^{(t-G_2)\sqrt{(\tau-1)^2\frac{A_{(i,b)}\log_2(1+\chi \text{SINR}_{(i,b)}(\mathbf{P}))}{D_i}-2A_{(i,b)}\sum_{j\in\mathcal{N}\backslash i}\frac{\log_2(1+\chi \cdot \text{SINR}_{(j,b)}(\mathbf{P}))}{D_j}}}-1\right)}. \tag{37}$$

Let G_3 be a constant number. Based on (31) and (29), the optimal data rate $r_i^{\#}(t)$ and the optimal state variable $x^{\#}(t)$ associated with $r_i^{\#}(t)$ can be described as:

$$r_i^{\#}(t) = \frac{Y_i(t)}{2B_{(i,b)}}\left(e^{\left(1-0.5\sum_{i=1}^{N}\frac{Y_i(t)}{B_{(i,b)}}\right)t}+G_3\right), \tag{38}$$

$$x^{\#}(t) = e^{\left(1-0.5\sum_{i=1}^{N}\frac{Y_i(t)}{B_{(i,b)}}\right)t}+G_3. \tag{39}$$

From (38), we can see that the optimal data rate $r_i^{\#}(t)$ is determined by both $B_{(i,b)}$ and auxiliary variable $Y_i(t)$ under the received SINR $\text{SINR}_{(i,b)}(\mathbf{P})$ of link $l_{(i,b)}$. Unfortunately, it is difficult to directly obtain the relationship between $\text{SINR}_{(i,b)}(\mathbf{P})$ and $r_i^{\#}(t)$ through an analytical derivation because $Y_i(t)$ cannot be further simplified into a concise structure. Thereby, given the channel gain $G_{(i,b)}$ of link $l_{(i,b)}$ under the large-scale slow-fading channel model, we use the numerical simulations to validate the effectiveness of the optimal data rate $r_i^{\#}(t)$ of upstream SU v_i. We further remark that Theorem 2 and Proposition 2 characterize the existence of the Nash equilibrium point to $\mathbb{P}2$. It should be also admitted that the optimal data rate $r_i^{\#}(t)$ has been formulated as a fixed and unique value in (38) by using auxiliary variable $Y_i(t)$ in (37). Correspondingly, Theorem 2 mathematically ensures the convergence of $r_i^{\#}(t)$ to a Nash equilibrium point. The key point to derive the optimal solution to the differential game model Γ_{HHCC} is also illustrated with a block diagram

depicted in Fig. 4b. With the help of vectors $\mathbf{R}^{\#}$ and $\mathbf{P}^{\#}$, shadow price $\aleph_i$ can be calculated as:

$$\aleph_i = \left(\frac{1}{\log_2\left(1+\chi \cdot \text{SINR}_{(i,b)}\left(\mathbf{P}^{\#}\right)\right)}\right)\frac{r_i^{\#}(t)}{x^{\#}(t)}. \tag{40}$$

4.2.0.3 Proposition 3 *For the given vector $\mathbf{R}^{\#}$ and vector $\mathbf{P}^{\#}$, a strict lower bound of shadow price $\aleph_i$ can be approximately calculated as follows:*

$$\aleph_i \geq \left(\log_2\left(\chi \cdot \frac{p_i^{\#}(t)}{\sum_{j\in\mathcal{N}\backslash i}p_j^{\#}(t)}\right)\right)^{-1}\frac{r_i^{\#}(t)}{x^{\#}(t)}. \tag{41}$$

Proof: See Appendix 5. ∎

Let $r_{(i,b)}^{\text{base}}$ represent the baseline rate along link $l_{(i,b)}$ from upstream SU v_i to bottleneck SU v_b. To guarantee the constraint of the saturation value of the buffer of bottleneck SU v_b, we design a distributed algorithm to obtain the optimal data rate update (ODRU) for upstream SUs as summarized in Algorithm 3, where $\mathbf{R} = \{r_1(t), r_2(t), \cdots, r_N(t)\}$ denotes the data rate vector of N upstream SUs at time t. A simple yet effective way to locally adjust the instant data rate $r_i^{\circ}(t)$ of upstream SU v_i is to employ the optimal data rate $r_i^{\#}(t)$ under the condition of buffer constraint $\phi_b(t)\leq \hat{L}_b$. Also, baseline rate $r_{(i,b)}^{\text{base}}$ in Algorithm 3 should be carefully chosen to ensure the effectiveness of instant data rate.

Algorithm 3 Distributed Optimal Data Rate Update Algorithm

1: **Input**: $\mathbf{R}$, $\mathbf{R}^{\#}$, $\mathbf{P}^{\#}$, $x^{\#}(t)$.

2: **Output**: Updated instant data rate vector $\mathbf{R}^{\odot} = \left\{ r_1^{\odot}(t), r_2^{\odot}(t), \cdots, r_N^{\odot}(t) \right\}$.

3: **Initialization**: $\mathrm{SINR}_{(i,b)}(\mathbf{P}^{\#})$, $\chi = -\varphi_1 / \log_2(\varphi_2 \cdot \mathrm{BER})$, D_i, G_2, G_3, τ, ε.

4: Update shadow price $\aleph_i$ using (40) for given $\mathbf{R}^{\#}$, $\mathbf{P}^{\#}$, and $x^{\#}(t)$.

5: **for** $i = 1 \rightarrow N$ **do**

6: $\quad$ $r_i^{\odot}(t) \leftarrow r_i^{\#}(t)$ using (38).

7: **end for**

8: **if** $\varphi_b(t) > \hat{L}_b$ **then**

9: $\quad$ **repeat**

10: $\quad\quad$ **for** $i = 1 \rightarrow N$ **do**

11: $\quad\quad\quad$ $r_i^{\odot}(t) \leftarrow r_i^{\odot}(t) - \aleph_i \times \mathrm{r}_{(i,b)}^{\mathrm{base}}$.

12: $\quad\quad$ **end for**

13: $\quad$ **until** $\varphi_b(t) \leq \hat{L}_b$

14: **end if**

4.3 Distributed implementation

So far, we have devised Algorithms 2 and 3 to satisfy the interference power constraint along with buffer constraint by locally adjusting the optimal transmit power and the optimal data rate, respectively. In what follows, we would like to describe the distributed implementation strategy to realize the cross-layer optimization framework for congestion and power control by jointly optimizing PLPC-HHCC simultaneously. In conclusion, the cross-layer optimization scheme for joint PLPC-HHCC design is implemented in a distributed manner as follows:

4.3.1 Shadow price

Update shadow price $\aleph_i$ using (28) and shadow price $\aleph_i$ using (40), respectively.

4.3.2 Power controller in the physical layer

For each upstream SU, we initially assign optimal transmit power $p_i^{\#}(t)$ using (25) to update instant transmit power $p_i^{\odot}(t)$ at power controller. Due to the interference power constraint in (6) to protect PUs, $p_i^{\odot}(t)$ should satisfy the following distributed power-update function when $\sum_{i \in \mathcal{N}} p_i(t) G_{(i,b)} > \varpi_{\mathrm{PBS}}$ via OTPU algorithm:

$$p_i^{\odot}(t) = \begin{cases} \left| p_i^{\odot}(t) - \aleph_i \times \overline{p}_i \right|, & \text{for } 0 < \aleph_i < 1 \\ \left| p_i^{\odot}(t) - (\aleph_i)^{-1} \times \overline{p}_i \right|, & \text{for } \aleph_i \geq 1 \end{cases} \tag{42}$$

4.3.3 Rate controller in the transport layer

For each upstream SU, we also initially assign optimal data rate $r_i^{\#}(t)$ using (38) to update instant data rate $r_i^{\odot}(t)$ at rate controller. Owing to the buffer constraint $\phi_b(t) \leq \hat{L}_b$ to guarantee that bottleneck SU v_b will not become congested, $r_i^{\odot}(t)$ should be subject to the following distributed rate update function when $\phi_b(t) > \hat{L}_b$ via ODRU algorithm:

$$r_i^{\odot}(t) \leftarrow r_i^{\odot}(t) - \aleph_i \times \mathrm{r}_{(i,b)}^{\mathrm{base}}. \tag{43}$$

4.3.4 Cross-layer coordination mechanism

With the aid of the updated power $p_i^{\odot}(t)$, the link capacity supply $C_{(i,b)}(\mathbf{P}^{\odot})$ with respect to each upstream SU is regulated by power controller by using (4) as shown in Fig. 3. The rate demand depends on instant data rate $r_i^{\odot}(t)$ regulated by rate controller, which is nonlinear function of instant transmit power vector $\mathbf{P}^{\odot}$ according to (37) and (38).

5 Simulation results

5.1 Simulation settings

The simulation scenario is shown in Fig. 5, which consists of one bottleneck SU and $N = 6$ randomly distributed upstream SUs transmitting data packets with elastic traffic towards bottleneck SU within a range of 100 m × 100 m. The scenario is easily extendable to a general case which involves much

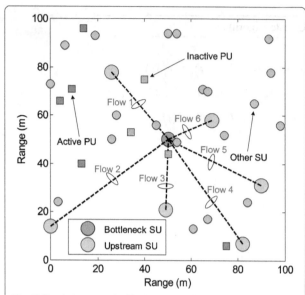

Fig. 5 Simulation scenario: $N = 6$ upstream SUs transmitting data packets with elastic traffic towards bottleneck SU within a range of 100 m × 100 m. The dotted lines correspond to the successful transmission links from upstream SUs to bottleneck SU

more randomly distributed upstream SUs. Our simulations pay more attention to evaluate the effect of the cross-layer optimization framework for congestion and power control on both optimal data rate and optimal transmit power of six different per-link flows. Different from the channel with carrier frequency of 890.4 MHz used by SUs, we assume that active PUs in simulation scenario occupy other uplink channels from set $\mathcal{H}$. This can make possible the successful transmissions of data packets from upstream SUs to bottleneck SU. The probability δ_i of uplink channels used by each SU is assumed to be 0.65. The channel gain of link $l_{(i,b)}$ is defined with large-scale slow-fading model, given by $G_{(i,b)} = 100 g_0 \| \vartheta_i(t) - \vartheta_b(t) \|^{-4}$ [34], where the reference channel gain g_0 is set to 9.7×10^{-4} [44]. We adopt a processing gain factor $\chi = -1.5 / \log_2(5 \mathrm{BER})$ where the target bit error rate is $\mathrm{BER} = 10^{-3}$ for multiple quadrature amplitude modulation with a symbol period of $T_s = 52.5$ μs. The thermal noise power at bottleneck SU and interference caused by PBS are assumed to be $n_0 = -50$ dBm and $I_P = 10$ dBm, respectively. In addition, the receiving reference power at bottleneck SU is chosen as $p_{(i,b)}^{ref} = -37$ dBm for each upstream SU. With regard to the elastic traffic modeled by Poisson process [32], the flow arrival intensity is set to a normalized value $\lambda_i = 125$ bps for each upstream SU, and the mean of flow size is given as $\mathbb{E}[\lambda_i] = 2$ Mbits. Under our differential game models Γ_{PLPC} and Γ_{HHCC}, we choose the

pricing factor $\omega = 22$, the penalty factor $\gamma = 0.7$, two constant numbers $G_2 = 5.5$ and $G_3 = 360$.

Due to the lack of empirical data about available contact or all encounters between a pair of SUs, to evaluate the uncertainty of contact $\mathcal{C}^{(i,b)}$ along link $l_{(i,b)}$, we assume that the minimum and maximum values of encounter duration for all encounters within contact $\mathcal{C}^{(i,b)}$ is offered in Fig. 6. Note that Flow i in Fig. 6 corresponds to contact $\mathcal{C}^{(i,b)}$ along link $l_{(i,b)}$, for $i = 1, 2, \cdots, 6$. We set the number of subintervals in Algorithm 1 to $M = 8$ for all upstream SUs in that the number of encounters K seems to be a lower value because of the short time interval in the simulations. In fact, game time or time interval is just set to $[0, 5]$ s in the following simulations. Under this setting, the probability distribution $\Upsilon^{(i,b)}$ of contact $\mathcal{C}^{(i,b)}$ derived from Algorithm 1 is assumed to comply with the contact distribution as provided by Fig. 7.

The proposed OTPU algorithm for the PLPC problem is compared with the existing classical distributed constrained power control (DCPC) algorithm in [45]. The DCPC algorithm is a SINR-constrained power control algorithm which distributively and iteratively searches for transmit power updated from the ς th iteration to the $(\varsigma + 1)$ th iteration. Let SINR_i^{tar} denote the target SINR for upstream SU v_i to maintain a certain QoS requirement. In the simulations, the target SINR can be set as $\mathrm{SINR}_i^{tar} = 8\ dB$. Therefore, the iterative function of transmit power update in the DCPC algorithm with number of iteration $\varsigma = 0, 1, 2, \cdots$ is specifically given as [45]:

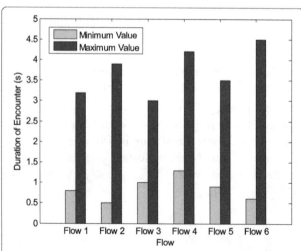

Fig. 6 Comparison between minimum value and maximum value of encounter duration among different per-link flows

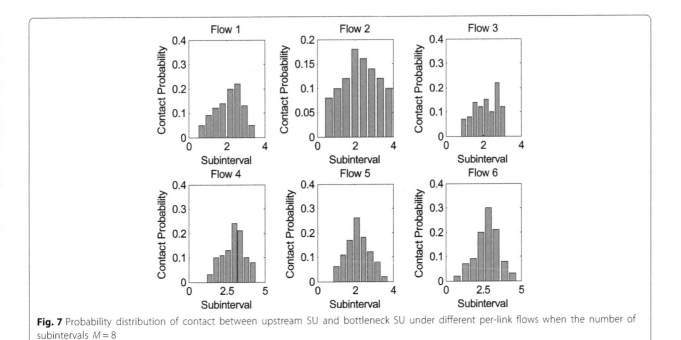

Fig. 7 Probability distribution of contact between upstream SU and bottleneck SU under different per-link flows when the number of subintervals $M = 8$

$$p_i^{(\varsigma+1)} = \min\left\{\bar{p}_i, \frac{\text{SINR}_i^{tar}}{\text{SINR}_{(i,b)}^{(\varsigma)}(\mathbf{P})} \cdot p_i^{(\varsigma)}\right\}. \qquad (44)$$

6 Results

6.1 Performance of OTPU algorithm

Figure 8 shows the optimal transmit power comparison for six per-link flows between the OTPU algorithm with the evolution of discount factor a and the DCPC algorithm with $\varsigma = 300$ iterations. This figure clearly depicts that an increased discount factor from 0.1 to 0.9 will increase the optimal transmit power of each flow under the OTPU algorithm. Apparently, this is a direct consequence of discount factor a on the optimal transmit power according to (25). However, it is observed that the optimal transmit power of each flow via the DCPC algorithm presents a fixed constant value. This is due to the fact that the optimal transmit power of each flow via the DCPC algorithm converges to an expected equilibrium point after 300 iterations. It is worth mentioning that Theorem 1 mathematically makes the optimal transmit power of each upstream SU converge to a Nash equilibrium point distributively. From the results, we can also see that the optimal transmit power of each flow by the OTPU algorithm is obviously lower than that of the DCPC algorithm. This can be explained by the fact that DCPC algorithm gives rise to more power

consumption for maintaining a certain SINR for each upstream SU. However, the optimal transmit power of each flow based on the OTPU algorithm mainly depends upon the maximum transmit power threshold of upstream SU. On the other hand, the instant power level can be further reduced via the change of discount factor a.

6.2 The impact of discount factor on optimal transmit power

Figure 9 illustrates the optimal transmit power comparison for six per-link flows via the OTPU algorithm under different discount factors. It is noted that the total interference caused by six upstream SUs satisfies the interference temperature limit $\varpi_{\text{PBS}} = -10$ dBm according to the constraint of (6). As the discount factor increases, the optimal transmit power of six flows obtained by the OTPU algorithm will raise as well. As expected, the optimal transmit power of flow 6 can achieve the minimum transmit power level with approximately 50 mW, and the optimal transmit power of flow 2 can obtain higher transmit power level with the maximum value nearly 570 mW. However, the increasing rate of the optimal transmit power in regard to flows 6, 3, and 1 flattens out after discount factor $a = 0.6$. The reason is as follows. Firstly, based on (25), the discount factor effect is in direct proportion to the optimal transmit power. One the other hand, with even higher Euclidean distance between upstream

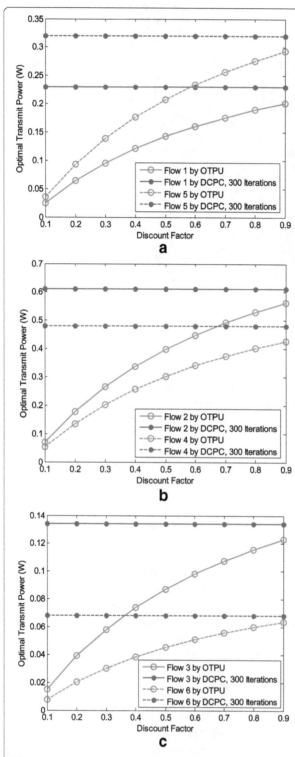

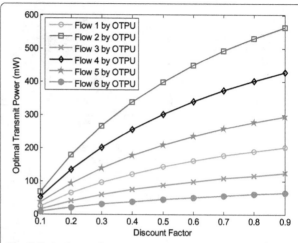

Fig. 9 Optimal transmit power comparison among six per-link flows through our proposed OTPU algorithm under interference temperature limit $\varpi_{PBS} = -10$ dbm

Fig. 8 Optimal transmit power comparison between our proposed OTPU algorithm and DCPC algorithm with 300 iterations under different per-link flows: **a** Flows 1 and 5, **b** flows 2 and 4, and **c** flows 3 and 6

SU and bottleneck SU, the maximum transmit power threshold will increase as well according to (11). Under the simulation scenario, the higher Euclidean distance of the successful transmission link of flow 2 leads to a higher transmit power level accordingly.

6.3 Optimal data rate performance of ODRU algorithm

Figure 10 exhibits the evolution of the optimal data rate for six per-link flows obtained by the ODRU algorithm versus game time $t \in [0, 5]$s under the condition of discount factor $\tau = 0.2$ and saturation value $\hat{L}_b = 1$ *Mbps*. From the results, we can see the optimal data rate for six flows gradually increase with the growth of game time t. Meanwhile, the optimal data rate levels of six flows are very close from 0 to 4 s. When game time t is more than 4 s, the gaps among the optimal data rate levels will be enlarged. This demonstrates that the optimal data rate has large values during the game time of the end interval of the game. Under discount factor $\tau = 0.2$, the optimal data rate value of flow 2 is much larger than those of other flows with maximum value of 300 kbps, and the optimal rate of flow 6 has the lowest level within 50 kbps. It can also be observed that the optimal data rate of flow 2 yields significant performance gains than other flows under the condition of the fixed discount factor. According to saturation value $\hat{L}_b = 1$ *Mbps*, we can observe that the total data rate generated by six upstream SUs is subject to the buffer constraint $\phi_b(t) \leq \hat{L}_b$ such that the instant data rate levels should not be adjusted through the ODRU algorithm.

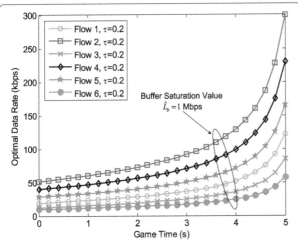

Fig. 10 Optimal data rate comparison among six per-link flows through our proposed ODRU algorithm under saturation value $\hat{L}_b = 1$ Mbps

Figure 11 depicts the optimal data rate update comparison for six per-link flows with the aid of the ODRU algorithm on the condition of discount factor $\tau = 0.2$ and saturation value $\hat{L}_b = 490$ kbps. It is implicitly revealed that the total data rate caused by six upstream SUs fail to guarantee the buffer constraint $\phi_b(t) \leq \hat{L}_b$ such that the instant data rate levels must be updated according to the ODRU algorithm. Hence, the evolution of the optimal data rate levels of six upstream SUs will enter the rate update zone (i.e., shadow area in Fig. 11) when the constraint $\phi_b(t) > \hat{L}_b$. From the results, we can see that the large values of the optimal data rate have been considerably dwindled according to the distributed rate update function in (41) when instant game time $t = 4.3$ s in order to meet the buffer constraint of bottleneck SU.

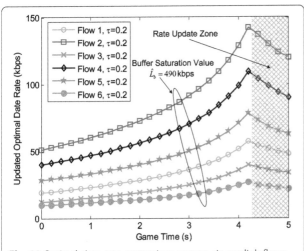

Fig. 11 Optimal data rate comparison among six per-link flows through our proposed ODRU algorithm under saturation value $\hat{L}_b = 490$ kbps

6.4 The impact of discount factor on optimal data rate

Figure 12 displays the evolution of the optimal data rate for six per-link flows via the ODRU algorithm versus discount factor τ under the condition of two fixed instant game time point t (i.e., $t = 3$ and $t = 4$) and saturation value $\hat{L}_b = 950$ kbps. From the results, we can see the total data rate generated by six upstream SUs accommodates for the buffer constraint $\phi_b(t) \leq \hat{L}_b$. It can be also observed that as the discount factor increases from 0.1 to 0.9, the optimal data rate of six flows obtained by the ODRU algorithm will decrease accordingly. The reason for this is that the utility function of each upstream SU must be discounted by the factor $e^{-\tau t}$ at time t under the differential game structure Γ_{HHCC}. As we expected, the optimal data rate of flow 6 can obtain the minimum rate level within approximately interval [6, 26] kbps, and the optimal data rate of flow 2 can gain the maximum value of data rate with nearly interval [30, 140] kbps. This can be explained by the fact that the higher Euclidean distance of the successful transmission link of flow 2 will result in a higher transmit power level accordingly. This result of higher transmit power level of flow 2 will lead to the more link capacity supply. It implies that the upstream SU has the enough link capacity supply to achieve higher data rate in the proposed cross-layer optimization framework. Essentially, this signifies the importance of cross-layer coordination mechanism on the coupling between rate demand regulated by rate controller and capacity supply regulated by power controller.

7 Conclusions

In this paper, a distributed cross-layer optimization framework for congestion and power control for CRANETs under predictable contact has been proposed. Particularly, we introduced a predictable contact model by achieving the probability distribution of contact between any pair of SUs, aiming to measure the uncertainty of contact. Also, an entropy paradigm was presented to characterize quantitatively the probability distribution of contact. We employed a differential game theoretic approach to formulate the PLPC problem and the HHCC problem, and obtained the optimal transmit power and the optimal data rate of upstream SUs via dynamic programming and maximum principle. To guarantee the interference power constraint for active PUs and the buffer constraint of bottleneck SU, we developed two distributed update algorithms to locally adjust optimal transmit power and optimal data rate of upstream SUs. Finally, we presented a distributed implementation strategy to construct the cross-layer optimization framework for

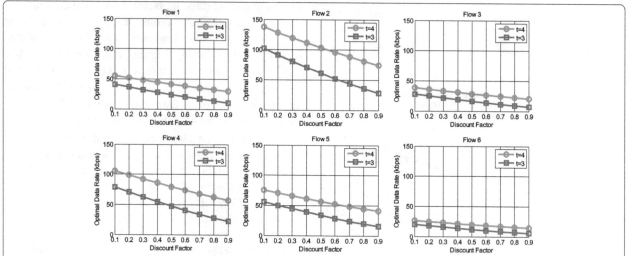

Fig. 12 Optimal data rate comparison among six per-link flows through our proposed ODRU algorithm for different discount factors under saturation value $\hat{L}_b = 950$ kbps

congestion and power control by jointly optimizing PLPC-HHCC simultaneously and validated its performance with simulations. What we have discussed in this paper is the portion of foundation for the cross-layer optimization framework in CRANETs. In the future work, a joint objective function to achieve congestion and power control will be considered. Moreover, it will be interesting and important to investigate a trade-off parameter as a whole to reflect the benefits of the proposed framework.

8 Appendix 1

8.1 Proof of Theorem 1

According to the dynamic optimization problem $\mathbb{P}1$, performing the maximization operation of the right hand side of (24) with respect to $p_i(t)$ yields the following optimal solution:

$$p_i^\#(t) = \bar{p}_i + \frac{C_{(i,b)}^*(\overline{\mathbf{P}})}{2} \frac{\partial V^i(p_i, I)}{\partial I}. \qquad (45)$$

Substituting $p_i^\#(t)$ in (45) into (24), we obtain:

$$aV^i\left(p_i^\#, I\right) = \frac{C_{(i,b)}^*(\overline{\mathbf{P}})}{4}\left(\frac{\partial V^i\left(p_i^\#, I\right)}{\partial I}\right)^2 - \omega I(t)$$
$$+ \frac{\partial V^i\left(p_i^\#, I\right)}{\partial I}\left(\sum_{i \in \mathcal{N}}\left(\bar{p}_i + \frac{C_{(i,b)}^*(\overline{\mathbf{P}})}{2}\frac{\partial V^i\left(p_i^\#, I\right)}{\partial I}\right) - \gamma \cdot I(t)\right). \qquad (46)$$

Upon solving the differential equation in (46), $V^i(p_i^\#, I)$ can be easily shown to be equivalent to the following equation:

$$V^i\left(p_i^\#, I\right) = \frac{\omega}{a(a+\gamma)}\left(\frac{\omega C_{(i,b)}^*(\overline{\mathbf{P}})(1+2N)}{4(a+\gamma)} - aI - \sum_{i \in \mathcal{N}}\bar{p}_i\right). \qquad (47)$$

Thus, an optimal transmit power $p_i^\#(t)$ which constitutes a Nash equilibrium solution to $\mathbb{P}1$ is given by:

$$p_i^\#(t) = \bar{p}_i - \frac{\omega C_{(i,b)}^*(\overline{\mathbf{P}})}{2(a+\gamma)}. \qquad (48)$$

Then, we can obtain the expression of $p_i^\#(t)$ and $V^i(p_i, I)$ as given by Theorem 1.

9 Appendix 2

9.1 Proof of Proposition 1

By substituting the vector of optimal transmit power $\mathbf{P}^\# = \{p_1^\#(t), p_2^\#(t), \cdots, p_N^\#(t)\}$ into the interference power constraint inequality in (6), we can obtain:

$$\sum_{i \in \mathcal{N}} p_i^\#(t)G_{(i,b)} \leq \varpi_{\mathrm{PBS}}. \qquad (49)$$

After taking the logarithm of both sides of (49), we have:

$$\sum_{i \in \mathcal{N}} \log_2\left(p_i^\#(t)G_{(i,b)}\right) \leq \log_2 \varpi_{\mathrm{PBS}}. \qquad (50)$$

Through rearranging terms, we have:

$$\log_2 \prod_{i \in \mathcal{N}} p_i^\#(t) + \log_2 \prod_{i \in \mathcal{N}} G_{(i,b)} \leq \log_2 \varpi_{\mathrm{PBS}}. \qquad (51)$$

By taking into account the large-scale slow-fading channel model to describe the wireless transmission environment, the channel gain of link from upstream SU v_i to bottleneck SU v_b can be formulated as [30]:

$$G_{(i,b)} = g_0 \left(\frac{\|\vartheta_i(t) - \vartheta_b(t)\|}{100} \right)^{-4}, \qquad (52)$$

where g_0 is a reference channel gain at a distance of 100 m [34]. We substitute $G_{(i,b)}$ in (52) into (51) and then we have:

$$\log_2 \prod_{i \in \mathcal{N}} p_i^{\#}(t) \le \log_2 \frac{\varpi_{\mathrm{PBS}}}{\prod_{i \in \mathcal{N}} g_0 \left(\frac{\|\vartheta_i(t) - \vartheta_b(t)\|}{100} \right)^{-4}}. \qquad (53)$$

Thus, (53) can be rewritten as follows:

$$\prod_{i \in \mathcal{N}} p_i^{\#}(t) \le \frac{\varpi_{\mathrm{PBS}}}{(10^6 g_0)^N} \prod_{i \in \mathcal{N}} \|\vartheta_i(t) - \vartheta_b(t)\|^4. \qquad (54)$$

Through defining $G_1 = \varpi_{\mathrm{PBS}}/(10^6 g_0)^N$, we can easily have the solution of (27).

10 Appendix 3
10.1 Proof of Theorem 2
According to the dynamic optimization problem P2, by performing the maximization operation of the right hand side of (30) with respect to $r_i(t)$, we can obtain:

$$r_i^{\#}(t) = \frac{e^{\tau t}}{2B_{(i,b)}} \frac{\partial W^i(r_i, x)}{\partial x}. \qquad (55)$$

Substituting $W^i(r_i, x) \triangleq (Y_i(t)x(t) + J_i(t))e^{-\tau t}$ and $r_i^{\#}(t)$ in (55) into (30), we have:

$$r_i^{\#}(t) = \frac{Y_i(t)x(t)}{2B_{(i,b)}}, \qquad (56)$$

$$\frac{dY_i(t)}{dt} = \frac{1}{4B_{(i,b)}} Y_i^2(t) + (\tau-1)Y_i(t) - A_{(i,b)} + \frac{Y_i(t)}{2} \sum_{j \in \mathcal{N} \backslash i} \frac{Y_j(t)}{B_{(j,b)}}, \qquad (57)$$

$$\frac{dJ_i(t)}{dt} = \tau \cdot J_i(t). \qquad (58)$$

Hence, this completes the proof.

11 Appendix 4
11.1 Proof of Proposition 2
Owing to the symmetric form of $Y_i(t)$ and $Y_j(t)$ in (57), we can immediately denote (57) by Riccati equation:

$$\frac{dY_i(t)}{dt} = \left(\frac{1}{4B_{(i,b)}} + \frac{1}{2} \sum_{j \in \mathcal{N} \backslash i} \frac{1}{B_{(j,b)}} \right) Y_i^2(t) + (\tau-1)Y_i(t) - A_{(i,b)}. \qquad (59)$$

As such, the form of (59) can be rearranged by differential equation as:

$$\wp_1(Y_i(t), t)dt + \wp_2(Y_i(t), t)dY_i(t) = 0, \qquad (60)$$

$$\wp_1(Y_i(t), t) = \left(\frac{1}{4B_{(i,b)}} + \frac{1}{2} \sum_{j \in \mathcal{N} \backslash i} \frac{1}{B_{(j,b)}} \right) Y_i^2(t) + (\tau-1)Y_i(t) - A_{(i,b)}, \qquad (61)$$

$$\wp_2(Y_i(t), t) = -1. \qquad (62)$$

Recall that we define $\Omega_{(i,b)} = 1/(4B_{(i,b)}) + 0.5\sum_{j \in \mathcal{N} \backslash i} 1/B_{(j,b)}$, for $4\Omega_{(i,b)}A_{(i,b)} + (\tau-1)^2 > 0$. Next, we turn to present a non-zero integrating factor $\Im(Y_i(t), t)$ that can make equation in (61) an exact form by multiplying it on both sides of (61). Here, we can easily obtain:

$$\Im(Y_i(t), t) = \frac{1}{\left| \Omega_{(i,b)} Y_i^2(t) + (\tau-1)Y_i(t) - A_{(i,b)} \right|}. \qquad (63)$$

So, (63) multiplied by $\Im(Y_i(t), t)$ is exact, and then we obtain:

$$\frac{\partial \Im(Y_i(t), t)}{\partial Y_i(t)} = -\frac{1}{\left| \Omega_{(i,b)} Y_i^2(t) + (\tau-1)Y_i(t) - A_{(i,b)} \right|}, \qquad (64)$$

$$\frac{\partial \Im(Y_i(t), t)}{\partial t} = 1. \qquad (65)$$

When $\Omega_{(i,b)} Y_i^2(t) + (\tau-1)Y_i(t) > A_{(i,b)}$, we can easily define $\varepsilon = \sqrt{(\tau-1)^2 + 4\Omega_{(i,b)}A_{(i,b)}}$. By integrating (64) and (65) with respect to $Y_i(t)$, we have:

$$\Im(Y_i(t), t) = -\frac{1}{\varepsilon} \ln \left| \frac{2\Omega_{(i,b)} Y_i(t) + (\tau-1) - \varepsilon}{2\Omega_{(i,b)} Y_i(t) + (\tau-1) + \varepsilon} \right| + h(t), \qquad (66)$$

$$\frac{\partial \Im(Y_i(t), t)}{\partial t} = \frac{dh(t)}{dt}. \qquad (67)$$

We can easily obtain $h(t) = t$. Let $\Im(Y_i(t), t) = G_2$, where G_2 is a constant number. Upon solving (66) as follows:

$$t - \frac{1}{\varepsilon} \ln \left| \frac{2\Omega_{(i,b)} Y_i(t) + (\tau-1) - \varepsilon}{2\Omega_{(i,b)} Y_i(t) + (\tau-1) + \varepsilon} \right| = G_2. \qquad (68)$$

Solving the above equation in (68) with respect to $Y_i(t)$, yields the desired result in (35).

12 Appendix 5

12.1 Proof of Proposition 3

For vector $\mathbf{P}^{\#}$, it is clear that the received SINR of link $l_{(i,b)}$ at bottleneck SU v_b satisfies the following inequality when all the upstream SUs with the equal Euclidean distance to bottleneck SU v_b:

$$
\begin{aligned}
\mathrm{SINR}_{(i,b)}\left(\mathbf{P}^{\#}\right) &\leq \frac{p_i^{\#}(t) G_{(i,b)}}{\sum_{j \in \mathcal{N} \setminus i} p_j^{\#}(t) G_{(j,b)}} \\
&= \frac{p_i^{\#}(t)}{\sum_{j \in \mathcal{N} \setminus i} p_j^{\#}(t)}.
\end{aligned} \tag{69}
$$

We can also approximate logarithmic function $\log_2(\cdot)$ by:

$$
\log_2\left(1 + \chi \cdot \mathrm{SINR}_{(i,b)}\left(\mathbf{P}^{\#}\right)\right) \approx \log_2\left(\chi \cdot \mathrm{SINR}_{(i,b)}\left(\mathbf{P}^{\#}\right)\right). \tag{70}
$$

Thus, shadow price $\aleph_i$ should be subject to a strict lower bound:

$$
\begin{aligned}
\aleph_i &\approx \left(\frac{1}{\log_2\left(\chi \cdot \mathrm{SINR}_{(i,b)}\left(\mathbf{P}^{\#}\right)\right)}\right) \frac{r_i^{\#}(t)}{x^{\#}(t)} \\
&\geq \left(\log_2\left(\chi \cdot \frac{p_i^{\#}(t)}{\sum_{j \in \mathcal{N} \setminus i} p_j^{\#}(t)}\right)\right)^{-1} \frac{r_i^{\#}(t)}{x^{\#}(t)}.
\end{aligned} \tag{71}
$$

which coincides with (41).

Abbreviations

BER: Bit error rate; CR: Cognitive radio; CRANET: Cognitive radio ad hoc network; CRN: CR network; DCPC: Distributed constrained power control; HHCC: Hop-by-hop congestion control; ODRU: Optimal data rate update; OTPU: Optimal transmit power update; PBS: Primary base station; PLPC: Per-link power control; PU: Primary user; QoS: Quality of service; RTT: Round trip time; SINR: Signal-to-interference-plus-noise ratio; SU: Secondary user; TCP: Transmission control protocol; UDP: User datagram protocol

Acknowledgements

We are grateful to the anonymous reviewers for their valuable comments and suggestions that have improved the paper.

Funding

This work was supported in part by the National Natural Science Foundation of China under Grants 61402147, 61402001, and 61501406; the Research Program for Top-notch Young Talents in Higher Education Institutions of Hebei Province, China, under Grant BJ2017037; the Research and Development Program for Science and Technology of Handan, China, under Grant 1621203037; and the Natural Science Foundation of Hebei Province of China under Grant F2017402068.

Authors' contributions

LZ conceived the idea of this work and wrote the paper. HX provided valuable insights for the scheme and reviewed the manuscript. LZ and FZ performed the simulations and revised the paper. All authors read and approved the final manuscript.

Competing interests

The authors declare that they have no competing interests.

Author details

[1]School of Information and Electrical Engineering, Hebei University of Engineering, Handan 056038, China. [2]School of Computer and Communication Engineering, University of Science and Technology Beijing, Beijing 100083, China.

References

1. S Haykin, Cognitive radio: brain-empowered wireless communications. IEEE J. Sel. Areas Commun. 23(2), 201–220 (2005)
2. Z Ding, Y Liu, J Choi, Q Sun, M Elkashlan, I Chih-Lin, HV Poor, Application of non-orthogonal multiple access in LTE and 5G networks. IEEE Commun. Mag. 55(2), 185–191 (2017)
3. V Rakovic, D Denkovski, V Atanasovski, P Mahonen, L Gavrilovska, Capacity-aware cooperative spectrum sensing based on noise power estimation. IEEE Trans. Commun. 63(7), 2428–2441 (2015)
4. N Zhang, H Liang, N Cheng, Dynamic spectrum access in multi-channel cognitive radio networks. IEEE J. Sel. Areas Commun. 32(11), 2053–2064 (2014)
5. PK Sharma, PK Upadhyay, Performance analysis of cooperative spectrum sharing with multiuser two-way relaying over fading channels. IEEE Trans. Veh. Technol. 66(2), 1324–1333 (2017)
6. IF Akyildiz, W-Y Lee, KR Chowdhury, CRAHNs: cognitive radio ad hoc networks. Ad Hoc Netw. 7(5), 810–836 (2009)
7. A Goldsmith, SA Jafar, I Maric, S Srinivasa, Breaking spectrum gridlock with cognitive radios: an information theoretic perspective. Proc. IEEE 97(5), 894–914 (2009)
8. S Gong, P Wang, Y Liu, W Zhuang, Robust power control with distribution uncertainty in cognitive radio networks. IEEE J. Sel. Areas Commun. 31(11), 2397–2408 (2013)
9. G Ozcan, MC Gursoy, Optimal power control for underlay cognitive radio systems with arbitrary input distributions. IEEE Trans. Wirel. Commun. 14(8), 4219–4233 (2015)
10. S Parsaeefard, AR Sharafat, Robust distributed power control in cognitive radio networks. IEEE Trans. Mobile Comput. 12(4), 609–620 (2013)
11. S-C Lin, K-C Chen, Spectrum-map-empowered opportunistic routing for cognitive radio ad hoc networks. IEEE Trans. Veh. Technol. 63(6), 2848–2861 (2014)
12. J Wang, H Yue, L Hai, Y Fang, Spectrum-aware anypath routing in multi-hop cognitive radio networks. IEEE Trans. Mobile Comput. 16(4), 1176–1187 (2017)
13. J Zhao, G Cao, Spectrum-aware data replication in intermittently connected cognitive radio networks (Proc. IEEE INFOCOM, Toronto, 2014), pp. 2238–2246
14. V Cerf, S Burleigh, L Torgerson, R Durst, K Scott, K Fall, H Weiss, Delay-tolerant network architecture: The evolving Interplanetary Internet (Internet-Draft, IPN Research Group, 2002). https://tools.ietf.org/html/rfc4838
15. KR Chowdhury, M Di Felice, IF Akyildiz, TCP CRAHN: a transport control protocol for cognitive radio ad hoc networks. IEEE Trans. Mobile Comput. 12(4), 790–803 (2013)
16. D Sarkar, H Narayan, Transport layer protocols for cognitive networks (Proc. IEEE INFOCOM, San Diego, 2010), pp. 1–6
17. B Soelistijanto, MP Howarth, Transfer reliability and congestion control strategies in opportunistic networks: a survey. IEEE Commun. Surveys & Tutorials 16(1), 538–555, First Quarter (2014)
18. AK Al-Ali, K Chowdhury, TFRC-CR: an equation-based transport protocol for cognitive radio networks. Ad Hoc Netw. 11(6), 1836–1847 (2013)
19. K Tsukamoto, S Koba, M Tsuru, Y Oie, Cognitive radio-aware transport protocol for mobile ad hoc networks. IEEE Trans. Mobile Comput. 14(2), 288–301 (2015)
20. M Chiang, Balancing transport and physical layers in wireless multihop networks: jointly optimal congestion control and power control. IEEE J. Sel. Areas Commun. 23(1), 104–116 (2005)
21. HJ Lee, JT Lim, Cross-layer congestion control for power efficiency over wireless multihop networks. IEEE Trans. Veh. Technol. 58(9), 5274–5278 (2009)
22. Q-V Pham, W-J Hwang, Network utility maximization-based congestion control over wireless networks: a survey and potential directives. IEEE Commun. Surveys & Tutorials 19(2), 1173–1200, Second Quarter (2017)
23. Q Gao, J Zhang, SV Hanly, Cross-layer rate control in wireless networks with lossy links: leaky-pipe flow, effective network utility maximization and hop-by-hop algorithms. IEEE Trans. Wirel. Commun. 8(6), 3068–3076 (2009)
24. Q-V Pham, W-J Hwang, Network utility maximization in multipath lossy wireless networks. International J. of Commun. Syst. 30(5), e3094 (2017)
25. S Guo, C Dang, Y Yang, Joint optimal data rate and power allocation in lossy mobile ad hoc networks with delay-constrained traffics. IEEE Trans. Comput. 64(3), 747–762 (2015)

26. K Xiao, S Mao, JK Tugnait, MAQ: a multiple model predictive congestion control scheme for cognitive radio networks. IEEE Trans. Wirel. Commun. **16**(4), 2614–2626 (2017)

27. Y Song, J Xie, *End-to-end congestion control in multi-hop cognitive radio ad hoc networks: to timeout or not to timeout?* (Proc. IEEE Globecom Workshops, Atlanta, 2013), pp. 1–6

28. X Zhong, Y Qin, L Li, TCPNC-DGSA: efficient network coding scheme for TCP in multi-hop cognitive radio networks. Wireless Pers. Commun. **84**(2), 1243–1263 (2015)

29. A Cammarano, FL Presti, G Maselli, L Pescosolido, C Petrioli, Throughput-optimal cross-layer design for cognitive radio ad hoc networks. IEEE Trans. Parallel Distrib. Syst. **26**(9), 2599–2609 (2015)

30. MV Nguyen, CS Hong, S Lee, Cross-layer optimization for congestion and power control in OFDM-based multi-hop cognitive radio networks. IEEE Trans. Commun. **60**(8), 2101–2112 (Aug. 2012)

31. MV Nguyen, S Lee, S You, CS Hong, LB Le, Cross-layer design for congestion, contention, and power control in CRAHNs under packet collision constraints. IEEE Trans. Wirel. Commun. **12**(11), 5557–5571 (2013)

32. A Alshamrani, X Shen, L-L Xie, QoS provisioning for heterogeneous services in cooperative cognitive radio networks. IEEE J. Sel. Areas Commun. **29**(4), 819–830 (2011)

33. SB Fredj, T Bonald, A Proutiere, G Regnie, JW Roberts, *Statistical bandwidth sharing: A study of congestion at flow level* (Proc. ACM SIGCOMM, San Diego, 2001), pp. 111–122

34. NH Tran, CS Hong, S Lee, Cross-layer design of congestion control and power control in fast-fading wireless networks. IEEE Trans. Parallel Distrib. Syst. **24**(2), 260–274 (2013)

35. H Zhang, S Huang, C Jiang, K Long, VCM Leung, HV Poor, Energy efficient user association and power allocation in millimeter-wave-based ultra dense networks with energy harvesting base stations. IEEE J. Sel. Areas Commun. **35**(9), 1936–1947 (2017)

36. H Zhang, Y Nie, J Cheng, VCM Leung, A Nallanathan, Sensing time optimization and power control for energy efficient cognitive small cell with imperfect hybrid spectrum sensing. IEEE Trans. Wirel. Commun. **16**(2), 730–743 (2017)

37. S Kandukuri, S Boyd, Optimal power control in interference-limited fading wireless channels with outage-probability specifications. IEEE Trans. Wirel. Commun. **1**(1), 46–55 (2002)

38. A Khelil, PJ Marron, K Rothermel, *Contact-based mobility metrics for delay-tolerant ad hoc networking* (Proc. IEEE MASCOTS, Atlanta, 2005), pp. 1–10

39. JC Principe, *Information theoretic learning: Renyi's entropy and Kernel perspectives* (Springer, New York, 2010)

40. F Heliot, MA Imran, R Tafazolli, On the energy efficiency-spectral efficiency trade-off over the MIMO Rayleigh fading channel. IEEE Trans. Commun. **60**(5), 1345–1356 (2012)

41. H Zhang, C Jiang, NC Beaulieu, X Chu, X Wang, TQS Quek, Resource allocation for cognitive small cell networks: a cooperative bargaining game theoretic approach. IEEE Trans. Wirel. Commun. **14**(6), 3481–3493 (2015)

42. DWK Yeung, LA Petrosyan, *Cooperative stochastic differential games* (Springer, New York, 2005)

43. H Zhang, H Liu, J Cheng, VCM Leung, Downlink energy efficiency of power allocation and wireless backhaul bandwidth allocation in heterogeneous small cell networks. IEEE Trans. Commun. **PP**(99), 1–1 (2017)

44. C-G Yang, J-D Li, Z Tian, Optimal power control for cognitive radio networks under coupled interference constraints: a cooperative game-theoretic perspective. IEEE Trans. Veh. Technol. **59**(4), 1696–1706 (2010)

45. Y Xing, CN Mathur, MA Haleem, R Chandramouli, KP Subbalakshmi, Dynamic spectrum access with QoS and interference temperature constraints. IEEE Trans. Mobile Comput. **6**(4), 423–433 (2007)

12

Analysis of hidden terminal's effect on the performance of vehicular ad-hoc networks

Saurabh Kumar[1] (ID), Sunghyun Choi[2] and HyungWon Kim[1*]

Abstract

Vehicular ad-hoc networks (VANETs) based on the IEEE 802.11p standard are receiving increasing attention for road safety provisioning. Hidden terminals, however, demonstrate a serious challenge in the performance of VANETs. In this paper, we investigate the effect of hidden terminals on the performance of one hop broadcast communication. The paper formulates an analytical model to analyze the effect of hidden terminals on the performance metrics such as packet reception probability (PRP), packet reception delay (PRD), and packet reception interval (PRI) for the 2-dimensional (2-D) VANET. To verify the accuracy of the proposed model, the analytical model-based results are compared with NS3 simulation results using 2-D highway scenarios. We also compare the analytical results with those from real vehicular network implemented using the commercial vehicle-to-everything (V2X) devices. The analytical results show high correlation with the results of both simulation and real network.

Keywords: Vehicular ad-hoc networks, Medium access control (MAC), Broadcast communication, Hidden terminal

1 Introduction

Vehicular communication known as vehicle-to-everything (V2X) is an integral part of the intelligent transportation system (ITS) for vehicle safety applications. For wireless communication among the vehicles and vehicle to the roadside unit (RSU), this paper considers communications based on dedicated short-range communication (DSRC) [1]. DSRC uses a band at 5.9 GHz with a bandwidth of 75 MHz. It transmits the basic safety message (BSM) at the center frequency of 5.89 GHz, which is the control channel (CCH) with a bandwidth of 10 MHz. DSRC allows a large number of vehicles and RSUs to communicate among each other, which construct a highly dynamic vehicular ad-hoc network (VANET).

DSRC utilizes IEEE 802.11p, which is an amendment of the physical and medium access control (MAC) layers of IEEE 802.11a [2]. The physical layer is based on the orthogonal frequency division multiplexing (OFDM) modulation, while the MAC layer is ameliorated for the low overhead communication. Vehicles in 802.11p communicate using the independent basic service set (IBSS) architecture. Thus, it does not require the initial setup phase

for the vehicle's association [2]. To obtain optimum performance, 802.11p is optimized for the highly mobile environment, fast-changing multi-path reflection, and Doppler shift (due to high relative speed).

The MAC protocol of IEEE 802.11p uses carrier sense multiple access with collision avoidance (CSMA/CA) random access mechanism as a basic access scheme. To avoid collisions, the CSMA/CA mechanism uses distributed co-ordination function (DCF) to access the channel [2].

In VANETs or V2X based on DSRC, each vehicle transmits BSM packets periodically [1]. BSM is a beacon message that contains the status, position, and movement information of the vehicle [1]. Since the transmitter broadcasts its BSM to all its neighbor vehicles, the transmitter cannot get the confirmation of the correct reception from the receivers. Because unlike unicast, which can efficiently utilize acknowledgment (ACK) packet, the regular broadcast communication does not support ACK.

The data received in the BSM packets are utilized by the safety applications, thus making the timely packet delivery an utmost performance goal. One of the leading causes of performance degradation in the IEEE 802.11 networks is hidden terminal collisions [3]. Thus, IEEE 802.11 networks use request to send / clear to send (RTS/CTS) mechanism to alleviate the hidden terminal

* Correspondence: hwkim@cbnu.ac.kr
[1]Department of Electronics Engineering, Chungbuk National University, Cheongju, Chungcheongbuk-do, South Korea
Full list of author information is available at the end of the article

problem [4]. However, the RTS/CTS mechanism is inapplicable in broadcast communication because unlike the unicast, the transmitter of the broadcast communication cannot receive CTS (same as an ACK) from all the receivers, individually. The nodes located within the channel sensing range from a receiver but are out of the channel sensing range from the transmitter, are called hidden terminals for the receiver-transmitter pair. For example, in Fig. 1, node Tx1 transmits a BSM data packet. Both the receivers Rx1 and Rx2 in the communication range of the node Tx1 are expected to receive the transmitted packet. However, if the nodes in the shaded area start transmitting their packets at the same time, the receiver Rx1 may not receive the packet correctly. Hence, the nodes in the shaded area are the hidden terminals for the node Rx1 (which can be different from the hidden terminals for the node Rx2). Therefore, all the receivers of the broadcast packet independently experience the hidden terminal problem. Therefore, the total region of hidden terminals is sizable. Moreover, in the absence of ACK for the broadcast packet, hidden terminals cannot deduce the transmission[1] even from the receiver nodes; hence, the vulnerable period (the time at which the hidden terminals' transmission can result in a collision) can be longer than the unicast communication [3].

The packet delivery performance in the case of BSM is determined by the ability of 1-hop receivers to receive the generated packets with high probability within allowed time.

Literature in the hidden terminal analysis can be classified in two types based on the communication category: hidden terminal analysis in unicast communication and hidden terminal analysis in broadcast communication. Many researchers have investigated the former. Firstly, Tobagi et al. investigated the effect of hidden terminals on the performance of the network with multiple transmitters and a single receiver in a saturated traffic case [5]. Subsequently, Ray et al. derived the analytical expressions for the packet collision probability, average packet delay, and maximum throughput for the saturated traffic condition [6]. The authors have used the queuing theoretical analysis in a 4-node segment network. Afterward, in [7], Ekici et al. estimated the performance under the hidden node problem in the unsaturated traffic conditions for a 3-node symmetric network. Similarly, Tsertou et al. presented the performance modeling of the hidden terminal problem in a 3-node symmetric network using fixed length timeslot [3].

The authors of [8–16] analyzed the performance of hidden terminals for IEEE 802.11p broadcast communication.

Ma et al. derived the performance metrics for one hop broadcast communication of VANET comprising hidden nodes in [8, 9]. They also analyzed the channel capacity using signal-to-interference ratio (SIR) in VANET under the hidden terminals, access collisions, as well as channel propagation in [10]. However, the authors assumed that the nodes travel in a 1-D highway. Similarly, Fallah et al. analyzed the hidden node interference for the performance of cooperative vehicle safety system in 1-D network [11]. Rathee et al. analyzed the throughput of VANET with hidden terminals in smaller networks of 5 or 10 nodes [12]. Furthermore, Song presented an analytical model for the performance analysis of multichannel MAC in 1-D VANET with hidden terminals [13]. Whereas, authors in [14–16] analyzed the hidden terminal problem in 2-D VANET, Ma et al. analyzed the effect of hidden terminals for 2-D networks in rural intersections in [14] and for a general 2-D VANET in [15]. In addition, Wang et al. analyzed the performance of enhanced distributed channel access (EDCA) mechanism for 2-D wireless networks [16].

Moreover, the authors in [17–20] presented the effect of hidden terminals via only simulations. Sjoberg et al. [17] and Tomar et al. [18] defined the hidden terminal problem and simulated the effect on both packet reception rate and throughput. The authors in [19, 20] simulated the networks for the received interference power from the hidden terminals. They calculated the effect of the interference power on the safety messages' reachable distance and hidden terminals radius. In contrast, the proposed protocol has the following advantages.

- Our protocol uses multi-lane 2-D VANET to derive the analytical model, which increases accuracy.
- Our new Markov chain accurately emulates multiple backoff counter freezing due to sequential transmission from more than one vehicle in the channel sensing range.
- We consider both the communication range and channel sensing range to effectively differentiate the communicating nodes from interfering nodes.
- Our proposed analytical model uses the fixed length timeslots; hence, can accurately quantify the collisions from the unsynchronized hidden terminals.

2 Methodology

The analytical model that can evaluate the performance of VANETs under various configurations of network parameters is derived. We then evaluate the accuracy of the proposed model through a comparison with the simulation results based on NS3 using multi-lane highway scenarios. Additionally, the model's accuracy is also compared with the results of a real network.

---

[1]Hidden terminals detect the ongoing unicast transmission by ACK and CTS transmitted by receiver nodes in the case of basic access and RTS/CTS access mechanism, respectively.

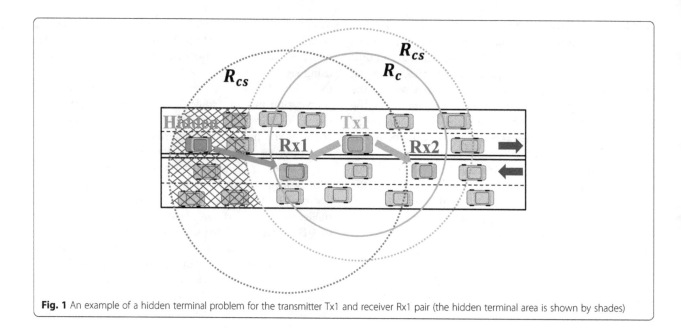

Fig. 1 An example of a hidden terminal problem for the transmitter Tx1 and receiver Rx1 pair (the hidden terminal area is shown by shades)

The analysis presented in [8–10] assumes the variable size timeslots, which is an extension of the prominent work done by Bianchi [4]. Bianchi assumed that the state transition for all the nodes occurs altogether. This assumption is valid for the connected networks since the nodes are synchronized. In the IBSS network with hidden terminals, nodes are unsynchronized; hence, we cannot assume that the nodes start transmission, simultaneously. In this paper, we consider the fixed-size slots, which are also assumed in [3], where the authors have analyzed the hidden terminals but only in the case of saturated unicast communication with a 3-node network. Additionally, in [3], a transmitter node can freeze its counter only once, i.e., there is only one other transmitter in the wireless sensing range. In the VANET, however, multiple nodes in the wireless sensing range can transmit the packets one after another. Thus, the counter can freeze multiple times for the same backoff counter value. We design a new Markov chain that can accurately emulate this behavior. We first define the hidden terminal problem and then model the effect of hidden terminals on three performance metrics: (1) packet reception probability (PRP), (2) packet reception delay (PRD), and (3) packet reception interval (PRI).

The rest of the paper is organized as follows: Section 3 depicts the operation of DSRC MAC for BSM broadcast. Section 4 introduces the system model and outlines the assumptions made in the analytical model. Then, Section 5 defines the hidden terminal problem in 2-D VANET and derives the model for performance metrics. Section 6 evaluates the accuracy of the proposed model in comparison with the results from NS3 simulator and real network. Section 7 presents the conclusions.

3 Dedicated short-range communication MAC

This section presents the salient facets of the IEEE 802.11p DCF MAC used by the IBSS architecture [2]. Here, only the broadcast communication is considered for simplicity.

Once the MAC receives a new packet for transmission from BSM application. It starts the transmission process by sensing the channel (using only physical carrier sensing, since virtual carrier sensing is inapplicable in broadcast communication) for the distributed interframe space (DIFS) time. If the channel is idle, the MAC transmits the packet. In contrast, if the sensing result gives channel busy during the DIFS, it continues to wait until the channel becomes idle for the DIFS time. Afterward, the MAC chooses a random backoff counter and delays the transmission for the backoff time as shown in Fig. 2. The backoff time is calculated by the backoff counter multiplied by the timeslot length σ. The counter is drawn from the uniform random distribution between 0 and contention window value. Contention window varies from the minimum value to the maximum value, according to the binary exponential backoff (BEB) procedure [2]. In the case of broadcast communication, however, the transmitter is not notified for the unsuccessful transmissions (no retransmission). Hence, MAC uses the constant contention window (CCW), i.e., the minimum contention window size (W_0). The random backoff time is used to mitigate the access collisions with the nodes in the channel sensing range. Here, an access collision is defined as the collision that occurs when two (or more) nodes in the channel sensing range transmit in the same timeslot.

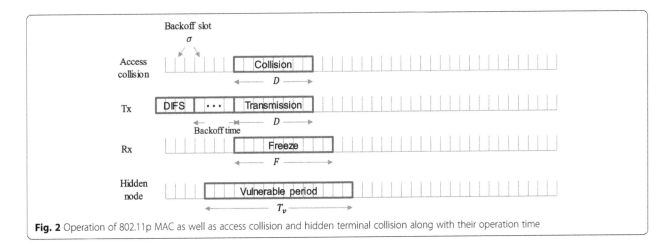

Fig. 2 Operation of 802.11p MAC as well as access collision and hidden terminal collision along with their operation time

Backoff counter is decremented at the start of each timeslot (also called backoff slot) of length σ, only if the channel is sensed idle. If the channel is busy, on the other hand, the backoff counter freezes for the sensed packet transmission plus DIFS time. Afterward, the MAC resumes decrementing the counter. Once the counter reaches zero, the node transmits the packet at the start of the next timeslot. The timeslot length σ is the sum of the time needed for the channel assessment, MAC processing delay, time to switch the transceiver from receiving to transmitting mode, and propagation delay [2]. After finishing the transmission, the node delays the next packet transmission for the random backoff time. This strategy allows a fair use of channel among all the neighbor nodes.

4 System model and modeling assumptions

To derive the system model, we assume that the vehicular nodes are randomly distributed in a multi-lane highway having a total of ξ lanes (including both directions). The width of each lane is ω. Vehicle's initial placement in the highway follows the poison point process (see Fig. 1) and its density is α vehicles/lane/km. However, their movement (vehicle speed with respect to time) exhibits exponential distribution [21] and constructs a highly dynamic 2-D VANET. The notation of node, terminal, or vehicle is used interchangeably. Additionally, the slot and timeslot both mean one timeslot of length σ. Frequently used notations in the model are specified in Table 1, and the descriptions of important notations are given as follows.

- BSM application in each node generates safety data packet periodically at λ Hz rate. The packet is generated at the start of the new period and stored in the MAC queue. Subsequently, DCF MAC starts the transmission process. Thus, the packet arrival to the MAC queue follows the deterministic

Table 1 Most frequently used modeling notations

Notation	Description
ξ	Total number of lanes
ω	Width of the lane (m)
a	Vehicle density (vehicles/lane/km)
D	BSM application data packet size (bytes)
$\lceil x \rceil$	Ceil(x): smallest integer greater than or equals to x
SD	Symbol duration (µs)
R	Data rate (bits/s)
T_{DATA}	Time required to transmit a data packet
S_{DATA}	Number of timeslots needed to transmit the data packet
σ	Size of the timeslot (µs)
R_c	Communication range (m)
R_{cs}	Channel (carrier) sensing range (m) ($1.5 \times R_c$)
N_c	Number of nodes in the communication range
N_{cs}	Number of nodes in the channel sensing range
λ	Packet generation rate in each node (Hz)
V	Set of vehicular nodes
C	Set of communication sets of the nodes
$H(v_i, v_j)$	Set of hidden terminals for the transmitter v_i and receiver v_j pair
P_h	Probability of collision due to hidden terminals
P_{Tx}	Probability of transmission in a slot
N_h	Average number of hidden terminal nodes for 2-D VANET
T_{vul}	Vulnerable time
S_{vul}	Number of slots in the vulnerable time
P_d	Probability of data availability in the MAC queue
$b(s)$	Value of backoff counter at state s in the Markov Chain
$f(s)$	Value of freezing slot at state s in the Markov Chain
τ	Stationary probability of data transmission in a slot
$\overline{M}_D$	Average (mean) MAC delay
W_0	Constant contention window size
F	Number of continuous slots in the freezing state

distribution. For modeling simplicity, this paper assumes an infinite size MAC queue because none of the packets are dropped due to queue full condition. Each packet is transmitted after the backoff time, which follows semi-Markov service time. Therefore, the MAC queue follows ($D/M/1$) queuing model ([22], p377), where 1 indicates the number of channels used (only CCH).

- Figure 3 shows the details of the transmitted BSM packet. Let the size of BSM packet generated by the safety application is D bytes. The network and the MAC layers add H_{Net} and H_{MAC} bytes, respectively, for the network header and MAC header. On top of that, the MAC layer also adds H_{Trail} byte frame check sequence (FCS) trailer information. Finally, the physical layer adds physical layer convergence procedure (PLCP) preamble (4 symbols), PLCP header (1 symbol), 16 bits service detail and 6 tail bits [2].

- The PLCP preamble and PLCP header are transmitted using the binary phase shift keying (BPSK) with $1/2$ code rate. It requires constant duration irrespective of the employed data rate. However, the rest of the packet is transmitted using physical layer data rate, which is R (bits/s).

- If one symbol duration is SD. The number of data bits transmitted per symbol are $R \times SD$. Therefore, the number of symbols (N_S) in a packet are,

$$N_S = 5 + \left\lceil \frac{16 + (D + H_{\text{Net}} + H_{\text{MAC}} + H_{\text{trail}}) \times 8 + 6}{R \times SD} \right\rceil \tag{1}$$

Hence, the total time needed to transmit the data is,

$$T_{\text{DATA}} = \left(5 + \left\lceil \frac{16 + (D + H_{\text{Net}} + H_{\text{MAC}} + H_{\text{trail}}) \times 8 + 6}{R \times SD} \right\rceil \right) \times SD \tag{2}$$

The number of timeslots needed to transmit the data packet can be calculated from Eq. (3).

$$S_{\text{DATA}} = \left\lceil \frac{T_{\text{DATA}}}{\sigma} \right\rceil \tag{3}$$

- If node v_i senses transmission from the nodes within v_i's channel sensing range, v_i freezes its backoff counter for the packet transmission plus DIFS time. Hence, the freezing time is $T_{\text{DATA}} +$ DIFS. Thus, the number of freezing slots is given as follows.

$$F = S_{\text{DATA}} + \left\lceil \frac{\text{DIFS}}{\sigma} \right\rceil \tag{4}$$

For tractability and simplicity of the model, the following assumptions are made:

- All the nodes are identical and have a uniform circular communication range R_c and channel (carrier) sensing range R_{cs} ($R_c < R_{cs} < 2 \times R_c$). All the nodes in the communication range of the transmitter can receive the packet correctly. However, the nodes in the channel sensing range can detect the transmission but might not be able to decode the packet contents correctly. The number of nodes in the communication range (N_c) and the channel sensing range (N_{cs}) are derived in the Appendix.

- We consider only BSM safety data transmission over continuous mode 802.11p MAC. In the continuous mode, the radio transceiver of the vehicle is always in operation at CCH for whole 100 ms [2].

- We do not consider shadowing, fading, and channel capturing in the modeling for simplicity. The focus of the paper is on the effect of hidden terminals in the 1-hop broadcast communication.

5 Modeling of hidden terminal problem

Unlike unicast, the safety data in the DSRC is broadcasted so that each node in the communication range of the transmitter is a prospective receiver. The hidden terminal problem in broadcast communication is different from that in unicast communication. For the broadcast of a transmitter, a set of nodes can be hidden terminals

	4 symbols	1 symbol	16 bits	24 bytes	36 bytes	D Bytes	4 bytes	6 bits
	PLCP P	PLCP H	Service	MAC H	Net H	BSM DATA	FCS	Phy tail
		Physical header →		MAC Header H_{MAC}	Network Header H_{Net}	BSM data contents	MAC Trailer H_{Trail}	Physical Trailer

Fig. 3 The detailed frame format of the BSM packet

for a receiver and might not be for others. For example, in Fig. 1, for the transmission of Tx1, the nodes in the shaded area are the hidden terminals for the receiver Rx1, whereas they are not the hidden terminals for the receiver Rx2. Although the authors in [17, 18] defined the hidden terminal problem, they had not considered channel sensing range, which in general, is larger than the communication range.

We represent a 2-D VANET by an undirected graph $G = (V, C)$, where V is the set of all vehicular nodes in the network. We can ignore the effect of mobility during the packet transmission since it takes a very short time (< 1 ms) for the vehicle to transmit a short packet like BSM [9]. Therefore, a vehicular node $v_i \in V$ is positioned at (x_{v_i}, y_{v_i}), and assumed to be stationary for the transmission time.

We define C as a set of communication set nodes, expressed by Eq. (5).

$$C = \{c_{v_i} | v_i \in V\} \qquad (5)$$

Here, communication set c_{v_i} for the node v_i is the set of all the nodes in the communication range of the node v_i for a transmission instant. In other words, c_{v_i} is the set of prospective receivers for the current transmission from the node v_i, which is denoted by Eq. (6).

$$c_{v_i} = \{v_j | v_j \in V, \neq v_i, d(v_i, v_j) \leq R_C\} \qquad (6)$$

where $d(v_i, v_j)$ denotes the Euclidean distance between the nodes v_i and v_j.

$$\begin{aligned} d(v_i, v_j) &= \|v_i, v_j\| \\ &= \sqrt{(x_{v_i} - x_{v_j})^2 + (y_{v_i} - y_{v_j})^2} \end{aligned} \qquad (7)$$

If node v_i is transmitting a packet and node v_j is one of the intended receivers among c_{v_i}. The set of hidden terminals for the transmitter-receiver pair is expressed by Eq. (8).

$$H(v_i, v_j) = \{v_h | v_h \in V, v_j \in (c_{v_i} \cap cs_{v_h}), R_{CS} < d(v_i, v_h) < 2 \times R_C\} \qquad (8)$$

Here, cs_{v_h} is the set of nodes in v_h's carrier sensing range and other notations are described in Table 1.

During the transmission from the node v_i, if any node in the set $H(v_i, v_j)$ transmit at the vulnerable period, v_j experiences collision thus, cannot receive the packet, correctly. This is called the hidden terminal problem in the broadcast communication for the vehicular ad-hoc networks and such collisions are called hidden terminal collisions. The probability of hidden terminal collision P_h is the probability of at least one of the hidden

terminal nodes transmitting at the vulnerable period. P_h can also be derived from the probability of none of the hidden terminal nodes transmit in the vulnerable timeslots.

$$P_h = 1 - \left((1 - P_{Tx})^{N_h}\right)^{S_{vul}} \qquad (9)$$

Here, P_{Tx} is the probability of transmission in a timeslot, N_h represent the average number of hidden terminals for the transmitter-receiver pair, and S_{vul} is the number of timeslots in the vulnerable period. To analyze the effect of hidden terminals using Eq. (9), we derive P_{Tx}, N_h (in the Appendix) and S_{vul}.

5.1 Vulnerable time period (T_{vul})

Unlike [4], the vehicular nodes in VANET are not connected with the common base station, and thus they are unsynchronized. Therefore, the start of transmission in the nodes is also unsynchronized. The time period when the hidden terminals' transmission can collide with the ongoing transmission is called vulnerable period T_{vul} [13]. Unicast communications exert ACK, which is received by the hidden terminals, and thus, they can suppress transmission. Hence, the vulnerable period in unicast communication is smaller [3]. For broadcast communications, T_{vul} can be expressed by Eq. (10).

$$T_{vul} = 3 \times T_{DATA} \qquad (10)$$

Here, T_{DATA} is the data packet transmission time.

As shown in Fig. 4, we assume the node v_i starts transmission at $t = 0$. Since nodes are not synchronized in the VANET, we can infer from Fig. 4 that any transmission from the hidden terminals (v_h) in the interval ($-T_{DATA}$, $2 \times T_{DATA}$) results in a collision at the receiver v_j. However, if the hidden terminal node v_h transmits before

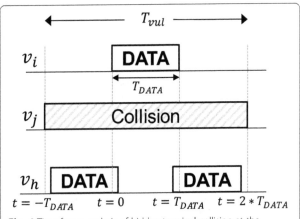

Fig. 4 Time frame analysis of hidden terminal collision at the receiver node to calculate the vulnerable period

$(-T_{\text{DATA}})$ or after $(2 \times T_{\text{DATA}})$, the receiver v_j receives the transmitted packet from the transmitter v_i without hidden terminal collisions.

The number of timeslots in a vulnerable period can be calculated by Eq. (11) using the vulnerable time period from Eq. (10).

$$S_{\text{vul}} = \left\lceil \frac{T_{\text{vul}}}{\sigma} \right\rceil \qquad (11)$$

5.2 Probability of transmission in a slot (P_{Tx})

The hidden terminals for the pair of transmitter and receiver lie within the channel sensing range of the receiver. Therefore, while calculating the probability of transmission, the cascading effect of the hidden terminals is not envisaged.

The probability of transmission in a timeslot for the unsaturated traffic in a node depends on the following two independent probabilities: (1) data arrival probability or data availability in the MAC queue (P_d), (2) the stationary transmission probability in a timeslot (τ). Hence, the probability of transmission in a slot can be written as Eq. (12).

$$P_{Tx} = P_d \times \tau \qquad (12)$$

However, two or more hidden terminals transmitting in the same slot can result in access collision among hidden terminals. Thus, the probability of successful packet transmission in a slot is expressed by Eq. (13).

$$P_S = P_d \times \tau \times (1-p) \qquad (13)$$

Here, p is the conditional collision probability (access collisions) given the node transmits in a timeslot.

5.3 Probability of data availability (P_d)

Data packet arrival at the MAC queue is deterministic with the rate λ Hz and service time is exponential (semi-Markov) with the mean MAC delay $\overline{M}_D$. Hence, the data availability in the MAC is a stochastic process with the Poisson distribution. As a result, the probability mass function (PMF) of the number of packets in time t can be written by Eq. (14) ([22], p 307).

$$p_n(t) = e^{-\lambda t} \frac{(\lambda t)^n}{n!}, \quad \text{where } n > 0 \qquad (14)$$

If the application generates a new packet before the current packet is transmitted, the current one becomes obsolete. Hence, after the generation of the next packet existing packet is preempted from the queue. Thus, the MAC queue always contains either one packet or zero packets. Hence, the probability of data availability is the

PMF of one packet in the queue in the mean MAC delay $\overline{M}_D$, which is calculated using Eq. (15).

$$P_d = e^{-\lambda \overline{M}_D} \times \lambda \overline{M}_D \qquad (15)$$

Section 5.6 derives the mean MAC delay $\overline{M}_D$, for the BSM packet. The network is unsaturated if the traffic follows $\lambda \overline{M}_D < 1$ condition. However, it becomes saturated if $\lambda \overline{M}_D \geq 1$, and the MAC queue start preempting non-transmitted packets.

5.4 Stationary transmission probability in a slot (τ)

We derive the stationary probability of transmission in a given slot (τ) for a node by using the Markov chain model of backoff counter transition. The timeframe is divided into discrete timeslots of a fixed length σ. Let s, and $s + 1$ are two consecutive timeslots. Initially, a vehicle v_i chooses its backoff counter from the uniform random distribution of $[0, W_0)$. The backoff counter of v_i is decremented at the start of the next timeslot $s + 1$ if the channel is sensed idle in the timeslot s. However, if any other node in the carrier sensing range starts transmitting with probability p_f, the channel becomes busy, and hence, the backoff counter of v_i freezes with the current value. The counter continues to freeze for the next F timeslots. Afterward, the counter starts to decrement, if the channel becomes idle, which occurs with probability $(1 - p_f)$. However, if any other node starts transmitting after F slots, the backoff counter of v_i remains the same and continues to freeze for the next F slots. The transition process of the Markov chain is described in Fig. 5. The non-null transition probabilities of individual steps in the Markov chain are given by Eq. (16)[2].

$$\begin{cases} P\{b, f+1|b, f\} = 1 & b \in (1, W_0-1), f \in (1, F-1) \\ P\{b, 1|b, 0\} = p_f & b \in (1, W_0-1) \\ P\{b, 1|b, F\} = p_f & b \in (1, W_0-1) \\ P\{b, 0|b+1, 0\} = 1-p_f & b \in (0, W_0-2) \\ P\{b, 0|b+1, F\} = 1-p_f & b \in (0, W_0-2) \\ P\{b, 0|0, 0\} = 1 \Big/ W_0 & b \in (0, W_0-1) \end{cases}$$

$$(16)$$

The first three equations in Eq. (16) account for transition related to freezing slots. The first equation represents the transition from the first freezing slot until the last (F^{th} slot). Once the node v_i starts sensing the channel as busy, the channel continues to be busy for the next F slots. In the Markov chain, the state of the next slot depends not only on the previous slot but also on the number of elapsed freezing slots. Thus, the state transition exhibits semi-Markov behavior [3].

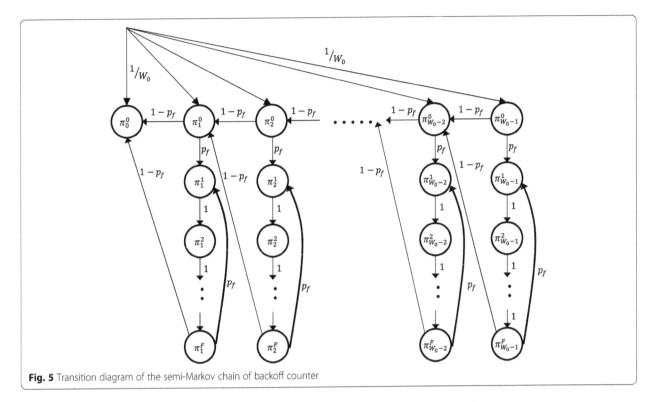

Fig. 5 Transition diagram of the semi-Markov chain of backoff counter

The second equation accounts for the node v_i entering the freezing process. Once v_i senses channel as busy with probability p_f, which occurs when other nodes in the carrier sensing range of v_i start the transmission, the counter of v_i freezes for the next F slots.

Equation three accounts for the continuation of the freezing process of the backoff counter for the next F timeslots, if any other nodes start transmitting after the current node finishes. The 4th and 5th equation account for decrementing the backoff counter; the former without going to freezing states and the latter after completing the freezing states, respectively. The last equation accounts for the initial backoff counter selection, once a new packet arrives in the MAC queue for the transmission.

Let $\pi_b^f = \lim_{t \to \infty} P\{k(s) = b, l(s) = f\}, b \in (0, W_0-1), f \in (0, F)$ be the stationary probability of the states in the semi-Markov chain. From the closed form solution, we can derive Eqs. (17) and (18).

$$\pi_b^1 = \pi_b^0 \times p_f + \pi_b^F \times p_f \qquad (17)$$

$$\pi_b^f = \pi_b^1 \quad 1 \leq f \leq F \qquad (18)$$

Solving Eqs. (17) and (18) gives Eq. (19).

$$\pi_b^f = \frac{p_f}{1-p_f} \pi_b^0 \quad 1 \leq f \leq F \qquad (19)$$

Another closed form solution from the backoff counter decrement states can be expressed by Eq. (20).

$$\pi_b^0 = \left(1 - p_f^2\right) \pi_{b+1}^0 + \frac{1}{W_0} \pi_0^0 \qquad (20)$$

Eq. (20) can be simplified to Eq. (21).

$$\pi_b^0 = \left(1 + \left(1 - p_f^2\right)(W_0 - b - 1)\right) \times \frac{1}{W_0} \pi_0^0 \qquad (21)$$

Eqs. (19) and (21) illustrate the values of the stationary probabilities π_b^f of Markov states in terms of π_0^0 and the freezing probability p_f. By using the Markov chain normalization condition, we determine π_0^0 as follows.

$$1 = \pi_0^0 + \sum_{b=1}^{W_0-1} \sum_{f=0}^{F} \pi_b^f = \pi_0^0 + \sum_{b=1}^{W_0-1}\left(1 + \frac{Fp_f}{1-p_f}\right)\pi_b^0 = \pi_0^0$$
$$+ \sum_{b=1}^{W_0-1}\left(1 + \frac{Fp_f}{1-p_f}\right)\left(1 + \left(1 - p_f^2\right)(W_0 - b - 1)\frac{1}{W_0}\pi_0^0\right)$$

$$(22)$$

Eq. (22) can be simplified for π_0^0 to Eq. (23).

$$\pi_0^0 = \frac{2W_0\left(1-p_f\right)}{2W_0\left(1-p_f\right) + \left(1 - p_f + Fp_f\right)\left(W_0-1\right)\left(2 + \left(1 - p_f^2\right)(W_0-2)\right)} \qquad (23)$$

The vehicular node transmits in the next slot after the backoff counter becomes zero. Hence, the stationary transmission probability in a slot can be calculated by Eq. (24).

$$\tau = \pi_0^0 = \frac{2W_0\left(1-p_f\right)}{2W_0\left(1-p_f\right) + \left(1-p_f + Fp_f\right)(W_0-1)\left(2 + \left(1-p_f^2\right)(W_0-2)\right)}$$

(24)

If we consider backoff counter transition without freezing ($p_f = 0$), a solution can be obtained by using a Markov chain based on classical constant contention window [4]. The stationary probability of transmission in a slot with $p_f = 0$ can be expressed by Eq. (25).

$$\tau = \frac{2}{W_0 + 1}$$

(25)

5.5 Freezing probability (p_f)

xTo derive τ from Eq. (24), we require p_f, freezing probability of the backoff counter for the node v_i in a slot. The counter freezes, when at least one of the nodes ($N_{cs} - 1$) within the channel sensing range from v_i transmit in a backoff slot. The backoff counter is chosen from [0, $W_0 - 1$] based on the uniform random distribution. Hence, the probability of a backoff slot selection is $1/W_0$. As a result, the probability that at least one of the nodes choose the given backoff slot is $(1-(1-1/W_0)^{N_{cs}-1})$. Additionally, as shown in Fig. 6, if the slot s_{n+1} is the first freezing slot, then the backoff counter continuously freezes for the next F slots, till slot S_{n+F}. As a result, the nodes in the sensing range do not transmit in the next F slots, which means that only one out of F slots has a freezing likelihood. Hence, the freezing probability is calculated as Eq. (26).

$$p_f = \frac{1}{F} \times \left(1-\left(1-1/w_0\right)^{N_{cs}-1}\right)$$

(26)

5.6 Mean MAC delay ($\overline{M}_D$)

Once the MAC queue receives a new packet (the previous packet is flushed if not transmitted), it starts the transmission process by choosing a backoff counter. Based on the channel state (idle or busy), the MAC decrements the backoff counter from the chosen value until zero. As soon as the counter becomes zero, the MAC transmits the packet in the next slot. Henceforth, the mean MAC delay for the packet is the average time spent in the Markov chain of the backoff counter transition. Markov chain consumes time in two states: (1) time in the freezing states in case of transmissions from the other nodes and (2) time consumed in decrementing the backoff counter until zero in the idle channel state. Let B be the random variable for the backoff counter selection, then the MAC delay for a packet is written by Eq. (27).

$$M_D = p_f \times F \times \sigma \times B + \sigma \times B$$

(27)

To obtain the mean MAC delay, we take the expectation on both sides of Eq. (27), which leads to Eq. (28).

$$E[M_D] = \left(p_f \times F + 1\right) \times \sigma \times E[B]$$

(28)

The expected value of the MAC delay is $\overline{M}_D$ and the expected value of the random variable of CCW backoff counter with range [0, $W_0 - 1$] is $W_0/2$. After substituting values in Eq. (28), we can obtain $\overline{M}_D$ expressed by Eq. (29).

$$\overline{M}_D = \left(p_f \times F + 1\right) \times \sigma \times \frac{W_0}{2}$$

(29)

The effect of hidden terminal collisions on the performance metrics PRP, PRD, and PRI are expressed in the following subsections using the probability of hidden terminal collision (P_h) obtained by Eq. (9).

5.7 Packet reception probability (PRP)

Packet reception probability is the ability of 1-hop receiver nodes to receive the BSM packet generated at the transmitter vehicle v_i successfully. Hence, it is measured as the probability of packets received with respect to the packets generated by the vehicles in the communication range. The receiver does not receive the transmitted packets if lost due to collision. Hence, the hidden terminal collisions directly affect the PRP. The loss in the PRP due to hidden terminals is the same as the probability P_h of hidden terminal collision at the receiver. As a result, the effect of hidden terminals on PRP can be calculated using Eq. (30).

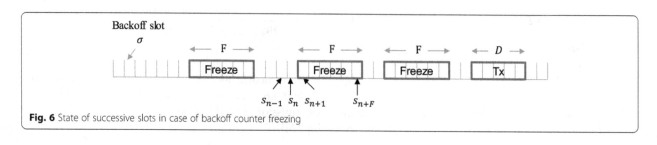

Fig. 6 State of successive slots in case of backoff counter freezing

$$PRP = 1 - P_h \qquad (30)$$

5.8 Packet reception delay (PRD)

Packet reception delay is defined as the time lapse between the packet arrival at the MAC queue of the transmitter and the packet reception in the MAC queue of the receiver. PRD can also be represented as MAC-to-MAC delay. For the sake of simplicity, in this paper, we ignore the time delay between the packet generation in the application and the start of the transmission process in the MAC. Thus, the MAC-to-MAC delay can be simplified as the sum of DIFS, mean MAC delay ($\overline{M}_D$), data packet transmission time (T_{DATA}), packet propagation, and physical layer processing time (σ). Hence, the average packet delay can be calculated by Eq. (31).

$$\mathrm{PRD} = \mathrm{DIFS} + \overline{M}_D + T_{\mathrm{DATA}} + \sigma \qquad (31)$$

5.9 Packet reception interval (PRI)

The efficiency of the safety applications depends on the freshness of the BSM data received from the neighbor nodes. The freshness of the BSM data depends on the rate of packet reception at the receiver, which in turn is affected by the probability of the transmission in a slot, access collision, and hidden terminal collision. We define the packet reception interval at the receiver as the average time between the reception of two consecutive packets from the same transmitter. The value of the packet reception interval should be $1\big/\lambda$ for the up-to-date data. Let a vehicle v_i receive λ packets/s from a transmitter, which is calculated using Eq. (32).

$$\lambda' = \lambda \times (1 - P_h) \qquad (32)$$

Eq. (32) signifies that the packets lost due to hidden terminal collisions decrease the rate of packet reception. PRI is, therefore, calculated as $1\big/\lambda'$ by Eq. (32).

6 Model verification

This section analyzes the accuracy of the analytical model and evaluates the effect of hidden terminals in 2-D VANET. Extensive simulation results are exhibited for the performance metrics PRP, PRD, and PRI and compared with the results of the analytical model derived in Eqs. (30), (31), and (32), respectively. Furthermore, the accuracy of the analytical model is compared with the measured result of a real network of V2X hardware devices.

6.1 Simulation setup

The simulations are conducted using the network simulator 3 (NS3) [23]. As shown in Fig. 7, the simulator is implemented using a realistic multi-lane highway mobility model. Simulations have been conducted in a 10 km highway segment. The highway model has a total of 10 lanes (5 in each direction) with a lane width of 4 m [24]. Each vehicle is randomly assigned to one of the lanes and kept in that lane throughout the simulation. The speed of the vehicle is assigned based on the lane number: 40 km/h for lane 1, 70 km/h for lane 2, 100 km/h for lane 3, 120 km/h for lane 4, and 140 km/h for lane 5. As a result, vehicles maintain the same speed during the simulation and do not collide or cross each other. Unless specified, the default data rate for BSM safety data is 6 Mbps and uses quadrature phase shift keying (QPSK) modulation and code rate $1\big/2$ for the robust performance.

Other simulation parameters are summarized in Table 2. The channel is simulated using the Nakagami-m propagation model, which is best suited for vehicular communication in highway scenarios [25]. Nakagami-m computes two distance dependent parameters: fading factor (m) and average power (Ω). Torrent-Moreno et al. computed the values of m and Ω for the highway scenario using maximum likelihood estimation [25]. The authors have shown that the average received power (Ω) is inversely proportional to the square of the distance between the transmitter and receiver ($\propto 1\big/d^2$). In addition, fading parameter m varies on the basis of distance range. For example, (1) $m = 3$ for a short distance between the transmitter and receiver ($d \le 100$), (2) $m = 1.5$ for an intermediate inter-distance ($100 < d \le 250$), and (3) $m = 1$ for the long distance ($d > 250$). Up to a distance of 250 m, the propagation follows the Racian distribution (incorporates the line of sight). Beyond that distance, the Rayleigh distribution is employed to calculate the average received power. We have chosen the threshold for communication range, and the threshold for channel sensing range as – 96 dBm and – 99 dBm, respectively. The simulations are conducted using BSM packets, which are transmitted in the CCH (Channel number 178) using continuous mode [1]. Each simulation result is computed by taking the average of 5 simulation readings by using different random seeds and the simulation time for each run is 50 s.

6.2 Real network with commercial V2X devices

We configured a real network using commercial V2X devices, MK5 V2X hardware from Cohda wireless [26]. We have tailored the application code to use link layer control (LLC) application programming interfaces (APIs) for controlling the detailed operations of the packet flow and avoid indeterministic delays. Hence, the network layer header H_{net} is zero, while the other parameters are configured the same as Table 2. Due to a limited number of commercial V2X devices, the results are exhibited for a network of four devices as shown in Fig. 8. The DSRC standard requires that each vehicle generates BSM packet every 100 ms (10 Hz). Thus, we have configured both the transmitter Tx1 and Tx2 to generate BSM packets every 1 ms to emulate 100 vehicles in each transmitter. Similar changes are made for other packet generation rates to emulate 100 vehicles per transmitter. Therefore, our network emulation technique can effectively measure the performance in a larger network using limited resources. Transmitters Tx1 and Tx2 are acting as hidden terminals for the receiver Rx1 since both transmitters are within the communication range of Rx1. In contrast, Rx2 can receive from Tx1 without collision, since the only transmitter, Tx1 is in the communication range of Rx2. The nodes are stationary during the experiment.

There are no access collisions because the two transmitters are hidden terminals to each other. The number of hidden terminals N_h for any transmission is always one (Tx1 when Tx2 is transmitting or Tx2 when Tx1 is transmitting). Additionally, for both transmitters, the number of nodes in the channel sensing range N_{cs} is also

one. By using this condition, the probability of freezing (p_f) is calculated from Eq. (26) as zero.

6.3 Effect on packet reception probability (PRP)

The collisions due to hidden terminals reduce the packet reception probability of the transmitted packets. In the case of the analytical model, the loss in the PRP due to hidden terminals (P_h) is calculated by Eq. (9). In case of simulation, on the other hand, the loss in the PRP due to hidden terminals (P_{h_s}) is calculated using Eq. (33).

$$P_{h_s} = \frac{\text{Hidden terminal collisions}}{\text{Expected receive packets}} \quad (33)$$

Figure 9 shows the loss of PRP due to hidden terminal collisions for both analytical model and simulation, and compares with all types of collisions for the simulation (note that analytical model derives the effect of only hidden terminal collisions). All types of collisions are comprised of both access collisions and hidden terminal collisions. The loss due to all types of collisions (P_{ALL}) is calculated by Eq. (34).

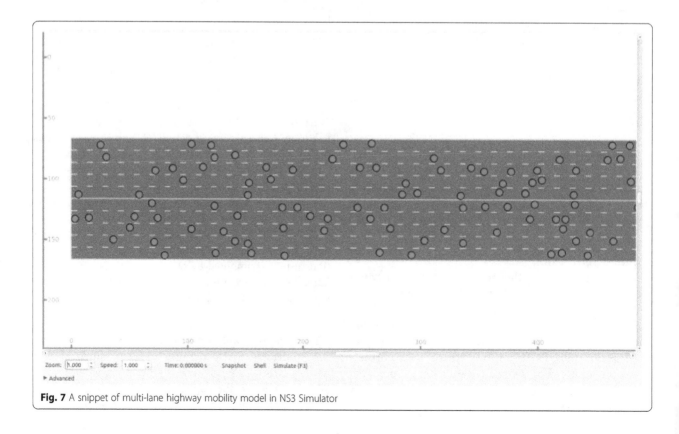

Fig. 7 A snippet of multi-lane highway mobility model in NS3 Simulator

Table 2 The parameters used in NS3 simulation and devices

Parameter	Value
Application	BSM wave
Physical and MAC layer	802.11p
Modulation and code rate	QPSK and 1/2
Timeslot size σ	13 μs
DIFS	58 μs
AISFN	6
SD	8 μs
H_{Net}	36 bytes
H_{MAC}	24 bytes
H_{Trail}	4 bytes
Antenna height (A)	$1 \leq A \leq 2$
Antenna gain	1 dBm

$$P_{ALL} = \frac{\text{Expected receive packets} - \text{actual received packets}}{\text{Expected receive packets}}$$

$$(34)$$

The comparison results from Fig. 9 manifest that the result of the analytical model closely matches the simulation result because we have considered only hidden terminal collisions. The results in Fig. 9 also show that the PRP loss due to hidden terminals increases more rapidly than the access collisions as the vehicle density increases. This result is also evident from Fig. 10, which shows that the PRP (only considering the hidden terminal collisions) decreases as vehicle density increases.

Figure 10 also shows that the reason for the decrease in PRP is due to an increase in the average number of hidden terminals per receiver as density increases. Furthermore, Figs. 11, 12, 13, 14, and 15 demonstrate the similarity in the PRP of the simulations and analytical model (only hidden terminal collisions are considered).

The results in Fig. 11 exhibit that the increase in the minimum contention window size (W_0) increases the PRP, although the increase is minuscule. The increase is due to an increased number of backoff slots. The results in Fig. 12 show that the PRP can be increased by decreasing the communication range (transmission power) of the transmitter. The reduction in the transmit power decreases the number of hidden terminal nodes, which leads to increased PRP. The results in Fig. 13 manifest that reducing the BSM data packet size increases the PRP in case of hidden terminals. This is attributed to the reduction in the vulnerable period because of the short packet transmission time.

Figure 14 presents the PRP measured by changing the packet generation rate of BSM application. The results show that reducing the packet generation rate reduces the contention (packet flow in the wireless channel), which in turn increases PRP. This increase can be explained as, even if the density increases, the total number of packets requested to be transmitted in the channel are not increasing as fast. Hence, the probability that hidden terminals transmit during the vulnerable period decreases.

The results in Fig. 15 show that by increasing the data rate of the transmitted packet, the effect of hidden terminals on PRP can be reduced. Higher data rate reduces the packet transmission time. Due to the reduction in the BSM packet transmission time, the vulnerable period decreases, which has a similar effect as reducing the BSM data size.

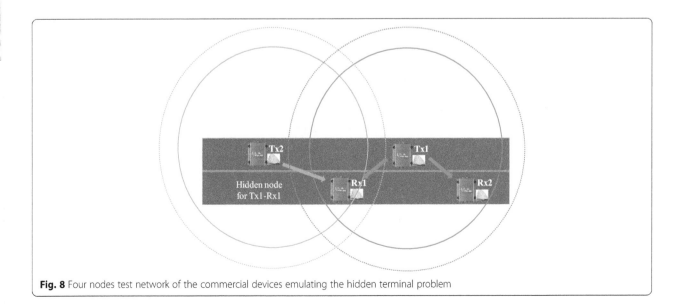

Fig. 8 Four nodes test network of the commercial devices emulating the hidden terminal problem

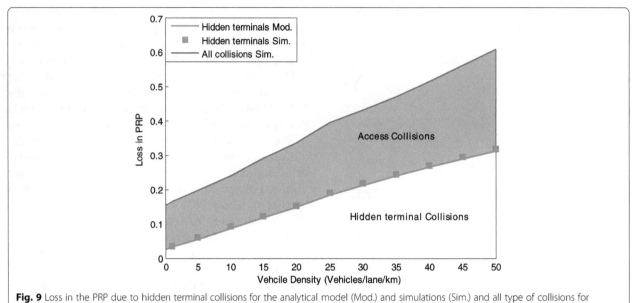

Fig. 9 Loss in the PRP due to hidden terminal collisions for the analytical model (Mod.) and simulations (Sim.) and all type of collisions for simulation. The network parameters are $W_0 = 16$, $R = 6$ Mbps, $R_c = 250$ m, $\lambda = 10$ Hz, $D = 186$ B

For the experiments with the real network, we calculate the loss in PRP (P_{h_N}) by Eq. (35).

$$P_{h_N} = \frac{\text{packets received at Rx2} - \text{packets received at Rx1}}{\text{packets received at Rx2}}$$

(35)

The result of the analytical model for the loss in PRP is calculated using Eq. (9). Figure 16 compares the results measured from the real network with the results of the analytical model. The results from the real network show the small difference due to non-uniform (circular) communication range of the transmitter as well as multi-path fading and shadowing. However, the difference is less than 1% compared with the analytical values. The results also exhibit that the density is fixed in the real network. Hence, Fig. 16 has been plotted for various values of minimum contention window sizes.

Since the MAC layer of the V2X device uses 8-bit unsigned integer for the minimum contention window size; hence, the results can be obtained only up to $W_0 = 128$. The results for the data rate, BSM data size, and packet generation rate also show similarity with the analytical results.

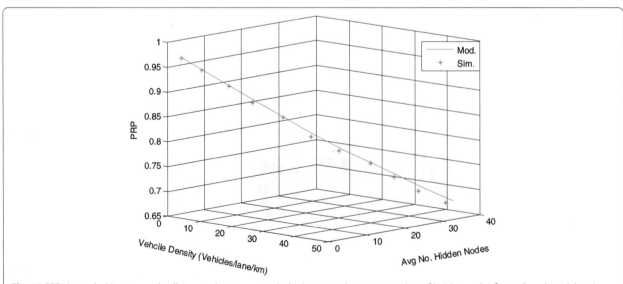

Fig. 10 PRP due to hidden terminal collisions with respect to vehicle density and average number of hidden nodes for analytical model and simulations, $W_0 = 16$, $R = 6$ Mbps, $R_c = 250$ m, $\lambda = 10$ Hz, $D = 186$ B

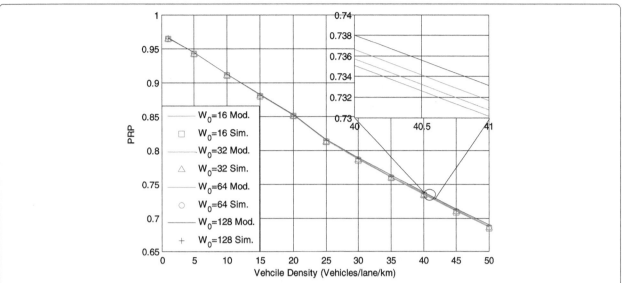

Fig. 11 Comparison of the analytic model and simulation results of PRP with hidden terminal collisions for various minimum contention window sizes with respect to vehicle density, $R_c = 250$ m, $R = 6$ Mbps, $\lambda = 10$ Hz, $D = 186$ B

6.4 Effect on packet reception delay (PRD)

As described in Section 5. *H*, the packet reception delay for the broadcast packet is defined as MAC-to-MAC delay for the packet. The analytical model calculates the delay by using Eq. (31), whereas the simulation obtains the delay by measuring the average delay between the packet reception time at the receiver and the packet generation time at the transmitter. The time difference is measured using the timestamp added by the transmitter in the BSM packet. Figure 17 shows the PRD with and without considering hidden terminal collisions. It shows that the results from the simulation and analytical model closely match, which confirms the high accuracy of the proposed analytical model. Additionally, results also exhibit that the hidden terminals have a negligible effect on PRD. This small change is attributed to the fact that the PRD also depends on the time consumed by the MAC queue before transmission. Hence, PRD remains nearly constant irrespective of the hidden collisions. When there is no hidden terminal collision, on the

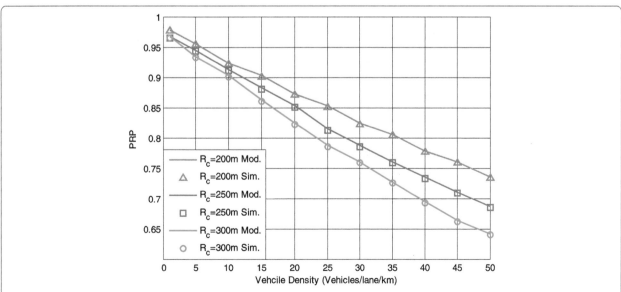

Fig. 12 Comparison of results for the analytic model and simulations of PRP with hidden terminals for various communication ranges with respect to vehicle density, $W_0 = 16$, $R = 6$ Mbps, $\lambda = 10$ Hz, $D = 186$ B

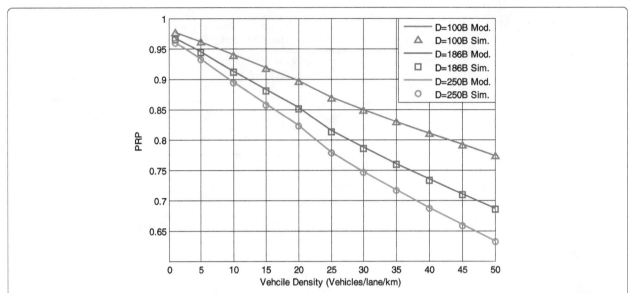

Fig. 13 Comparison of analytic model and simulation results of PRP with hidden terminal collisions for various BSM data sizes with respect to vehicle density, $W_0 = 16$, $R = 6$ Mbps, $R_c = 250$ m, $\lambda = 10$ Hz

other hand, more packets are received correctly, which smooths out the average PRD value (negligible difference).

Figure 18 compares the PRD results of the real network experiment (Fig. 8) and the analytical model. The PRD results are measured at Rx1 and Rx2 for the effect of hidden terminal collisions and without collisions, respectively. The results show that the hidden terminals have no noticeable effect on PRD.

6.5 Effect on packet reception interval (PRI)

The packet reception interval is defined as the average time between the reception of two packets at a receiver from the same transmitter. In the analytical model, the PRI is calculated by Eq. (32). In the simulation, the receiver measures the average arrival time difference between two successive packets from the same transmitter. Figure 19 compares the PRI calculated by the analytical model and measured by the simulation. The results

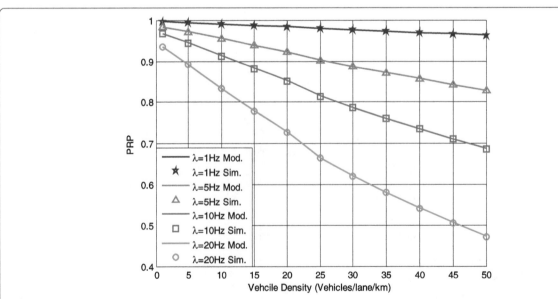

Fig. 14 Comparison of analytic model and simulation results of PRP with hidden terminal collisions for various packet generation rates with respect to vehicle density, $W_0 = 16$, $R = 6$ Mbps, $R_c = 250$ m, $D = 186$ B

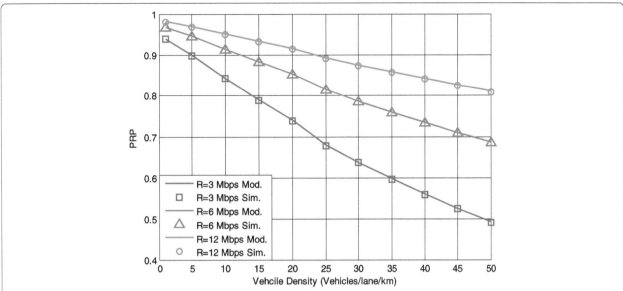

Fig. 15 Comparison of analytic model and simulation results of PRP with the hidden terminal collisions for data rates with respect to vehicle density, $W_0 = 16$, $R_c = 250$ m, $D = 186$ B, $\lambda = 10$ Hz

show that the PRI from the analytical model match with the simulation results. It also exhibits that the PRI increases with the increase in the density, which incurs due to increased number of hidden terminal collisions. However, increasing the packet generation rate reduces the PRI.

For the real network of Fig. 8, the PRI is calculated as the average inter-packet interval for the packets received from the device Tx1 at Rx1 for the hidden terminal case and Rx2 for no collisions case. Figure 20 compares the analytical results with the results measured from the real

network. Figure 20 exhibits that the results from the analytical model closely match the results from the real network. It also shows that the PRI is higher in case of hidden terminals. Additionally, it deduces that the value of the minimum contention window size does not affect the PRI, although increasing the packet generation rate decreases the PRI.

7 Conclusion and future work

In this paper, we proposed an analytical model to evaluate the effect of the hidden terminals in 2-D VANET.

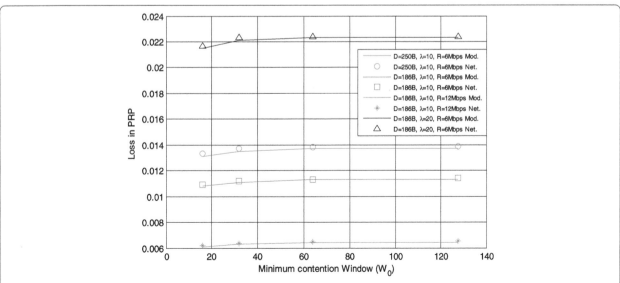

Fig. 16 Loss in the PRP due to hidden terminals for the analytical model (Mod.) and real network in Fig. 8 (Net.) with respect to minimum contention window size for the various network parameters, $R_c = 250$ m

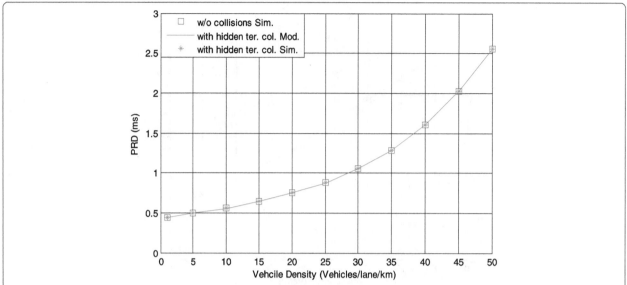

Fig. 17 Comparison of analytical model and simulation results of PRD with the hidden terminal collisions and without collisions with respect to vehicle density, $W_0 = 16$, $R = 6$ Mbps, $R_c = 250$ m, $\lambda = 10$ Hz, $D = 186$ B

The proposed model accurately estimates the performance metrics; packet reception probability (PRP), packet reception delay (PRD), and packet reception interval (PRI) for BSM safety data broadcast for a wide range of realistic vehicular networks. In order to verify the accuracy of the proposed analytical model, we have used NS-3 simulator with realistic vehicle mobility in various highway scenarios. Furthermore, we implemented a real vehicular network using commercial V2X devices. Our extensive simulations and experiments demonstrated that the performance estimated by the analytical model

accurately matches the performance measured from the NS3 simulator and real hardware network. The analytical model shows only minuscule discrepancy due to variance in the vehicle movement, channel fading and shadowing in the model. The results exhibit a decline in the PRP and PRI metrics' performance, while the PRD is not affected by the hidden terminals.

The metrics PRP and PRI depend on the network parameters; hence, the model can be used in adaptive congestion control to reduce the effect of hidden terminal problem. In future work, we plan to use this model as a

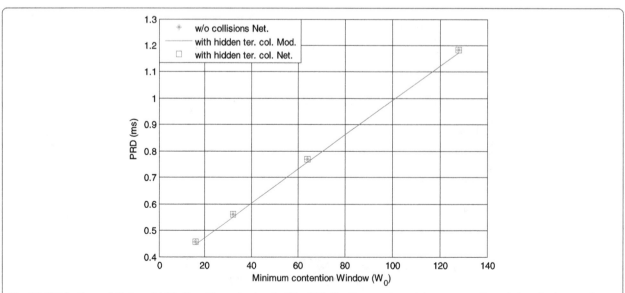

Fig. 18 PRD for the analytical model (Mod.) and for real network of Fig. 8 (Net.) with respect to minimum contention window size in case of hidden terminals and without, $R = 6$ Mbps, $\lambda = 10$ Hz, $D = 186$ B, $R_c = 250$ m

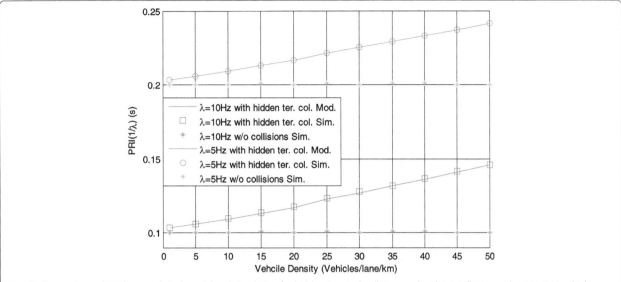

Fig. 19 Comparison of PRI from analytical model and simulation for hidden terminal collisions and without collisions with respect to vehicle density for different packet generation frequencies, $W_0 = 16$, $R = 6$ Mbps, $R_c = 250$ m, $D = 186$ B

cost metric to optimize the performance of 2-D VANET under hidden terminals.

8 Appendix

In the Appendix, we first derive the average number of nodes in the communication range (N_c) and channel sensing range (N_{cs}). We then calculate the average number of hidden terminal nodes (N_h) experienced by a receiver for a 2-D multi-lane highway.

8.1 Number of nodes in the communication range

Let X be a random variable representing the current lane of the vehicle. As shown in Fig. 21, for a vehicle in the xth lane, the number of vehicles in the communication range for each lane varies. Suppose, the vehicle in the xth lane is placed at the origin of 2-D coordinates, and each lane is spaced by one step of lane width ω in the coordinate system. Let m be the y-axis value for the distance vector for each lane from the origin. The distance to each lane from the origin can be expressed as $|m \times \omega|$.

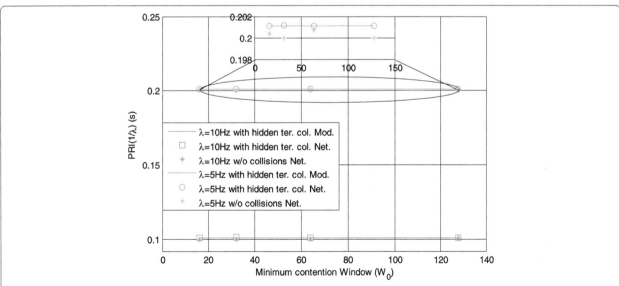

Fig. 20 PRI for analytical model and real network in Fig. 8 (Net.) for hidden terminal collisions and without collisions with respect to minimum contention window sizes for different packet generation frequencies, $R = 6$ Mbps, $D = 186$ B, $R_c = 250$ m

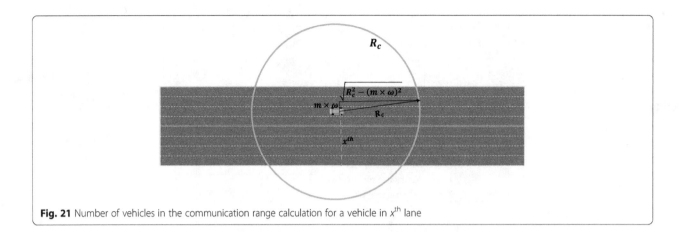

Fig. 21 Number of vehicles in the communication range calculation for a vehicle in x^{th} lane

As a result, the average number of neighbor nodes for a vehicle in the xth lane can be expressed by Eq. (36).

$$(N_c)_x = 2\alpha \sum_{m=1-x}^{\xi-x} \sqrt{R_c^2-(m \times \omega)^2} \qquad (36)$$

The average number of nodes in the communication range for whole network can be calculated by taking the average of the nodes in the communication range for the transmitters in each lane. Thus, the average number of nodes in the communication range can be calculated by Eq. (37).

$$N_c = \frac{2\alpha \sum_{x=1}^{\xi} \sum_{m=1-x}^{\xi-x} \sqrt{R_c^2-(m \times \omega)^2}}{\xi} \qquad (37)$$

The average number of nodes in the channel sensing range can also be calculated using Eq. (37) by replacing R_c with R_{cs}.

8.2 Number of hidden terminals

Suppose X is the random variable representing the lane of the transmitter node, and Y is the random variable representing the lane of the receiver with respect to the transmitter. In Fig. 22, a transmitter in x^{th} lane broadcasts a packet, while a node in y^{th} lane is a prospective receiver. Assume that the transmitter node is placed at the origin of 2-D coordinates and that the receiver is positioned at $(b, y \times \omega)$. Where ω is the lane width. The number of hidden terminal nodes in the mth lane from the receiver is calculated using Eq. (38) for the transmitter-receiver pair.

$$h_m = \alpha \left(b + \sqrt{R_{cs}^2-(m \times \omega)^2} - \sqrt{R_{cs}^2-((y+m) \times \omega)^2} \right) \qquad (38)$$

The total number of hidden terminal nodes for the transmitter-receiver pair can be calculated by summing the hidden nodes in all the lanes (see Eq. (39)).

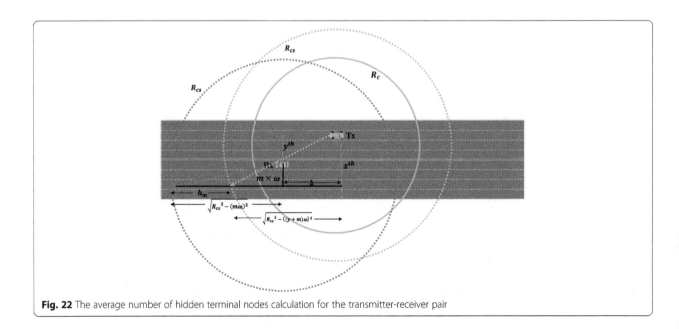

Fig. 22 The average number of hidden terminal nodes calculation for the transmitter-receiver pair

$$H = \sum_{m=1-y-x}^{\xi-y-x} \alpha \left(b + \sqrt{R_{cs}^2 - (m \times \omega)^2} - \sqrt{R_{cs}^2 - ((y+m) \times \omega)^2} \right) \tag{39}$$

To calculate the average number of hidden terminals in the network, we first calculate the average number of hidden terminals H_y for the receivers in y^{th} lane. H_y can be calculated by taking the average of the hidden nodes for all receivers in y^{th} lane. The positions of receivers in y^{th} lane can be obtained by changing the coordinate b for all the receivers in the communication range of the transmitter. As a result, Eq. (40) expresses the average number of hidden nodes for the receivers in the yth lane.

$$H_y = \frac{1}{2 \times \sqrt{R_c^2 - (y \times \omega)^2}} \sum_{i=1}^{\alpha\sqrt{R_c^2 - (y\times\omega)^2}} \sum_{m=1-y-x}^{\xi-y-x} \left(\frac{i}{\alpha} + \sqrt{R_{cs}^2 - (m \times \omega)^2} - \sqrt{R_{cs}^2 - ((y+m) \times \omega)^2} \right) \tag{40}$$

Eq. (40) can be simplified to Eq. (41).

$$H_y = \frac{1}{2} \sum_{m=1-y-x}^{\xi-y-x} \left(\frac{1 + \alpha\sqrt{R_c^2 - (y \times \omega)^2}}{2} + \alpha\sqrt{R_{cs}^2 - (m \times \omega)^2} - \alpha\sqrt{R_{cs}^2 - ((y+m) \times \omega)^2} \right) \tag{41}$$

From Eq. (41), We can derive Eq. (42), which calculates the average number of hidden terminal nodes for all the receivers of the transmitter in x^{th} lane.

$$H_x = \frac{\sum_{y=1-x}^{\xi-x} H_y}{\xi} \tag{42}$$

By extending Eq. (42), we can derive Eq. (43), which gives the average number of hidden terminals in the network. Here, N_h is obtained by taking the average of the number of hidden terminals for the transmitters in all the lanes.

$$N_h = \frac{\sum_{x=1}^{\xi} H_x}{\xi} \tag{43}$$

Abbreviations
ACK: Acknowledgment; BSM: Basic safety message; CCH: Control channel; DSRC: Dedicated short-range communication; MAC: Medium access control; PRD: Packet reception delay; PRI: Packet reception interval; PRP: Packet reception probability; RSU: Roadside unit; RTS/CTS: Request to send/clear to send; V2X: Vehicle to everything; VANET: Vehicular ad-hoc network

Authors' contributions
All three authors contributed equally. SK formulated the analytical model and implemented the simulation and real network. Afterward, together SK, HK, and SC verified the model, generated results, and wrote the manuscript. All authors read and approved the final manuscript.

Funding
This work was supported in part by IITP Grant through the Korean Government, under the development of wide area driving environment awareness and cooperative driving technology which are based on V2X wireless communication under grant R7117-19- 0164 and in part by the Center for Integrated Smart Sensors funded by the Ministry of Science, ICT & Future Planning as Global Frontier Project, South Korea (CISS-2019).

Competing interests
The authors declare that they have no competing interests.

Author details
[1]Department of Electronics Engineering, Chungbuk National University, Cheongju, Chungcheongbuk-do, South Korea. [2]Department of Electrical and Computer Engineering, Seoul National University, Seoul, South Korea.

References
1. US Federal Communications Commission. *Standard specification for telecommunications and information exchange between roadside and vehicle systems—5 ghz band dedicated short range communications (DSRC) medium access control (MAC) and physical layer (PHY) specifications*. Washington, DC (2003)
2. IEEE 802.11 Working Group, *IEEE Standard for Information Technology–Telecommunications and Information Exchange Between Systems–Local and Metropolitan Area Networks–Specific Requirements–Part 11: Wireless LAN Medium Access Control (MAC) and Physical Layer (PHY) Specifications Amendment 6: Wireless Access in Vehicular Environments*. IEEE Std **802**(11) (2010).
3. A. Tsertou, D.I. Laurenson, Revisiting the hidden terminal problem in a CSMA/CA wireless network. IEEE Trans. Mob. Comput. **7**, 817–831 (2007)
4. G. Bianchi, Performance analysis of the IEEE 802.11 distributed coordination function. IEEE J. Selected Areas Commun. **18**(3), 535–547 (2000)
5. F. Tobagi, L. Kleinrock, Packet switching in radio channels: Part 2—the hidden terminal problem in carrier sense multiple access and the busy-tone solution. IEEE Trans. Comm. **23**(12), 1417–1433 (1975)
6. S. Ray, D. Starobinski, J.B. Carruthers, Performance of wireless networks with hidden nodes: a queuing-theoretic analysis. Comput. Commun. **28**(10), 1179–1192 (2005)
7. O. Ekici, A. Yongacoglu, IEEE 802.11 a throughput performance with hidden nodes. IEEE Commun. Lett. **12**(6), 465–467 (2008)
8. X. Ma, X. Chen, H.H. Refai, Performance and reliability of DSRC vehicular safety communication: a formal analysis. EURASIP J. Wirel. Commun. Netw. **2009**, 3 (2009)
9. X. Ma, J. Zhang, W. Tong, Reliability analysis of one-hop safety-critical broadcast services in VANETs. IEEE Trans. Veh. Technol. **60**(8), 3933–3946 (2011)
10. X. Ma et al., Comments on "interference-based capacity analysis of vehicular ad-hoc networks". IEEE Commun. Lett. **21**(10), 2322–2325 (2017)
11. Y.P. Fallah, C.L. Huang, R. Sengupta, H. Krishnan, Analysis of information dissemination in vehicular ad-hoc networks with application to cooperative vehicle safety systems. IEEE Trans. Veh. Technol. **60**(1), 233–247 (2011)
12. P. Rathee, R. Singh, S. Kumar, Performance analysis of IEEE 802.11 p in the presence of hidden terminals. Wirel. Pers. Commun. **89**(1), 61–78 (2016)
13. C. Song, Performance analysis of the IEEE 802.11 p multichannel MAC protocol in Vehicular Ad-Hoc Networks. Sensors **17**(12), 2890 (2017)
14. X. Ma et al., Performance of VANET safety message broadcast at rural intersections. In *2013 9th International Wireless Communications and Mobile Computing Conference (IWCMC)* (IWCMC) (pp. 1617–1622). IEEE.
15. X. Ma, K.S. Trivedi, Reliability and performance of general two-dimensional broadcast wireless network. Perform. Eval. **95**, 41–59 (2016)
16. P. Wang et al., Performance analysis of EDCA with strict priorities broadcast in IEEE802. 11p VANETs. In *2014 International Conference on Computing, Networking and Communications (ICNC)* (pp. 403–407). IEEE.
17. K. Sjoberg, E. Uhlemann, E.G. Strom, How severe is the hidden terminal problem in VANETs when using CSMA and STDMA?. In 2011 IEEE *Vehicular Technology Conference (VTC Fall)* (pp. 1–5). IEEE.
18. R.S. Tomar et al., in *Harmony Search and Nature Inspired Optimization Algorithms*. Performance analysis of hidden terminal problem in VANET for safe transportation system (Springer, Singapore, 2019), pp. 1199–1208

19. S. Bastani, B. Landfeldt, The effect of hidden terminal interference on safety-critical traffic in vehicular ad-hoc networks. In *Proceedings of the 6th ACM symposium on development and analysis of intelligent vehicular networks and applications.* (pp. 75–82) (ACM, 2016)

20. E. Bozkaya, K. Chowdhury, B. Canberk, in *Proceedings of the 12th ACM Symposium on QoS and Security for Wireless and Mobile Networks.* SINR and reliability based hidden terminal estimation for next generation vehicular Networks. (pp. 69–76) (ACM, 2016)

21. M.I. Serra, B.I. Hillier, in *Proceedings of the 11th Space Syntax Symposium.* Spatial configuration and vehicular movement (2017)

22. K.S. Trivedi, *Probability & statistics with reliability, queuing and computer science applications* PHI Learning Pvt. Limited (2011)

23. G.F. Riley, T.R. Henderson, in *Modeling and tools for network simulation.* The ns-3 network simulator (Springer, Berlin, Heidelberg, 2010), pp. 15–34

24. S. Mecheri, F. Rosey, R. Lobjois, The effects of lane width, shoulder width, and road cross-sectional reallocation on drivers' behavioral adaptations. Accid. Anal. Prev. **104**, 65–73 (2017)

25. M. Torrent-Moreno et al., in *Proceedings of the 9th ACM International Symposium on Modeling Analysis and Simulation of Wireless and Mobile Systems.* IEEE 802.11-based one-hop broadcast communications: understanding transmission success and failure under different radio propagation environments (pp. 68–77) (ACM, 2006)

26. Cohda wireless MK5 device: description, Available At: https://cohdawireless.com/solutions/hardware/mk5-obu/. Accessed 18 Feb 2019

Permissions

The contributors of this book come from diverse backgrounds, making this book a truly international effort. This book will bring forth new frontiers with its revolutionizing research information and detailed analysis of the nascent developments around the world.

We would like to thank all the contributing authors for lending their expertise to make the book truly unique. They have played a crucial role in the development of this book. Without their invaluable contributions this book wouldn't have been possible. They have made vital efforts to compile up to date information on the varied aspects of this subject to make this book a valuable addition to the collection of many professionals and students.

This book was conceptualized with the vision of imparting up-to-date information and advanced data in this field. To ensure the same, a matchless editorial board was set up. Every individual on the board went through rigorous rounds of assessment to prove their worth. After which they invested a large part of their time researching and compiling the most relevant data for our readers.

The editorial board has been involved in producing this book since its inception. They have spent rigorous hours researching and exploring the diverse topics which have resulted in the successful publishing of this book. They have passed on their knowledge of decades through this book. To expedite this challenging task, the publisher supported the team at every step. A small team of assistant editors was also appointed to further simplify the editing procedure and attain best results for the readers.

Apart from the editorial board, the designing team has also invested a significant amount of their time in understanding the subject and creating the most relevant covers. They scrutinized every image to scout for the most suitable representation of the subject and create an appropriate cover for the book.

The publishing team has been an ardent support to the editorial, designing and production team. Their endless efforts to recruit the best for this project, has resulted in the accomplishment of this book. They are a veteran in the field of academics and their pool of knowledge is as vast as their experience in printing. Their expertise and guidance has proved useful at every step. Their uncompromising quality standards have made this book an exceptional effort. Their encouragement from time to time has been an inspiration for everyone.

The publisher and the editorial board hope that this book will prove to be a valuable piece of knowledge for researchers, students, practitioners and scholars across the globe.

List of Contributors

Janani V S and Manikandan M S K
Thiagarajar College of Engineering, Madurai-15, India

Sara Berri
Research Unit LaMOS (Modeling and Optimization of Systems), Faculty of Exact Sciences, University of Bejaia, 06000 Bejaia, Algeria
L2S (CNRS-CentraleSupelec-Univ. Paris-Saclay), 91192 Gif-sur-Yvette, France

Mohammed Said Radjef
Research Unit LaMOS (Modeling and Optimization of Systems), Faculty of Exact Sciences, University of Bejaia, 06000 Bejaia, Algeria

Samson Lasaulce
L2S (CNRS-CentraleSupelec-Univ. Paris-Saclay), 91192 Gif-sur-Yvette, France

Mohammad Faisal and Haseeb Ur Rahman
Department of Computer Science and IT, University of Malakand, Chakdara, KPK, Pakistan

Sohail Abbas
Department of Computer Science, College of Science, University of Sharjah, Sharjah, UAE

Jipeng Zhou, Liangwen Liu and Haisheng Tan
Department of Computer Science, Jinan University, Guangzhou 510632, People's Republic of China

Mohamed A. Abd El-Gawad
Department of Electronics Engineering, Chungbuk National University, Cheongju, South Korea
National Telecommunication Institute, Cairo, Egypt

Mahmoud Elsharief
Department of Electronics Engineering, Chungbuk National University, Cheongju, South Korea
Electrical Engineering Department, Al-Azhar University, Cairo, Egypt

Jaeshin Jang
Department of Electronic Telecommunications Mechanical and Automotive Engineering, Inje University, 197 Inje-ro, Gimhae, Gyeongnam 50834, South Korea

Balasubramaniam Natarajan
Department of Electrical and Computer Engineering, Kansas State University, 1701D Platt St., Manhattan, KS 66506, USA

Sung-Jeen Jang and Sang-Jo Yoo
Department of Information and Communication Engineering, Inha University, 253 YongHyun-dong, Nam-gu, Incheon, South Korea

Chul-Hee Han
Hanwha Systems 188, Pangyoyeok-Ro, Bundang-Gu, Seongname-Si, Gyeonggi-Do 13524, South Korea

Kwang-Eog Lee
Agency for Defense Development, Yuseong-Gu, Daejeon, South Korea

Kan Yu and Lina Ni
College of Computer Science and Engineering, Shandong University of Science and Technology, Qingdao, 266000 Shandong, China

Guangshun Li and Jiguo Yu
School of Information Science and Engineering, Qufu Normal University, Rizhao, 276826 Shandong, China

Shirin Rahnamaei Yahiabadi
Faculty of Computer Engineering, Najafabad Branch, Islamic Azad University, Najafabad, Iran

Behrang Barekatain
Faculty of Computer Engineering, Najafabad Branch, Islamic Azad University, Najafabad, Iran
Big Data Research Center, Najafabad Branch, Islamic Azad University, Najafabad, Iran

Kaamran Raahemifar
Electrical and Computer Engineering, Sultan Qaboos University, Al-Khoud 123, Sultanate of Oman
Chemical Engineering Department, University of Waterloo, 200 University Avenue West, Waterloo, Toronto, Ontario N2L 3G1, Canada

Adnan Shahid Khan
Department of Computer Science and Information Technology, Universiti Malaysia Sarawak (UNIMAS), 9300 Kota Samarahan, Malaysia

Irshad Ahmed Abbasi
Department of Computer Science and Information Technology, Universiti Malaysia Sarawak (UNIMAS), 9300 Kota Samarahan, Malaysia
Department of Computer Science, Faculty of Science and Arts at Balgarn, University of Bisha, Sabt Al-Alaya 61985, Kingdom of Saudi Arabia

Shahzad Ali
Department of Computer Science, COMSATS Institute of Information Technology, Abbottabad, Pakistan

Long Zhang and Fan Zhuo
School of Information and Electrical Engineering, Hebei University of Engineering, Handan 056038, China

Haitao Xu
School of Computer and Communication Engineering, University of Science and Technology Beijing, Beijing 100083, China

Saurabh Kumar and HyungWon Kim
Department of Electronics Engineering, Chungbuk National University, Cheongju, Chungcheongbuk-do, South Korea

Sunghyun Choi
Department of Electrical and Computer Engineering, Seoul National University, Seoul, South Korea

Index

CPSIA information can be obtained
at www.ICGtesting.com
Printed in the USA
BVHW012039300822
645853BV00002B/88